SURFACE AND COLLOID CHEMISTRY IN ADVANCED CERAMICS PROCESSING

SURFACTANT SCIENCE SERIES

CONSULTING EDITORS

MARTIN J. SCHICK
Consultant
New York, New York

FREDERICK M. FOWKES
(1915–1990)

ADDITIONAL VOLUMES IN PREPARATION

SURFACE AND COLLOID CHEMISTRY IN ADVANCED CERAMICS PROCESSING

edited by

Robert J. Pugh
Lennart Bergström

Institute for Surface Chemistry
Stockholm, Sweden

CRC Press
Taylor & Francis Group
Boca Raton London New York

CRC Press is an imprint of the
Taylor & Francis Group, an **informa** business

CRC Press
Taylor & Francis Group
6000 Broken Sound Parkway NW, Suite 300
Boca Raton, FL 33487-2742

First issued in paperback 2019

© 1994 by Taylor & Francis Group, LLC
CRC Press is an imprint of Taylor & Francis Group, an Informa business

No claim to original U.S. Government works

ISBN-13: 978-0-8247-9098-1 (hbk)
ISBN-13: 978-0-367-40223-5 (pbk)

Visit the Taylor & Francis Web site at
http://www.taylorandfrancis.com

and the CRC Press Web site at
http://www.crcpress.com

Library of Congress Cataloging-in-Publication Data

Surfactant and colloid chemistry in ceramic processing / edited by
  Robert J. Pugh, Lennart Bergström.
    p. cm. -- (Surfactant science series ; v. 51)
  Includes bibliographical references and index.
  ISBN 0-8247-9098-7
  1. Ceramics. 2. Surface active agents. 3. Colloids. I. Pugh,
Robert J. II. Bergström, Lennart.
III. Series.
TP815.S87 1994
666--dc20   93-32077
                                                                  CIP

Preface

Traditionally, ceramics processing has been regarded more as a conventional production process rather than as a science that involves the preparation and mixing of various components according to traditional formulations followed by compaction and heat treatment. However, with the advent of new types of ceramic materials with unique properties, ceramics processing has attained increasing interest. Processing has also become more complex, since there is a trend to (1) decrease the size of the ingoing powders to within the colloidal size range (typically < 1 μm) in order to enhance sinterability at lower temperatures and (2) reduce the scale of mixing homogeneity of the different components. In order to control and optimize the processing of such colloidal systems, it is often necessary to carry out investigations beginning at the powder stage and work through to the final microstructure. It is clear that defects in the final material very often can be traced back to the early stages of the processing.

In this book we attempt to cover the surface and colloidal chemical aspects of the processing of ceramic powders in depth. Although many ceramics scientists and engineers recognize the importance of surface and colloidal chemistry in processing, there are few texts available that cover the various engineering aspects of the discipline. For example, particle preparation and characterization, the dispersions and mixing of powders in liquids, rheology, slurry consolidation, pressure filtration, and the packing of powders all involve basic surface and colloidal chemical principles. With the control and manipulation of the interparticle forces in powder suspensions, heterogencities can be removed and the suspension properties optimized. This concept of improving the reliability is called colloidal processing.

We have recruited specialists in both surface and colloid chemistry and processing of ceramics with the aim of covering both the fundamentals of surface and colloid chemistry relevant to the processing of advanced ceramic powders and the surface and colloid chemistry aspects of the most widely used forming methods. The book focuses on the processing of high-strength, high-performance ceramics with some coverage of other types of advanced ceramics.

Chapter 1 presents an overview of the importance of surface and colloid chemistry in ceramics and discusses the need for microstructure control in order to improve high-strength and high-temperature properties. Chapter 2 describes various technologies for controlling powder characteristics. The surface characterization of ceramic powders is covered in Chapter 3 with an emphasis on characterization methods of the solid/liquid interface. Chapter 4 introduces the fundamentals of colloid science, with an application to the dispersion of powders in liquids, in both aqueous and organic media. Chapter 5 describes and discusses the rheological behavior of colloidal suspensions in relation to different types and magnitudes of the interparticle forces. Chapter 6 treats the dry pressing operation, including the importance of surface and colloid chemistry in granulation, spray-drying, and compaction. The surface and colloid chemistry aspects of the most common ceramic-forming methods are covered in Chapters 7 and 8. Chapter 7 focuses on ceramic casting operations such as slip casting, centrifugal casting, pressure filtration, and tape casting. Chapter 8 discusses the injection-molding process, with relevance to all types of plastic-forming operations.

It is hoped that the book is sufficiently current and well balanced to serve as a guideline and reference for ceramic engineers and for interdisciplinary research. We are grateful to a number of people who helped us during this project, including Martin Schick and Fred Fowkes, who urged us to start up the project, and Per Stenius, who helped us with the organization in the initial stages. Special thanks to all the authors whose time and effort made it possible

Robert J. Pugh
Lennart Bergström

Contents

Contributors

Lennart Bergström, Ph.D. Deputy Section Manager, Institute for Surface Chemistry, Stockholm, Sweden

Elis Carlström, Ph.D. Research Manager, Swedish Ceramic Institute, Gothenburg, Sweden

J. R. G. Evans, B.Sc., Ph.D., C. Eng., M.I.M. Reader, Department of Materials Technology, Brunel University, Uxbridge, Middlesex, United Kingdom

Michael Persson, M.Sc., Ph.D.[*] Project Manager, Swedish Ceramic Institute, Gothenburg, Sweden

Robert J. Pugh, D.Sc. Professor, Institute for Surface Chemistry, Stockholm, Sweden

Richard E. Riman, B.S., Ph.D. Associate Professor of Ceramics, Department of Ceramic Engineering, Rutgers University, Piscataway, New Jersey

Present affiliation: R&D Manager, Colloidal Silica Department, Eka Nobel AB, Bohus, Sweden

1

Surface and Colloid Chemistry in Ceramics: An Overview

ELIS CARLSTRÖM Swedish Ceramic Institute, Gothenburg, Sweden

By using techniques from colloid chemistry, defects can be removed and the strength of ceramic materials increased considerably. Colloid chemistry techniques can also be used to enhance homogeneity in the mixing of ceramic powders and to improve sinterability, and they play a crucial role in formation of ceramic powder compacts. Synthesis of ceramic powder is yet another area of application of surface chemistry. A number of these possibilities are discussed briefly in this overview and in greater detail in the following chapters.

I. HIGH-STRENGTH CERAMICS

Ceramics are brittle materials at moderate temperatures. This brittleness has been a major obstacle in the production of high-strength ceramics, but owing to their superior high-temperature and wear properties, great efforts have been made to produce high-strength ceramics.

A brittle material is not necessarily a low-strength material. Brittleness means that when the fracture stress is exceeded the material fails, with no extra energy required. The failure is instantaneous and is thus often called a catastrophic failure (Fig. 1). A stress and strain diagram can be used to describe this. As tensile stress is applied to the material, the strain increases linearly. If we apply the same stress to a metal, we observe a yield stress. Above the yield stress, the metal deforms plastically and only after a considerable amount of plastic deformation does the fracture occur (Fig. 2).

The strength of a brittle material is controlled by local stress concentrations. If no plastic deformation can occur, a local stress concentration can only be relieved by fracture. If we put a brittle material under tensile stress, we find local stress concentration around cracks or other defects in the material. Any defect in a brittle material will cause a local stress concentration when the material is subjected to stress (Fig. 3).

The strength of a ceramic material can be described by Griffith's equation [1]:

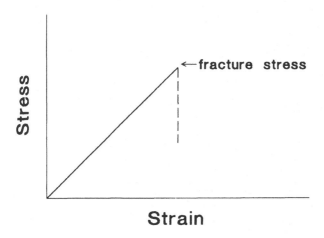

FIG. 1 Stress-strain curve of a brittle material.

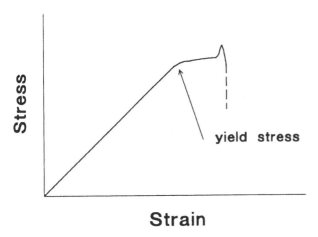

FIG. 2   Stress-strain curve of a metallic material.

$$\sigma = YK_{Ic}c^{-1/2} \tag{1}$$

where c is the defect size, K_{Ic} the fracture toughness, and σ the fracture strength. The Y factor is a constant that depends on the position and shape of the defect.

This provides a simple recipe for a high-strength ceramic. To achieve a strong material, we either have to increase the fracture toughness or decrease the size of the defects. In theory this is very simple; in practice there are a number of interrelated difficulties that have to be solved.

As one single defect can limit the strength, a considerable variation in strength from sample to sample will occur. Because of this, the measure of

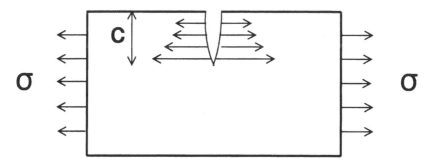

FIG. 3   A defect in a brittle material causes a stress concentration when the material is subject to tensile stress.

the strength of a ceramic material has only really a valid mean value when associated with a description of the variation in strength such as the Weibull modulus [2]. Processing of ceramics with colloid chemistry methods aims to increase both strength and reliability. The strength increase comes from reduction of the maximum defect size, while reliability is increased by reducing the statistical variation of defect sizes, i.e., reducing the spread in strength.

In order to describe the statistical variation of strength in brittle materials, Weibull used the analogy of a chain in which the weakest link determines the strength. In a ceramic material, the links are represented by defects such as cracks, inclusions, and pores, all of which give rise to stress concentrations and a consequent risk of brittle failure. The strength of a brittle material can be deduced from the fracture toughness and the size (and shape and position) of a critical defect. With a given fracture toughness, an increase in strength and reliability of a brittle material is achieved by decreasing the size and number of defects in the material.

There are some exceptions to this phenomenon. Transformation toughened ceramics and fiber/whisker reinforced ceramic materials can be relatively insensitive to flaw size. Examples of such materials are optimally aged transformation toughened ZrO_2 [3] and fiber composites with very weak fiber-matrix interfacial bond strengths [4].

The defects in a ceramic material are often caused by defects or impurities in the starting powders. Defects are also introduced during processing before firing, during firing, and during machining to final shape. A variation on the chain analogy theme can be used to describe the introduction of defects during fabrication of ceramics. Powder synthesis and preparation, powder processing and forming, sintering, and machining are all links in this chain. Additions of a new link to this chain can never increase the strength of the chain, only maintain it or decrease it. In the same way, each of these processing steps has a very limited if any ability to decrease the number of defects produced by a preceding step. Sintering, for example, can never remove foreign inclusions, but it can introduce new defects such as large grains owing to abnormal grain growth. Powder processing and forming of powder compacts are to a large extent controlled by surface and colloid chemistry mechanisms. This makes application of surface and colloid chemistry techniques among the most important steps in the production of high-strength ceramics.

Kendall, Alford, and Birchall [5] measure the variation in strength (Weibull modulus) of an alumina ceramic before and after sintering. Their experiments show that the strength increases, while the variation in strength (Weibull modulus) of the bodies remains unchanged. This indicates that the same defects limited the strength in the green and in the sintered state in this system. Moreover, the size of these defects remains the same during sintering. Hot isostatic pressing is capable of removing some types of processing flaws such as pores and cracks [6]. Unfortunately, certain other types of defects, such as pores at the surface or just below the surface, cannot be removed [7].

In order to increase the strength and reliability of ceramic materials, the links in the chain must be examined beginning from the starting materials, and efforts must be made to remove or restrict defects or defect formation at each step [8]. In doing this, the process can be simplified somewhat by examining the fracture surfaces of a material to find out what the dominating fracture origins consist of [9,10]. Efforts can then be directed to removing these defects rather than trying to remove every conceivable type of defect that might possibly occur.

Most ceramics are formed as powder compacts and made dense by sintering. Powder used for fabrication of high performance ceramics will only sinter to dense materials if the particle size is small enough. In practice this means particles of sizes 5–0.05 µm. In this size range colloidal forces will often be more important than the force of gravity. For this reason, techniques of surface and colloid chemistry are some of the most important tools in ceramic processing.

By using colloid chemistry techniques, defects can be removed and the strength of the ceramics increased considerably. As described below, there are a number of techniques available. There are also examples in the literature, where colloid approaches have indeed been used to increase the fracture strength of ceramics. Lange et al. [11] have shown that the strength of composites in the alumina-zirconia system can be increased using the colloidal approach. Linde et al. [12] have shown that this can also be done in a nonoxide ceramic system. Alford, Birchall, and Kendall [13] have shown that notable strength improvement in a pure alumina system can be achieved. They report strengths in the range of 1 GPa with a high Weibull modulus preserved. This should be compared with the typical strength of 300–400 MPa for commercial aluminas. These are some of the practical examples that show that both the strength and the reliability of ceramics can be improved using colloid chemistry methods.

II. INCREASING THE FRACTURE TOUGHNESS

Another way of increasing the strength is to increase the fracture toughness [14]. For a single phase ceramic, the chosen phase and grain size determine the fracture toughness. However, many single phase ceramics have a grain boundary interphase of special composition. By controlling the amount and properties of the grain boundary phase, the fracture toughness can be optimized. This gives a limited space for increase of the fracture toughness, often at the expense of the high-temperature properties.

A. Particle Reinforcement

More drastic ways of increasing fracture toughness include particle and fiber (whisker) reinforcement. Particle reinforcement works by crack deflection [15], microcracking, or transformation toughening [16]. Crack deflection means that cracks are deflected by the second phase particles and have to work their way around these particles, losing some of their propagation energy. A microcrack that intersects with a crack path often stops the crack propagation, as it tends to deflect and split the crack in two. Transformation toughening involves using a phase (second or major) of particles that can transform. When the transformable particle is subjected to the tensile field at the crack tip, it transforms and changes its volume or shape so that the crack tip is subjected to compressive stress. The transformation can be of either the martensitic or the ferroelastic type. A martensitic transformation is encountered in zirconia stabilized with yttria. The tetragonal form of zirconia is stabilized down to room temperature by the yttria doping. A tensile field destabilizes a tetragonal particle. The particle then transforms to the monoclinic type; this is a transformation that involves a significant volume expansion. This volume expansion causes a compressive stress field around the particle.

B. Fiber and Whisker Reinforcement

When fiber composites are made with ceramic matrix materials, the fiber only rarely has a significantly higher E-modulus than the matrix. In such cases the fiber can put the matrix in compression, and any local tensile stresses first have to overcome this compression before a crack can experience tensile stress. Most ceramic fiber and whisker composites are made with fibers of approximately the same E-modulus as the matrix. In this case the strengthening mechanism is crack deflection or crack bridging [17].

Crack bridging means that fibers or whiskers extend perpendicularly over a crack, making crack propagation more difficult. Only the matrix is broken by the extending crack, leaving the fibers/whiskers connecting the two parts. This can dramatically increase the fracture thoughness. Also, the material no longer exhibits linear elastic behavior at high stresses. A propagating crack encounters increasing resistance as it advances. This can be described with a resistance curve (often called an R-curve) [18] showing fracture toughness as a function of crack length. The physical explanation is that when the crack extends, the number of bridges across the crack increases. These bridges make the propagation of the crack increasingly difficult. Resistance curve behavior can be observed with most types of high-toughness ceramics.

If crack bridging or crack deflection is to take place, the interface between the fiber/whisker and the matrix has to be weak. This makes the interfacial properties of the whiskers very important in controlling possible reactions at the interface.

These means of increasing strength show very promising results but put our processing abilities to greater tests. Increased fracture toughness with a simultaneous increase in defect size is of limited use. Thus we have to process these materials just as thoroughly as other ceramics.

C. Processing Requirements for Fiber and Particle Reinforcement

We face several problems in trying to process fiber or particle reinforced ceramics. First, we have to disperse two (or sometimes more) types of materials at the same time. Sometimes these materials will have very different surface characteristics, making the dispersion we need, to form or mix, very difficult. Second, in the case of whiskers we want to pack the powder and whiskers just as efficiently as if we only had a powder. Anyone who has tried to put the matches back in a box after spilling them will appreciate this problem. One cannot align the whiskers before packing them, and generally one does not want to do so. Third, we want to do this without any unwanted changes in the surface characteristics of the whiskers.

Composites with long fibers are often formed by methods such as dipping the fibers in a powder slurry and then packing them, or weaving a cloth of fibers that is then infiltrated with a slurry or precursor of the matrix. Close control of the rheology of the powder slurry is necessary in both these cases. All of this demands an extra processing effort and efficient use of colloidal methods.

III. HIGH-TEMPERATURE PROPERTIES

Good high-temperature properties are one of the most important reasons for using ceramics. Performance at high temperatures is often controlled by a combination of phenomena such as creep, slow crack growth, oxidation of nonoxide phases, grain growth, and crystallization of glassy phases. The purity and composition of the starting materials are very important in achieving good high-temperature properties. Ceramic materials often have grain boundary phases. The composition of these phases and whether they are amorphous or crystalline will often determine the creep or slow crack growth tendencies.

In order to make good high-temperature materials it is important to be able to mix the components well, to avoid varying amounts and compositions of the secondary phases. It is also important that no contamination be introduced during processing. Even small amounts of contaminants can reduce high-temperature strength considerably. One example of this is small amounts of fluorine in silicon nitride, which decrease the high-temperature properties of the grain boundary glassy material that controls the mechanical properties of the material. Another example is alumina, in which impurities restrict high-temperature properties. Even small additions of alkaline components will significantly reduce the usable temperature of the material.

A. Processing Requirements for High-Temperature Properties

Many surfactants contain alkaline ions as counterions. Careful selection of additives for dispersion, including acids and/or bases used to control pH, is necessary. Halogens can have adverse effects on sintering and high-temperature properties. They can be introduced during processing as hydrochloric and hydrofluoric acid and remain adsorbed on the powder surfaces and only wash away gradually. Grinding is often the most effective means of mechanically dispersing powders that sometimes contain hard agglomerates, but grinding is always a source of contamination. Using the same material in the mill and milling media as in the produced ceramic decreases the risk of milling contamination, but this is not always possible. In order to reduce milling times efficient chemical dispersion methods can be used. Removal of large grains and agglomerates can also be done using colloid chemistry methods such as sedimenting.

IV. SINTERING OF HIGH-STRENGTH CERAMICS

A. Sintering Behavior

In order to make ceramic materials, powder compacts have to be sintered. The driving force for sintering is the reduction of surface energy. Sintering, in itself, does not necessarily involve densification, and densification is essential to making strong ceramic materials. In most cases, the high-strength ceramic material needs to be as dense as possible.

Sintering can take place by solid phase or liquid phase sintering. Liquid sintering is influenced by the viscosity of the liquid, its wetting characteristics, and the solubility of the solid phase in the liquid. Both solid and liquid phase sintering are influenced by the quality of the particle compact, as large pores will tend not to densify. In liquid phase sintering they might be filled with the liquid at high temperatures. In the final microstructure they will show up as large glassy areas, which is a potential defect. Inhomogeneous distribution of the glass forming components in a powder compact makes sintering more difficult. This is especially true in liquid phase sintering of ceramics for high-temperature applications, where the liquid must have a high viscosity.

1. Powder Characteristics Required for Sintering

What are the powder characteristics that control sintering and densification? Particle size is important; a smaller particle has a higher tendency to increase in size, and one of the ways of increasing in size is sintering with a neighbouring particle. Sintering stops at the thermodynamic equilibrium when $dA_s/dA_b = \gamma_b/\gamma_s$. In this equation dA_s is the change in surface area of the grain, while dA_b is the change in grain boundary area. γ_b and γ_s are the corresponding surface energies associated with the creation of new surface area (particle-pore) and new grain boundary area (grain-grain). It has been shown experimentally many times that particles with smaller sizes start sintering at lower temperatures [19]. Sintering cannot be equated with densification, which is influenced by particle packing as well as by particle size.

2. The Effect of Agglomeration

A number of experimental results show that powders with fewer hard agglomerates have improved sintering properties [20–24]. Lange [25–27] has shown that the difference in density between an agglomerate and its surroundings causes differential shrinkage. This differential shrinkage causes cracks or residual porosity and has a negative influence on both sintering and strength.

3. Particle Shape and Distribution

Particle shape can also be important. Unequiaxed particles are more difficult to pack than equiaxed particles. Bowen [28] states that the ideal sinterable powder to produce a theoretically dense single phase ceramic has the following properties:

1. Fine particle size (0.1–1.0 μm)
2. Narrow particle size distribution
3. Equiaxed shape
4. Nonagglomerated state

The reasons for wanting nonagglomerated powder and fine particle sizes have been described above. The equiaxed shape facilitates most forming processes. Viscosity increases both in a slurry and in a mix for injection-molding when the particles are rough or elongated.

A relatively narrow distribution of particle sizes is necessary to avoid differential shrinkage. A particle that is much larger than the surrounding matrix will shrink less than the matrix (because it is already dense), causing compressive stresses in the matrix. These will inhibit sintering or cause matrix cracking. A narrow or monodisperse particle size, on the other hand, might cause problems with particle packing if the powder is packed in the dispersed state. This phenomenon, called formation of second generation particle clusters, is described below.

The lower limit for particle size that can be used for sintering is set by the forming methods. It is very difficult to produce a concentrated slip with particles of sizes below 0.1 μm [29]. During injection molding, particles that are too small dramatically raise the viscosity of the injection molding batch. In pressing, smaller particles need higher pressures to be pressed to the same density as a powder with larger particles.

4. Can Colloid Chemistry Be Used to Obtain Better Powders?

Colloid chemistry has many uses in this context. As mentioned above, agglomerates can be removed by sedimentation rather than by milling. One advantage of this is less contamination; another is that *all* agglomerates above a certain size can be removed. With milling, a high percentage but not all can be removed.

Colloid chemistry is also used to synthesize powder with no agglomeration and equiaxed shapes. Sol-gel processing allows powders with very small particles to be synthesized and, at times, to form powder compacts directly with these small particles. Sol-gel processes can also be used to produce intimate mixes of several powder components.

Grinding of powders is influenced by colloidal processes. Addition of surfactants is commonly used in the grinding of minerals. They can also be used in the grinding of ceramic powder for high-performance ceramics. Surfactants can influence viscosity during grinding, but it is also possible that, through adsorption, they influence the grinding process as such.

5. Particle Packing

Particle packing is equally important as particle shape and size. The packed density is less important than the pore size distribution. Large pores do not disappear during sintering. Whether or not a pore shrinks during sintering depends on its pore coordination number according to Lange [30]. A high pore coordination number is the same as a large number of grains defining the pore. As the number of grains around a pore increases, the dihedral angle between the grains increases. Young's equation can be used to describe the stable angle configuration between two grains and a pore. If the angle is below the equilibrium angle, the pore shrinks and eventually disappears. If it is above this angle, it is stable and does not shrink in size during sintering.

If the powder is spherical and monosized, it can be packed in a regular hexagonal close packed structure [28]. This structure is achieved if the powder is allowed to settle from a diluted stable dispersion. However, it has not been possible to form large close packed structures. In practice, close packed domains separated by domain boundaries are formed. An analogy can be drawn with close packed atoms in a crystal structure, in which the domain boundaries would be the crystal boundaries. On sintering, these structures sinter perfectly within the close packed domains. At the domain boundaries, large pores develop and do not disappear [31]. Hierarchical clusters, such as ordered domains, always develop when compacts are formed from colloidal powders [29]. Consequently, the pore size distribution is multimodal even for a monodisperse powder. High concentrations of solids and shear deformation during forming (as in injection molding) can be used to reduce the width of the pore size distribution. A second approach is to use a multimodal particle system.

Sintering is a special kind of material transport. During the early stages of sintering, several paths for material transport compete. The sink of the transport always occurs at the particle necks. The source can be either the surface or the grain boundary. Transport mechanisms involve diffusion or evaporation-condensation. The transport paths for diffusion are lattice, surface, or grain boundary. Of these paths, only transfer of matter from the particle volume or grain boundary to the neck causes shrinkage and pore elimination.

High temperatures are necessary to increase the rate of diffusion. The diffusion rate is also influenced by particle size, as reduction of particle curvature is the driving force for diffusion. Smaller particle sizes decrease the temperature needed for sintering. Better particle packing facilitates sintering by shortening the distances for material transport [32]. Small pore sizes are necessary to avoid stable pores. When enough grain growth occurs around a stable pore it becomes unstable again. The reason for this is the reduction of the pore coordination number that takes place as a result of particle coarsening [33].

Thus if we have a poor powder compact with large pores we can often sinter it, using high temperatures and long sintering times. This however causes considerable grain growth during sintering, and large grains are not usually desired in the final microstructure. Large grains function as defects in the material and cause fracture at low stresses.

For this reason, it is important to be able to reduce sintering temperatures and sintering times to avoid excessive grain growth. In doing this, optimized particle packing is essential.

Monosized particles can be packed in an ordered formation. However, the packed density of a monosized powder will always be smaller than the density of a powder compact made from a powder with a wider distribution of sizes. This is because small particles will fit into the pores at a four grain junction of larger particles [34].

B. Homogeneity in Ceramics

A ceramic material is usually composed of more than one component. What we usually think of as single phase materials such as alumina and silicon carbide are fabricated from a mixture of materials. Alumina is often doped with small amounts of magnesia. Magnesia inhibits excessive grain growth during sintering of alumina. Silicon carbide cannot be densified without the use of pressure, unless sintering aids are added. Small amounts ($\approx$1 wt %) of boron or aluminum and some free carbon are added to promote densification during sintering. Excessive additions of these sintering aids will degrade the properties of the material. If we mix them well, we can minimize the amounts needed.

Other materials are deliberately made up of two or more phases. Silicon nitride needs a glassy interface between the grains to have a high fracture toughness. In practice, this is achieved by adding small amounts of oxides such as magnesia or yttria and alumina. Some silicon oxide will also be present on the silicon nitride surface, and this oxide is incorporated into the glassy phase. Sialons can be made with or without a glassy grain boundary phase. But even without a final second phase, several oxide and nitride phases have to be mixed at some stage of the preparation. If we cause two or more phases to react to form a stoichiometric phase, we need exact amounts of each. To decrease the time needed to complete the reaction, we also want the material transport paths to be short. Good mixing will do this, and only with good mixing do we have the exact amounts of the separate components locally. Poor mixing always results in long transport paths.

In order to fabricate composite materials such as particle or whisker reinforced materials, we also need good mixing. Alumina toughened by additions of zirconia will not achieve optimum fracture toughness if the zirconia particles are distributed unevenly. The same thing applies if we toughen the alumina with silicon carbide whiskers. In these cases the second phase also inhibits grain growth. If there is local depletion of the second phase, abnormal grain growth can occur at that spot.

An ideal mix would consist of particles evenly distributed, with particles of one kind alternating regularly with particles of others (Fig. 4). A mixture of this kind cannot be created by stirring. If an ideal mixture is stirred, it no longer remains ideal. By stirring we can, in the best case, achieve a "random homogeneous mix" [35]. Quantifying the degree of random mixing can be done by taking spot samples of composition and studying the varying values obtained for them. For particles of the same size, and small spot samples (compared with the total sample), a binomial distribution can be used to describe the probabilities of a certain composition. For particles of uneven sizes, the equations for the variance become more complex. A third type of mix, the "ordered mix," is also possible. Ordered mixing might occur spontaneously if the difference in particle size is large. Small particles pack spontaneously in the porosity created between the large particles, creating an ordered mix. In high-performance ceramics the differences in particle size are often not of this magnitude. It is, however, also possible to use colloid chemistry methods in order to create ordered mixes.

1. The Effect of the Degree of Dispersion on Mixing

Colloid chemistry influences mixing in a number of ways. Fine grained ceramic powders can seldom be mixed dry because of the clogging that takes place. If powders are mixed wet, the suspension of powder in liquid will be in either the dispersed or the flocced state. Well-dispersed particles will allow mixing with a minimum of shear force applied. Ceramics are often mixed using a ball mill, but this is only really necessary if hard

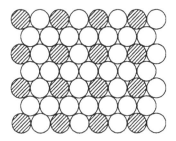　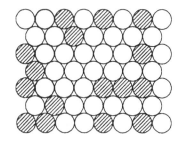

Ideal homogeneous mix　　Random homogeneous mix

FIG. 4 Ideal and random homogenous mixes.

agglomerates are present in the powder. Mixing in the dispersed state has one disadvantage, which is that segregation may take place instead of mixing. Segregation will occur if the components have different densities and are allowed to settle. Differences in particle size lead to a differential settling rate and can also lead to segregation. Segregation can also take place as a consequence of the liquid transport during slip casting.

Particles can be mixed in the flocced state if enough shear force is applied [36,37]. For high shear forces, the pseudoplastic mix decreases in viscosity and flows easily. This happens because the flocs are broken down in the applied shear field. When shearing is stopped, new flocs will form and the existing state of mixing can be preserved (Fig. 5).

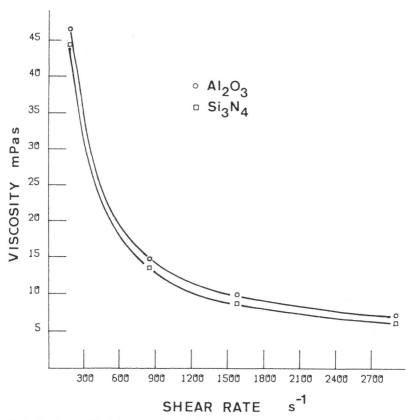

FIG. 5 Shear thinning behavior of flocced suspensions [37]. (E. Carlström and F. F. Lange, Mixing of flocced suspensions, *J. Am. Ceram. Soc. 67*(8): C169–C170 (1984), Figure 1. Reprinted by permission of the American Ceramic Society.)

2. Random Mixing and Segregation

When we try to describe the state of mixing, we always need to know the scale at which we observe our system. The scale at which the system starts to exhibit inhomogeneity can actually be experimentally defined. Using scanning electron microscopy with x-ray analysis, the homogeneity of a ceramic body can be quantified. The method consists of measuring the statistical variance of the chemical composition. When the chemical composition is measured for a large area of the sample, the measurement variance is small (usually <3%) and mainly consists of the statistical variation of the x-ray emission count. When small areas are examined, the variance is much greater, because the composition of each area does not constitute an average of the total sample. The smallest area with a composition equal to the average composition of the total sample can be defined as the scale of inhomogeneity of the sample [38,39].

3. Using Solid Solubility in Mixing

The smallest scale mixing that can be achieved is solid solution at the atomic level, i.e., the solids are dissolved into each other, as is necessary when we need to stabilize a high-temperature crystal structure at room temperature. One example of this is when yttria is added to zirconia to stabilize the cubic phase (or to partially stabilize the tetragonal phase). Solid solubility can also be utilized when sintering additives are mixed with silicon carbide. If boron is added during the synthesis of the powder, as can be done in plasma synthesis of silicon carbide, it dissolves in the silicon carbide forming a solid solution [40]. Mixing on an atomic scale can also be done using wet chemical methods such as sol-gel methods with alkoxides or solution precipitation of salts. One of the limitations of this method is the upper limit of solid solubility. It is also necessary to form the solid solubility during the synthesis of the powder. When we wish to create a grain boundary phase as in silicon nitride materials we want the secondary phase to stay at the grain boundary and not dissolve in the grains, in which case solid solubility has to be avoided.

4. Ordered Mixing

Another way of getting around the problems of mixing is to use ordered mixing. The homogeneity in this type of mixing is of the same scale as the largest particle size. There is no random variation in the degree of mixing. Usually we can prevent demixing as well by using the methods available for ordered mixing. Ordered mixing can be accomplished by adsorption of sols, adsorption of ions, and surface reactions with metal alkoxides (Fig. 6).

**Spontaneous ordered mix
by size difference**

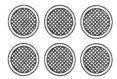   

**Ordered mix by reaction   Ordered mix by adsorption
with surface   of sol particles**

FIG. 6   Ordered mixing.

(a) Adsorption of Sols. A sol is a stable colloidal dispersion of particles in a liquid. The sol particles are much smaller than the particles of a normal ceramic powder. A sol is often stabilized by electrostatic stabilization. If the sol is added at a pH value where it is countercharged to the particles, it will adsorb on the particles. If we add a small quantity of sol to a powder and cause it to adsorb, we get an ordered mixture. Each powder particle will be covered with smaller sol particles. If the concentration of sol particles exceeds what is needed to cover the particle surfaces completely, the mixture will no longer be completely ordered. Flocculation of the sol must also be avoided throughout the mixing procedure. Flocculation of the sol can lead to larger inhomogeneities than normal mixing of powders.

Bostedt et al. [41] have shown how yttria could be added to silicon nitride as a sol. Positively charged yttria sol particles were adsorbed on the silicon nitride (Fig. 7). A maximum of 2 wt % yttria could be adsorbed on the silicon nitride. At the maximum yttria adsorption, the silicon nitride particles had a low absolute charge. The viscosity of the slip exhibited a minimum at this point. The homogeneity of the green bodies was better than in a body of the same composition fabricated by mixing powders.

(b) Adsorption of Ions. Ordered mixing can also be done by the adsorption of ions. Ions with an opposite charge adsorb on particles and tend to

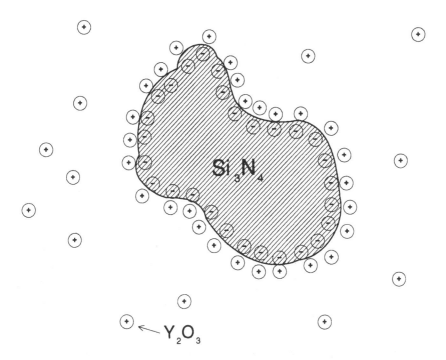

FIG. 7 Adsorption of yttria sol onto silicon nitride powder [41].

counsercharge them. In this way, small quantities of magnesium hydroxide can be adsorbed onto silicon carbide or alumina [42].

(c) Surface Reactions. A third way to achieve ordered mixing is to use surface reactions instead of adsorption. Alkoxides, such as aluminum iso-propylate, tend to react with most ceramic materials. The alkoxide reacts with the hydroxyl groups on the ceramic surface (Fig. 8). The hydroxyl group is present on oxides as well as on nitrides and carbides. This makes surface reaction possible with a number of alkoxides. Alkoxides can be synthesized from a large number of elements [43]. Many of the commercially available alkoxides, such as Si-, Mg-, Al-, Zr- and Ti-alkoxides, are interesting in a ceramic context.

Alkoxides can also be used to synthesize ceramic powder. Because the precursors can be mixed on a molecular scale, for example in a solution, the homogeneity of the final powders can be excellent [44].

Alkoxides can be used to add sintering aids to silicon nitride [45] or to manufacture sialons [46,47]. In such cases, the alkoxide is hydrolyzed and

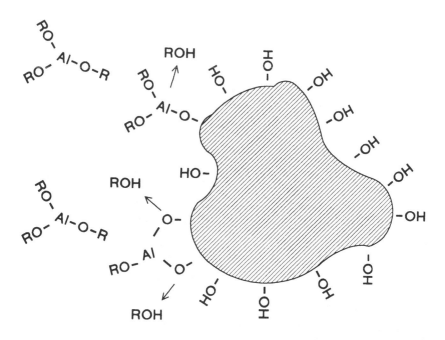

FIG. 8 Schematic drawing of the aluminum alkoxide reaction with powder surface.

reacts with the powder. By adding sintering aids in this manner, a homogeneous mixture can be formed thanks to the fine grained oxides that are produced during hydrolysis. If the alkoxide reacts with the particles and the excess is washed away, an ordered type of mixture is formed. If large amounts are added in this way, a thick porous layer with high specific surface area is created around each particle. Powders with these characteristics are very difficult to form and often require hot pressing to be sinterable.

By using small quantities of alkoxides, the surface of a powder can be covered with a single molecular layer of the preferred type. Controlled hydrolysis of the alkoxide can make the layer thicker than a monomolecular layer while preserving the ordered mixing [48]. The controlled hydrolysis can be performed at a level where the increase in specific surface area is marginal. This is also a method of modifying the surface characteristics of a ceramic powder. A powder treated with aluminum alkoxide will take on aluminalike surface properties.

V. FORMING

This part is an introduction to the forming methods for ceramics. These methods are described in greater detail in the respective chapters of this book, but they are summarized below.

A. Pressing

Two principal methods are used for pressing. In isostatic pressing the pressure is applied by a liquid through a flexible mold. In uniaxial pressing the pressure is applied through a rigid mold with one or two dies (Fig. 9). Powder is fed into the mold where pressure is applied and released. The compact is then removed from the mold.

Die pressing is often used for small components of relatively simple shapes. The major advantage is the fast forming cycle, and the major disadvantage is the relatively limited range of shapes that can be formed.

Isostatic pressing is used to generate rough shapes of small to large sizes. Green machining is often used to arrive at the final formed shape. Isostatic pressing together with green machining is well suited to prototype production, and it can also be automated for large scale production. One example of this is the production of insulators for spark plugs.

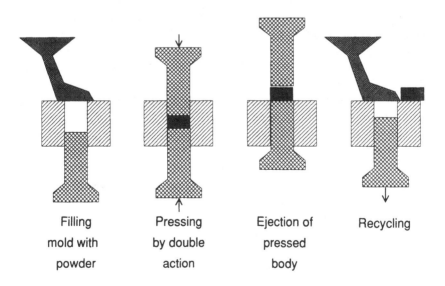

| Filling mold with powder | Pressing by double action | Ejection of pressed body | Recycling |

FIG. 9 Die pressing.

The quality of a pressed body depends greatly on the quality of the powder granulate. If granule boundaries remain after pressing, they cause defects in the sintered microstructure [49]. At the same time, the granules must be strong enough not to be broken down during storage, handling, and mold filling.

Production scale granulation is often done by spray-drying [50]. Prior to spray-drying the powder has to be dispersed in a well-concentrated slurry usually containing some pressing additives. The pressing additives function both as binder for the granules and, equally often, as lubricant during pressing. A low viscosity of the slurry at a high powder concentration is necessary to ensure high-density granules. Owing to the high shear forces in the nozzles used in spray drying, a shear thickening slurry cannot be used [51].

1. Granulation

To press a ceramic powder, a free flowing powder is preferred. It has to be free flowing to be able to be fed automatically through the feed shoe of the press. In less automatic types of pressing, such as isostatic pressing, a free flowing powder is still preferable if not necessary, because it is much easier to fill the mold to an even fill density with a free flowing powder. In order to make the powder free flowing it has to be granulated. The granule size usually has to be above 50 μm. Below this size the powder clogs and does not flow.

(a) Spray Drying. Ceramic powders can be granulated by spray drying in which the powder is dispersed in a liquid and then sprayed through an atomizer (often a small nozzle) into a hot air drying chamber. The droplets from the atomizer dry in the spraying chamber and form granules that are separated from the air in a cyclone.

(b) Sieve Granulation. Granulation can also be done by pressing a mixture of the powder and a small amount of liquid through a screen. This can be done without dispersing the powder and requires simpler equipment. The problems with this type of granulation involve uneven distribution of the moisture in the granules and contamination from wear of the screen.

(c) Spray Granulation. Spray granulation is not often used with high-performance ceramic materials as it requires a larger particle size. It involves spraying moisture into a fluidized bed of the powder.

(d) Freeze Drying. A powder suspension can also be sprayed into liquid nitrogen and frozen into granules. The frozen granules are then freeze dried. No segregation of material takes place during freeze drying, as the frozen liquid is transported in gas phase by sublimation. The granules do not shrink during drying, and a more porous granulate is formed.

2. Additives for Pressing

Additives are often used during pressing to give the pressed body higher green strength or to lubricate the powder and mold during pressing. When a dispersion is prepared for spray drying, these additives have to be present. A stable suspension with no tendencies toward shear thickening has to be prepared for spray drying. A shear thickening suspension clogs the nozzle at the high shear rates present in the atomizer. Pressing additives can sometimes complicate the making of a spray drying dispersion, as they tend to increase the viscosity of the dispersion and migrate during drying.

B. Slip Casting

Slip casting consists of several steps (Fig. 10). The first step is the preparation of a low-viscosity slip with a high solid content. The slip is then poured into a mold (usually a plaster mold). During slip casting, a solid body is formed along the walls of the mold. If drain casting is used, excess slip is then poured out of the mold. The cast body solidifies as water is removed from the slip. When the solid body is formed, it is initially dried in the mold to improve its green strength. It is then removed from the mold and the residual moisture is evaporated by slowly drying the formed body. Defects can be created in the slip cast body during any one of these steps.

A special form of slip casting is tape casting. Slip is made to form a thin tape with the help of a sharp blade. Organic binders are usually added to increase the green strength of the tape. Tape casting is used by the electronics industry to manufacture chip carriers and multilayer capacitors.

The same type of lost wax technique used for precision casting of metals can be used for slip casting. This, together with drain casting and molds made from several parts, makes it possible to cast an infinite variety of shapes using slip casting. The advantage of slip casting is the relative simplicity with which prototypes or small series can be made, since plaster molds are cheap and easy to manufacture.

The main disadvantage is imprecision in the formed shape, caused by corrosion of the plaster molds (in water systems). Another disadvantage is the relatively time-consuming casting of fine powders. These disadvantages can be overcome by pressure casting in polymer molds, but this also increases the cost of the procedure. The difficulties of finding proper dispersing agents and procedures for new slip casting systems can also be a disadvantage in some cases.

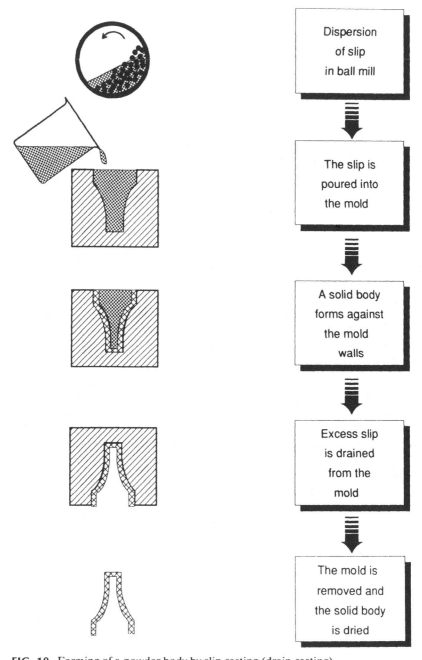

FIG. 10 Forming of a powder body by slip casting (drain casting).

Slip casting is based on the dispersion of a powder in a liquid. The quality of the green body formed depends on how well this dispersion is prepared [52]. Drying cracks, pores from air bubbles in the dispersion, segregation of mixed components, and density gradients are all controlled or at least partly controlled by the colloidal properties of the slip.

The powder can be treated prior to slip formation to remove agglomerates. One way of doing this is by using sedimentation. Sedimentation theoretically removes *all* large particles or agglomerates above a certain size. Sedimentation is performed at low concentration of solids, where the settling rate for spherical particles can be calculated using Stokes' law (Fig. 11).

When the concentrated slip is prepared, sedimentation can no longer be used. Sieving can still be performed at this stage. Even if only the larger sizes of agglomerates can be removed, sieving immediately before slip casting has the advantage of removing defects introduced during the slip preparation [53].

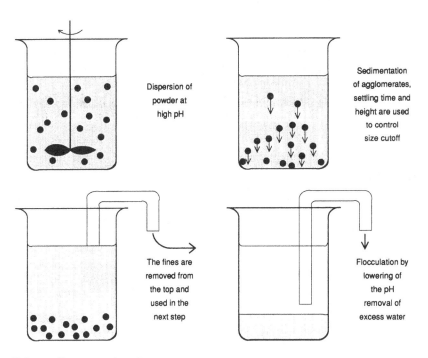

FIG. 11 Removal of hard agglomerates by dispersion and sedimentation.

During the first stage of drying, the constant rate period, the drying body shrinks. At the same time, the body transfers from a plastic to a brittle behavior. If the concentration of solids in the slip is low, the shrinkage will be excessive, and drying cracks are likely to occur. This can also take place if the drying shrinkage is obstructed by the shape of the mold. The drying rate outside the mold has to be kept low to ensure that no cracking occurs. The moisture has to be removed completely prior to sintering.

C. Injection Molding

Ceramics can also be formed using methods normally used for polymers. The ceramic powder is mixed with a polymer binder that gives the required plasticity. The major component of the binder is usually a thermoplastic polymer or a wax. After mixing, the powder-polymer mix is injection molded (Fig. 12). The injected part usually contains between 35 and 50 vol % organic binder. The polymer must be removed before sintering. This is done by heating in air or in an inert atmosphere. The heating must be done very slowly and controlled in order not to create cracks or blisters in the material.

Complicated shapes can be made with good precision by injection molding. It is also easy to automate the method for large scale production. The disadvantages are that molds and injection molding equipment are expensive and that binder removal is a comparatively slow process.

In order to be able to injection mold a ceramic powder, processing additives have to be added. These decrease the viscosity of the injection mold-

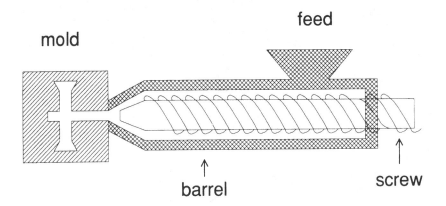

FIG. 12 Schematic drawing of an injection molding machine with reciprocating screw.

ing batch and enable higher solids contents. These additives interact with the surface of the powder and thus reduce the aggregation tendency or the viscosity of the polymer matrix.

Producing parts without defects with injection molding demands a complex interaction between the polymer-powder mix, the machine parameters, and the mold design [54–57]. Machine parameters include the temperatures of the different zones of the injection molder and mold, pressure and flow cycle during molding, clamp pressure of the mold, and holding time before demolding. Important parameters in mold design include location and diameter of the gate and mold opening angle.

The terms of colloid chemistry are not used often in the literature on the injection molding of ceramics. Nevertheless, a powder is dispersed in a molten polymer and has to be lubricated both in terms of particle-particle contact and the particle-mold surface contact. These processes are not yet as well understood as for example the processes that take place during slip casting, but they are nevertheless surface phenomena on particles with a large specific surface area. Thus colloidal forces are at play. As described above, the rheology during injection molding, to a large extent, controls the final properties of the injection molded part. Weld lines, density gradients, and internal voids can be caused by improper rheology.

A good polymer-ceramic powder mix has low viscosity, pseudoplastic behavior, viscosity that is insensitive to temperature variations, and high concentrations of solids. Dispersing or lubricating agents decrease viscosity and are thus useful even if perfect stability is not explicitly desired. The desired goal is low viscosity at high solids concentration rather than colloidal stability per se.

D. Extrusion

The same type of polymer-ceramic mixes used for injection molding can also be extruded. This method is used to form pipes or honeycomb structures. Extrusion can also be done in water systems. In this case a water soluble polymer binder is added to increase plasticity during extrusion and to improve the green strength of the extruded material.

VI. FUNCTIONAL CERAMICS

Many ceramics are used in applications where electric, piezoelectric, magnetic, and optical properties are the main reason for using the material. These types of ceramics are sometimes called functional ceramics. The main properties of functional ceramics are not controlled by single defects but rather by the average microstructure. Control of grain size, grain orientation, grain boundary composition, and homogeneity are necessary to obtain the desired properties. Many of these factors can be controlled using the same surface or colloid chemistry methods described here for use in the processing of high-strength ceramics.

REFERENCES

1. A. G. Evans, in *Fracture Mechanics of Ceramics*, Vol. 1 (R. C. Bradt, D. P. H. Hasselman, and F. F. Lange, eds.). Plenum Press, New York, 1974, pp. 17–48.
2. S. B. Batdorf, in *Fracture Mechanics of Ceramics*, Vol. 3 (R. C. Bradt, D. P. H. Hasselman, and F. F. Lange, eds.). Plenum Press, New York, 1978, pp. 1–30.
3. M. V. Swain, in *Fracture Mechanics of Ceramics*, Vol. 5 (R. C. Bradt, A. G. Evans, D. P. H. Hasselman, and F. F. Lange, eds.). Plenum Press, New York, 1983, pp. 355–370.
4. D. B. Marshall and A. G. Evans, *J. Am. Ceram. Soc. 68*(5): 225–231 (1985).
5. K. Kendall, N. McN. Alford, and J. D. Birchall, in *Special Ceramics 8, British Ceramic Proceedings 37* (S. P. Howlett and D. Taylor, eds.). Institute of Ceramics, Stoke-on-Trent, 1986, pp. 255–265.
6. H. T. Larker, in *Proceedings of International Symposium on Ceramic Components for Engine* (S. Somiya, E. Kanai, and K. Ando, eds.). KTK Scientific Publishers, Tokyo, 1983, pp. 304–310.
7. J. Kellet and F. F. Lange, *J. Am. Ceram. Soc. 71*(1): 7–12 (1988).
8. F. F. Lange, *J. Am. Ceram. Soc. 72*(1): 3–15 (1989).
9. R. W. Rice, in *Fracture Mechanics of Ceramics*, Vol. 1 (R. C. Bradt, D. P. H. Hasselman, and F. F. Lange, eds.). Plenum Press, New York, 1974, pp. 323–345.
10. J. J. Mecholsky, S. W. Freiman, and R. W. Rice, *J. Mater. Sci. 11*(7): 1310–1319 (1976).
11. F. F. Lange, B. I. Davis, and E. Wright, *J. Am. Ceram. Soc. 69*(1): 66–69 (1986).
12. K. Linde, M. Persson, R. Pompe, S. Karlsson, and R. Carlsson, in *EuroCeramics Vol. 3: Engineering Ceramics* (G. de With, R. A. Terpstra, and R. Metselaar, eds.). Elsevier, London, 1989, pp. 3.126–3.130.
13. N. McN. Alford, J. D. Birchall, and K. Kendall, *Nature 330*(6143): 51–53 (1987).
14. A. G. Evans, *J. Am. Ceram. Soc. 73*(2): 187–206 (1990).
15. K. T. Faber and A. G. Evans, *Acta Metall. 31*(4): 565–584 (1983).
16. A. G. Evans and A. H. Heuer, *J. Am. Ceram. Soc. 63*(5–6): 241–248 (1980).
17. P. L. Swanson, C. J. Fairbanks, B. R. Lawn, Y. W. Mai, and B. J. Hockey, *J. Am. Ceram. Soc. 79*(4): 279–289 (1987).
18. R. W. Steinbrech and A. H. Heuer, *Mater. Res. Soc. Symp. Proc. 60*: 469–481 (1986).
19. I. B. Cutler, in *Ceramic Processing Before Firing* (G. Y. Onoda, Jr., and L. L. Hench, eds.). John Wiley, New York, 1978, pp. 21–29.

20. K. Haberko, *Ceramics Int.* 5: 148 (1979).
21. D. E. Niesz and R. B. Bennett, in *Ceramic Processing Before Firing* (G. Y. Onoda, Jr., and L. L. Hench, eds.). John Wiley, New York, 1978, pp. 61–73.
22. J. S. Reed, T. Carbone, C. Scott, and S. Lukasiewicz, in *Processing of Crystalline Ceramics* (H. Palmour III, R. F. Davis, and T. M. Hare, eds.). Plenum Press, New York, 1978, pp. 171–180.
23. C. A. Bruch, *Am. Ceram. Soc. Bull.* 41(12): 799–806 (1962).
24. W. H. Rhodes, *J. Am. Ceram. Soc.* 64(1): 19–22 (1981).
25. F. F. Lange, *J. Am. Ceram. Soc.* 66(6): 396–398 (1983).
26. F. F. Lange and M. Metcalf, *J. Am. Ceram. Soc.* 66(6): 398–406 (1983).
27. F. F. Lange and B. I. Davis, *J. Am. Ceram. Soc.* 66(6): 407–408 (1983).
28. E. A. Barringer and H. K. Bowen, *J. Am. Ceram. Soc.* 65(12): C199–C201 (1982).
29. I. Aksay, in *Ceramic Transactions Vol. I: Ceramic Powder Science II, B* (G. L. Messing, E. R. Fuller, Jr., and H. Hausner, eds.). American Ceramic Society, Westerville, Ohio, 1987, pp. 663–674.
30. F. F. Lange, *J. Am. Ceram. Soc.* 67(2): 83–89 (1984).
31. I. A. Aksay, in *Advances in Ceramics Vol. 9: Forming of Ceramics* (J. A. Mangels and G. L. Messing, eds.). American Ceramic Society, Columbus, Ohio, 1983, pp. 94–104.
32. A. Roosen and H. K. Bowen, *J. Am. Ceram. Soc.* 71(11): 970–977 (1988).
33. F. F. Lange and B. J. Kellet, *J. Am. Ceram. Soc.* 72(5): 735–741 (1989).
34. D. V. Miller and J. S. Reed, in *Ceramic Transactions Vol. 1: Ceramics Powder Science II, B* (G. L. Messing, E. R. Fuller, Jr., and H. Hausner, eds.). American Ceramic Society, Westerville, Ohio, 1987, pp. 733–740.
35. P. F. Messer, *Trans. J. Br. Ceram. Soc.* 82: 156–162 (1983).
36. F. F. Lange, U.S. Patent 4,624,808, (1986).
37. E. Carlström and F. F. Lange, *J. Am. Ceram. Soc.* 67(8): C169–C170 (1984).
38. F. F. Lange and M. M. Hirlinger, *J. Mater. Sci. Lett.* 4: 1437–1441 (1985).
39. F. F. Lange and K. T. Miller, *J. Am. Ceram. Soc.* 70(12): 896–900 (1987).
40. H. R. Baumgarten and B. R. Rossing, in *Ceramic Transactions, Silicon Carbide '87*, Vol. 2. American Ceramic Society, Westerville, Ohio, 1989.
41. E. Bostedt, M. Persson, E. Carlström, and R. Carlsson, Paper VII in *Defect Minimisation in Silicon Carbide, Silicon Nitride and Alumina Ceramics.* Dissertation by Elis Carlström, Chalmers University of Technology, 1989.
42. R. J. Pugh and L. Bergström, *J. Colloid Interface Sci.* 124(2): (1988).
43. D. C. Bradley, R. C. Mehrotra, and D. P. Gaur, *Metal Alkoxides.* Academic Press, London, 1978, Chs. 1–4.
44. K. S. Mazdiazni, in *Better Ceramics Through Chemistry, Mat. Res. Soc. Symp. Proc.*, Vol. 32 (J. Brinker, D. Clark, and D. Ulrich, eds.). Elsevier, 1984, pp. 175–186.
45. T. Mah, K. S. Mazdiyasni, and R. Run, *Am. Ceram. Soc. Bull.* 54(9): 840–844 (1979).
46. K. Kishi, S. Umebayashi, E. Tani, and K. Kobayashi, *Yokyo-Kyukai-Shi* 93(10): 629–635 (1985).

47. K. Kishi, S. Umebayashi, E. Tani, and K. Kobayashi, *Yokyo-Kyukai-Shi* 94(1): 179–192 (1986).
48. E. Bostedt, M. Persson, and R. Carlsson, *Euro-Ceramics Vol. 1: Processing of Ceramics* (G. de With, R. A. Terpstra, and R. Metselaar, eds.). Elsevier, London, 1989, pp. 1.140–1.144.
49. S. J. Lukasiewicz and J. S. Reed, *Am. Ceram. Soc. Bull.* 57(9): 798–801, 805 (1978).
50. S. J. Lukasiewicz, *J. Am. Ceram. Soc.* 72(4): 617–624 (1989).
51. J. S. Reed, *Introduction to the Principles of Ceramics Processing.* John Wiley, New York, 1988, p. 24.
52. G. W. Phelps and M. G. McLaren, in *Ceramic Processing Before Firing* (G. Y. Onoda, Jr., and L. L. Hench, eds.). John Wiley, New York, 1978, pp. 211–225.
53. D. J. Shanefield and R. E. Mistler, *Am. Ceram. Soc. Bull.* 55: 213 (1976).
54. M. J. Edirisinghe and J. R. G. Evans, *Int. J. High. Tech. Ceramics* 2(1): 1–31 (1986).
55. M. J. Edirisinghe and J. R. G. Evans, *Int. J. High. Tech. Ceramics* 2: 249–278 (1986).
56. M. J. Edirisinghe and J. R. G. Evans, *Br. Ceram. Trans. J.* 86: 18–22 (1987).
57. J. G. Zhang, M. J. Edirisinghe, and J. R. G. Evans, *Industrial Ceramics* 9(2): 72–82 (1989).

2

The Chemical Synthesis of Ceramic Powders

RICHARD E. RIMAN   Rutgers University, Piscataway, New Jersey

I. INTRODUCTION

The purpose of this chapter is (1) to familiarize the reader with the benefits of processing with an ideal (chemically derived) powder versus a conventionally processed multicomponent mixture and (2) to describe various technologies for controlling powder characteristics for a spectrum of oxide and nonoxide ceramic materials.

II. THE NEED FOR CHEMICALLY DERIVED POWDERS

Ideal physicochemical characteristics of a ceramic powder include submicron particle size, controlled particle size distribution, uniform and equiaxed shape, minimum degree of agglomeration, high degree of chemical purity, controlled chemical and phase homogeneity, maximized bulk particle density, minimum weight loss upon heating, and controlled surface chemistry. The importance of these characteristics for the processing of controlled and reliable ceramic materials is discussed in other chapters. However, it is worthwhile to discuss here the differences between multicomponent powders prepared via conventional and synthetic routes in order better to appreciate the differences in multicomponent phase reaction kinetics as well as resultant homogeneity.

A. Mixing

Conventional processing to prepare multicomponent powders involves the three consecutive steps of *mixing*, *solid state reaction*, and *milling*. Powders can be mixed either in a structured (selective) fashion or randomly. In the structured case, mixing is achieved on a size scale similar to that of the mean diameter of the powders employed. Heterocoagulation approaches using electrostatic stabilization-destabilization have been successful in achieving this type of mixing [1-5].

In the random case, the composition of nearby particles coordinating a specific particle is based on probability. Using probability theory and chemical characterization tools, the frequency of observing a composition that coincides with the average composition can be predicted for a two-component mixture by the relation [6]

$$P(n_1, n) = \frac{n!}{n!(n-n_1)!} X^n (1-X)^{n-n_1} \tag{1}$$

where X is the number fraction of component 1 in the entire mixture, n_1 is the number of particles of component 1 in the sample, n is the number of particles in the sample, and $P(n_1, n)$ is the probability of seeing a sample with n_1. The randomness of a mixture can be determined with the use of sampling statistics that allow comparisons between experimental histograms and ones based on probability.

Assuming the mixture is random, is the resultant homogeneity scale acceptable? Envisioning the mixture as an equiaxed monodisperse packed bed consisting of two components of equal density, Hogg [6] has derived the relation

$$d = \sigma^{2/3} D \left[\frac{1-\varepsilon}{X(1-X)} \right]^{1/3} \tag{2}$$

where σ is the expected variance for the chosen homogeneity scale, D is the homogeneity scale, ε is the porosity of the powder bed, and d is the particle size. If powder components 1 and 2 having a particle size of 0.4 μm and X = 0.5 composition are randomly mixed such that the standard deviation is 0.5%, a homogeneity scale of 10 μm results. Smaller homogeneity scales require finer particle sizes. In order to achieve a homogeneity scale below 1 μm, nanometer-size particles are required. This task is challenging, since nanometer-size particles agglomerate due to significant adhesional forces. These attractive forces are difficult to overcome during a conventional mixing process.

The problems associated with random mixing can be compounded by segregation mechanisms whose prevalence depends on the distribution of powder characteristics in the mixture such as shape, density, and electrical, magnetic, or surface properties. For instance, on the basis of size, mixtures of particles larger than 50 μm demix easily due to gravity and percolation mechanisms. In contrast, particle sizes finer than 1 μm are less susceptible to these types of segregation mechanisms, since adhesional forces significantly hamper their motion.

B. Solid State Reactions

Solid state reaction is the next step in the formation of multicomponent phases. Salt decomposition and reactions between solids play a role in conventional processing, with reaction kinetics governing the rate at which these processes occur. For example, $Mg(OH)_2$ decomposes to MgO according to the expression [7]

$$(1-\alpha)^{1/2} = 1 - \frac{kt}{r_0} \tag{3}$$

where α is the fraction of $Mg(OH)_2$ reacted, k is the rate constant, t is the reaction time, and r_0 is the initial particle size. Particles of zinc oxide and alumina react to form $ZnAl_2O_4$, as described by the Carter equation

$$[1 + (Z-1)\alpha]^{2/3} + [Z-1] [1-\alpha]^{2/3} = Z + \frac{(1-Z)tkD}{r_0^2} \tag{4}$$

where Z is the ratio of volume formed to volume consumed and D is the diffusion coefficient of the rate-determining species. Thus these two models show that, for both decomposition and interdiffusion solid state reactions, fine particles accelerate the reaction kinetics. Further evidence for this phenomenon could be provided by examining many other solid state reaction models, but such examination is beyond the scope of this discussion.

As the above models imply, reaction mechanism and the rate-controlling step are important factors that are critical to understand when phase pure multicomponent powders are desired. For instance, in the preparation of conventionally derived barium titanate, TiO_2 (anatase) can be reacted with $BaCO_3$. These components undergo the reaction path [8]

$$2BaCO_3 + TiO_2 = 2BaCO_3 \cdot TiO_2 \tag{5}$$

$$2BaCO_3 \cdot TiO_2 = Ba_2TiO_4 + 2CO_2 \tag{6}$$

$$Ba_2TiO_4 + TiO_2 = 2BaTiO_3 \tag{7}$$

This is an illustration of how conventional solid state reactions form reactive intermediates. Elimination of these intermediates is dependent not only on kinetic factors (described earlier) and thermodynamic considerations but also on the homogeneity scale of the powder mixture. In conventional processing of multicomponent ceramics, this scale commonly is larger than a viable diffusion distance for the rate-determining species. Furthermore, this effect is magnified when multicomponent materials require several reaction steps in order to form.

C. Milling

Conventional solid state reactions to prepare single- and multicomponent powders typically result in the formation of aggregates (hard agglomerates) that require comminution processes to reduce the particle size to the micrometer level. This process not only produces asymmetric particles but also introduces impurities from milling media.

Given the many challenges associated with commercial powders (reviewed above), the importance of chemically derived uniform multicomponent submicron powders is more easily appreciated. Atomic mixing, inherent in chemically derived powders, avoids the large concentration gradients present in conventionally processed systems and facilitates precise and uniform compositional control of major and minor constituents of the powder. When this powder is an unstable metal salt, the fine particle sizes that are possible via chemical preparation promote rapid decomposition kinetics. In those cases where intermediate phases can form, the size scale accelerates reaction kinetics by minimizing the diffusion distance for solid state reaction. Chemical impurities are minimized not only by the highly purified state of the powders but also because the need for milling is minimized or eliminated.

III. POWDER SYNTHESIS TECHNOLOGY

Chemically derived ideal multicomponent powders can be synthesized in a variety of ways. This review encompasses technologies that utilize chemistry in the liquid phase, chemistry between heterogeneous phases, chemistry in a droplet, and chemistry in the vapor phase.

A. Chemistry in the Liquid Phase

Chemistry in the liquid phase is an approach in which the physical characteristics of a ceramic powder are controlled by nucleation, growth, and aging processes in solution. In these systems, precipitation is driven by chemical supersaturation using one of the following synthetic methods: direct strike precipitation, solvent removal, nonsolvent addition, or precipitation from homogeneous solution.

1. Direct Strike Precipitation

Direct strike precipitation, commonly employed for a spectrum of materials, involves the addition of a precipitating agent to a solution of soluble metal salts to form a sparingly soluble salt. When the soluble salts are added to a solution containing the precipitating agent, the procedure is referred to as a reverse strike process.

In direct strike precipitation, copious precipitation ensues at a rate faster than the time scale required for mixing of the reagents (Fig. 1). For example, in the preparation of lanthanum-modified lead zirconate titanate [9], the nitrates of lead and lanthanum, zirconium oxychloride, titanium tetrachloride, and hydrogen peroxide are dissolved in water at 40°C. Precipitation is induced by the addition of ammonium hydroxide to this solution to maintain a pH of ~9. The precipitate, most likely a mixture of various hydroxides, is filtered from suspension, washed free of soluble species with hot distilled water, and dried at 100°C. Calcination at a temperature of 650°C is necessary to form the oxide, followed by subsequent milling to break up aggregates formed during precipitation and calcination.

Nonoxides can be prepared by direct strike precipitation, but the methods require a good background in synthetic chemistry. Aluminum hydride can be reacted with ammonia at room temperature in tetrahydrofuran to yield a precipitate ($Al(NH)NH_2$) that can be calcined at 1100°C to form aluminum nitride [10]. Silicon diimide can be prepared by

$$SiCl_4 + 18NH_3 = [Si(NH)_2]_x + 4NH_4Cl \cdot 3NH_3 \qquad (8)$$

Submicron silicon nitride (95% α-Si_3N_4/5% β-Si_3N_4) has been prepared by pyrolysis of silicon diimide $[Si(NH)_2]_x$ at temperatures of 1200–1700°C [11–

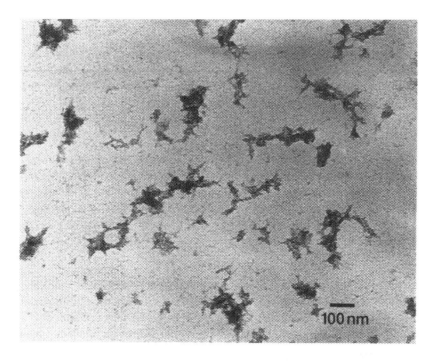

FIG. 1 Alkoxide-derived strontium titanium hydrous oxide prepared by a direct strike precipitation method.

13]. Boron trichloride (BCl$_3$) can be reacted with an ammonia-in-benzene solution at $-75°C$ via a direct strike reaction to form a precipitate that can be pyrolyzed in vacuo at $1200-1400°C$ to form BN [14,15]. With these latter two reactions, ammonium salts that coprecipitate with the desired product must be eliminated via washing processes prior to pyrolysis of the precipitate.

The common ion effect and differential precipitate solubilities are factors that can lead to chemical and phase heterogeneities in direct strike–derived precipitate. The common ion effect can cause solution homogeneity problems in multicomponent solutions when one of the soluble salts has an anion that renders one of the components insoluble. Differential solubilities of the various precipitating phases affect the precipitation kinetics of each component, causing the stoichiometry of the precipitate to vary with reaction time. The copious nature of a direct strike precipitation creates concentration gradients in solution that can further accentuate this effect.

Differential solubility problems can be reduced by introducing a mixture of anions that renders all cations highly insoluble. For instance, manganese-zinc ferrites can be prepared by additions of ammonium hydroxide (NH_4OH) and ammonium bicarbonate (NH_4CO_3) to metal sulfate solutions to prepare hydroxide-carbonate coprecipitates [16]. When ammonia is able to complex metal ions such as copper, as in the case of $Bi_{1.6}Pb_{0.3}Sb_{0.1}Ca_2Sr_2Cu_3O_{10}$ superconductors, alkyl ammonium hydroxide/alkyl ammonium carbonate solutions are used instead to form a homogeneous stoichiometric coprecipitate [17].

To minimize segregation during direct strike precipitation and optimize control of the stoichiometry, multicomponent salts of fixed stoichiometry can be precipitated. $BaTiO_3$ with a Ba:Ti molar ratio of 0.995 ± 0.005 has been prepared from calcination of $BaTi(C_6H_6O_7)_3 \cdot 6H_2O$ [18]. In order to prepare this salt, titanium tetrabutoxide, citric acid, and barium carbonate are mixed in aqueous solution in a pH range of 0.5 to 2.6. At a pH of 3.2, a $Ba_2Ti(C_6H_6O_7)_2 \cdot 7H_2O$ salt forms instead.

Multicomponent salts such as oxalates have been prepared for non-stoichiometric oxides such as $Sr_{0.33}Ba_{0.65}TiO_3$, $Ba_{0.91}Pb_{0.09}TiO_3$, or $Ba_{1.00}Ti_{0.87}Zr_{0.12}O_3$ [19]. However, stoichiometries are restricted to limited ranges, and many times the solution composition does not correspond to that of the precipitate. Saegusa, Rhine, and Bowen [20] have addressed this problem for lead-substituted barium titanyl oxalates by utilizing cation exchange of soluble $(NH_4)_2TiO(C_2O_4)_2$ with Ba^{2+} and Pb^{2+} acetate solutions to form an oxalate precipitate with the desired stoichiometry.

Other undesirable effects such as coprecipitated anion impurities (e.g., nitrates, chlorides, and sulfates) plague direct strike precipitation. These impurities can be incorporated into the particles as lattice defects or can adsorb onto powder surfaces. In addition, soluble species in solution can become occluded within agglomerates of particles. While such impurities are difficult to remove with a washing process, failure to remove these ions may affect later stages of ceramics processing such as sintering or resultant properties of the ceramic material.

Proper choice of the soluble anions in solution is important in order to minimize the above problems. Anions such as nitrates, which decompose at low temperatures, are more desirable since they can be eliminated from the powder during calcination. Anion contamination problems can be avoided with the use of anionless precursors such as metal alkoxides. Metal alkoxides react with water to precipitate *ceramic polymers* by hydrolysis:

$$M(OR)_4 + H_2O = (OR)_3M-OH + ROH \qquad (9)$$

and polymerization:

$$(OR)_3M-OH + HO-M(OR)_3 = (OR)_3M-O-M(OMe)_3 + H_2O \qquad (10)$$

$$(OR)_3M-OH + RO-M(OR)_3 = (OR)_3M-O-M(OR)_3 + ROH \qquad (11)$$

In many cases, a nearly fully cross-linked oxide (e.g., silica) can result, subsequently forming a hydrous oxide. However, formation of various types of hydrates such as AlO(OH) or $Sr(OH)_2$ that cannot fully or even partially polymerize according to reactions 10 and 11 is also possible.

Metal alkoxides are useful for preparing multicomponent solutions since they exhibit excellent solubility in organic solvents such as alcohol, toluene, or tetrahydrofuran. For instance, the double alkoxide $MgAl_2(O-i-Pr)_8$ can be hydrolytically decomposed to form a coprecipitate of hydroxides and calcined at 800°C to prepare submicron spinel $(MgAl_2O_4)$ [21,22]. This stable double alkoxide is prepared by complexation between a magnesium and an aluminum alkoxide [23,24]:

$$Mg(O-i-Pr)_2 + 2Al(O-i-Pr)_3 = MgAl_2(O-i-Pr)_8 \qquad (12)$$

Alkoxide complexation has been used to make a variety of oxides such as mullite [25] or yttria-stabilized zirconia [26]. When a multiphase coprecipitate is undesirable, condensation complexes can be prepared by partial hydrolysis routes [27]:

$$M(OR)_4 + H_2O = (OR)_3M-OH + ROH \qquad (13)$$

$$(OR)_3M-OH + RO-M'(OR)_3 = (OR)_3M-O-M'(OR)_3 + ROH \qquad (14)$$

This approach has been utilized widely in sol-gel processing [e.g., Ref. 28] and in the preparation of multicomponent ceramic powders such as cordierite $(2MgO \cdot 2Al_2O_3 \cdot 5SiO_2)$ with no evidence of segregation [29].

The literature contains many examples of alkoxides complexed with other precursors such as inorganic or metal organic salts to prepare useful precursors for powders [e.g., Ref. 30]. Dosch [31] reacted barium hydroxide with titanium isopropoxide in methanol to prepare a soluble complex as a precursor for barium titanate. Gurkovitch and Blum [32] have reacted lead acetate and titanium isopropoxide in methoxymethanol as a precursor for lead titanate.

2. Solvent Removal and Nonsolvent Addition

Sol-gel methods generally are useful for preparing powders, since the gels can be broken up into granules during the drying step or ground in the dry state to produce granules of high surface area material [33,34]. Supercritical CO_2 extraction, in which an overpressure of supercritical CO_2 is applied to extract the solvent from the gel, is another route for powder preparation [35]. A constant flow of supercritical CO_2 is maintained until alcohol can no longer be seen separating from the CO_2 exiting the autoclave. Solvent is removed from the autoclave by depressurizing the gel at the rate required to maintain the temperature above the critical temperature of the fluid while drying the gel. Single-component (silica, titania, etc.) and multicomponent powders (aluminum titanates, aluminosilicates, etc.) with surface areas as high as 950 m^2/g and particle sizes less than 1 μm have been prepared [36].

In addition to solvent removal techniques, there are also direct precipitation methods based on addition of a nonsolvent. In this case, an aqueous solution is prepared and precipitation is induced by dilution of this solution with a hydrophilic nonsolvent. Soluble salts employed include sulfates, citrates, and nitrates, depending on the assortment of cations desired. Alcohol (ethanol, isopropanol, etc.) or acetone typically is employed as the nonsolvent. Proper choice of the salt concentration, solution environment (pH, addition of complexing agents), and nonsolvent is essential toward achieving the proper stoichiometry and satisfactory homogeneity. Materials such as barium titanate, lead zirconate, magnesium aluminate, manganese ferrites, etc., can be prepared [37–39]. This method and other approaches such as freeze drying and evaporation are commonly employed in chemistry in a droplet approaches as discussed later in this chapter.

3. Drawbacks of the Above Technologies

In most liquid phase approaches (including all of the cases described above), the precipitation product must be calcined to form the oxide and thus is subjected to the same types of problems inherent in conventional processing. Reduced particle size and greater homogeneity accelerate the reaction kinetics, as mentioned earlier, and resultantly lower calcination temperatures are observed. However, the precipitate also becomes a great deal more aggregated (Fig. 2). Thus in many cases, a milling step is required, allowing impurities to be introduced into the powder.

In a few cases, direct strike precipitation can result in the direct formation of oxides. Kiss et al. [40] hydrolyzed titanium alkoxides with an aqueous solution of barium hydroxide, producing submicron agglomerates of ~10 nm $BaTiO_3$. Mazdiyasni et al. [41] accomplished the same result by hydrolyzing a barium-titanium double alkoxide solution. The use of anions such as the tungstate, chromate, or molybdate anions is useful for preparation of oxides directly from solution [42]. For instance, barium chromate can be precipitated from a solution of potassium chromate and barium chloride [43]. Oxides also can be directly precipitated from solution via oxidation reactions. Basic aqueous solutions of iron nitrates or sulfates can be reacted with oxygen to prepare polymorphs of Fe_2O_3 and other multicomponent ferrites (Mn-, Zn-, or Co-) at temperatures as low as 70°C.

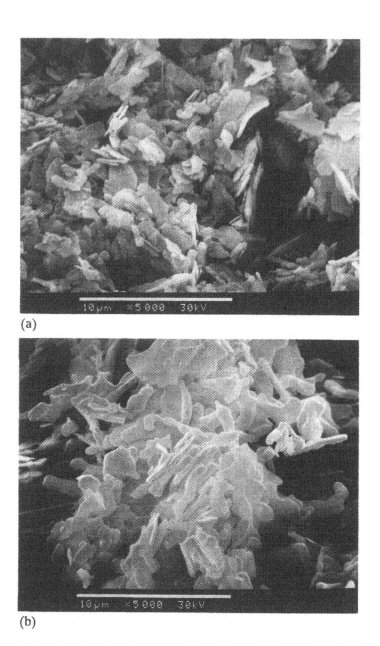

(a)

(b)

FIG. 2 Lead oxalate (a) as precipitated from solution and (b) calcined at 600°C in air for 2 h to form lead oxide.

Despite advances in direct strike precipitation technology, the inability to avoid nonuniformities in solution hinders control of physical characteristics of the powder. For example, high particle number densities and high electrolyte concentrations encountered during direct strike precipitation lead to a low degree of colloid stability and resultantly high degree of agglomeration. As growth and aging proceed, aggregates (hard agglomerates) are formed. Efforts are underway to determine how the degree of aggregation caused by direct strike precipitation methods can be minimized. Voigt [44] has used a population balance model to explain how agglomerate/aggregate properties are controlled by precipitation pH. Anzai [45] has reported the use of polyethylene glycol as an in situ dispersant to precipitate disperse 10 nm yttrium-stabilized zirconia using a direct strike approach.

4. Precipitation from Homogeneous Solution

Precise control of physical powder characteristics requires uniform supersaturation in solution. When this is achieved, nucleation precedes growth and aging, and these processes occur uniformly throughout the reaction medium. Controlled precipitation is best accomplished using a precipitation from homogeneous solution (PFHS) technique. This approach utilizes chemical reactions whose kinetics somehow rate limit the release of supersaturating species. A wide variety of techniques are available, as discussed in a text by Gordon et al. [46]. Only a small portion of the available techniques are discussed below.

When considering PFHS techniques for ceramics processing, it should be kept in mind that many of these methods have been developed to precipitate filterable coarse-grained solids with well-defined chemical compositions for quantitative analysis. When interest is shifted to control of submicron particle size, size distribution, and morphology, special attention must be given to concentration, pH, heating rate, temperature, corresponding anion of the soluble metal salt, choice of precipitating reaction, and aging conditions. For monodisperse powders (Fig. 3), precipitation must occur at concentrations typically less than 0.05 M and at a pH near the solubility limit of the precipitate. At higher concentrations, increased morphology variation and agglomeration result. These phenomena limit the applicability of this technology in conventional ceramics processing.

Furthermore, while the important role played by chemical complexation in achieving controlled supersaturation and colloid stability is well recognized [48–50], a detailed understanding of all the factors (e.g., solution nonidealities, adsorption phenomena, ionic strength) is lacking at present. Thus each system must be investigated empirically. For instance, the use of chloride versus nitrate solutions to precipitate α-Fe_2O_3 under otherwise

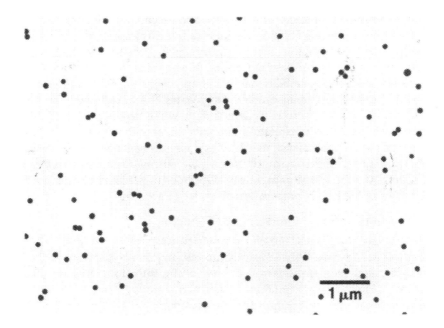

FIG. 3 Monodisperse lead hydroxycarbonate prepared by a precipitation from homogeneous solution method.

identical conditions yields elliptical versus star-shaped particles [50]. An understanding of the role of these *inert* anions in controlling the morphology might enable prediction of the morphologies of other systems.

There usually are a variety of ways to prepare the same precipitate. The most commonly used PFHS technique employs a soluble salt codissolved with a precursor that decomposes upon heating, releasing an anion that quantitatively precipitates the dissolved cation. For instance, zinc oxalate dihydrate can be prepared in aqueous solution by formamide decomposition (pH increase) in a zinc nitrate–nitric/oxalic acid solution:

$$Zn(NO_3)_2 = Zn^{2+} + 2NO_3^- \tag{15}$$

$$H_2C_2O_4 = HC_2O_4^- + H^+ \tag{16}$$

$$HC_2O_4^- = C_2O_4^{2-} + H^+ \tag{17}$$

$$HCONH_2 + H_2O = HCOOH + NH_3 \tag{18}$$

$$NH_3 + H^+ = NH_4^+ \tag{19}$$

$$Zn^{2+} + C_2O_4^{2-} + 2H_2O = ZnC_2O_4 \cdot 2H_2O(s) \tag{20}$$

Zinc oxalate dihydrate also can be precipitated from an aqueous oxalic acid–zinc nitrate solution via anion release by hydrolyzing diethyl oxalate in a 15 vol % water in a nonaqueous acetic acid solution:

$$Et_2C_2O_4 + 2H_2O = 2EtOH + 2H^+ + C_2O_4^{2-} \tag{21}$$

In both systems, the release of $C_2O_4^{2-}$ is being throttled in a solution of dissociated Zn^{2+}. However, in terms of precipitation kinetics, the systems are by no means similar. The diethyl oxalate system precipitates ~10 μm (~7 aspect ratio) needlelike particles, while the formamide system produces ~10 μm cube-shaped particles [47]. As the results of these and other PFHS techniques [48] show, choice of the particular PFHS system should be made very carefully.

PFHS techniques also can be used to prepare multicomponent powders with controlled stoichiometry and phase homogeneity. In order to obtain a phase homogeneous powder with the intended cation stoichiometry, careful attention must be paid to preparation of the homogeneous solution and the types of multicomponent complexes that form. Gherardi and Matijevic [51] thermally hydrolyzed a peroxidic complex, precipitating a Ba-Ti salt of controlled Ba:Ti ratio. The catalyst, $Ti_aM_bM'_cO_d$ (M=Zn, Cd, and Cu; M'=Ni and La), has been precipitated from a chloride solution by using urea to increase the pH [52]. Problems such as stoichiometry control that are associated with multicomponent monodisperse syntheses can be avoided when multilayer structures are prepared, since each component is added in a separate processing step. Gherardi and Matijevic [53] precipitated hematite on titania cores by hydrolysis of an acidified aqueous iron chloride solution. Kayima and Qutubuddin [54] coated basic yttrium–copper carbonate with barium carbonate using urea decomposition to prepare a composite carbonate powder suitable for $Ba_2YCu_3O_{(7-x)}$ superconductors. Blok and Geus [55] prepared a permanent magnetic material by precipitating Fe_2O_3-based compositions on a porous silica carrier in suspension to oxidize Fe^{2+} solutions gradually.

In nonaqueous media, controlled hydrolysis of metal alkoxides with dilute alcoholic solutions provides the simplest means of preparing spherical monodisperse powders and composite powders. Oxides such as hydrous titania [56], hydrous silica [57], and zinc oxide [58] have been prepared. Multicomponent systems such as Sr-Ti hydrous oxides [29], niobium or tantalum-doped titanium hydrous oxides [59], and yttria-doped zirconium hydrous oxide [60] can be prepared by hydrolysis of multicomponent alkoxide solutions. In addition, composite particles such as alumina coated with hydrous titania have been prepared as a precursor system for alumina-titania composites [61].

Most monodisperse systems can be prepared in alcoholic media. However, some metal alkoxide precursors form gels rather than powders. For instance, controlled hydrolysis of $SrTi(O\text{-}i\text{-}Pr)_6$ in isopropanol yields gels, while the same reaction conducted in acetonitrile results in spherical submicron powders [29,64], even though both systems appear to exhibit adequate colloidal stability. Solubility measurements indicate that hydrolysis polymer solubility should be low when forming powders with metal alkoxides but high when preparing gels.

To overcome the aggregation problems that can occur with precipitations, Jean and Ring [62] developed a technique utilizing hydroxypropyl cellulose, which is believed to provide steric stabilization during the precipitation process. Sherif and Via [63], examining postsynthetic approaches, developed a microexplosive chemical deagglomeration (MED) technique. After precipitation, a microexplosive such as ammonium nitrate (NH_4NO_3) is coated on the surface of the powders. Ammonium nitrate decomposes during calcination, yielding expanding gases that break up agglomerates.

Nonoxide powder preparation is possible via a small number of PFHS techniques. Gallagher et al. [65] prepared submicron diborides of titanium, zirconium, and hafnium by refluxing metal borohydrides in xylene at 140°C:

$$Zr(BH_4)_4 = ZrB_2 + B_2H_6 + 5H_2 \tag{22}$$

The powders were amorphous as precipitated but crystallized when annealed. By the addition of polybutadiene to the reaction mixture, agglomeration was reduced and excess carbon (a sintering aid) was evident. Ritter [66] dehalogenated $SiCl_4$ at 130°C in n-heptane to form 1–5 μm (70 m^2/g) amorphous silicon carbide:

$$SiCl_4 + CCl_4 + 8Na = SiC + 8NaCl \tag{23}$$

Using a high shear rate stirrer and controlled heating with an oil bath, the reaction began as the sodium metal melted (~100°C) and a fine precipitate appeared. Vacuum distillation and sublimation processes followed to purify the nonoxide powders. By utilizing the following chloride pairs, $TiCl_4$ and BCl_3, $TiCl_4$ and CCl_4, or BCl_3 and CCl_4, the nonoxides TiB_2, TiC, or B_4C, respectively, were obtained.

B. Chemistry Between Heterogeneous Phases

Chemistry between heterogeneous phases is an approach in which the physical powder characteristics are controlled by interaction of a discrete solid phase with a continuous one. Synthetic methods falling into this category include hydrothermal synthesis, molten salt synthesis, pyrolysis, and spark erosion.

1. Hydrothermal Synthesis

Hydrothermal synthesis provides a means for preparing anhydrous oxides in a single step [67,68]. In this process, a suspension or in some cases a homogeneous solution is subjected to elevated temperature ($<1000°C$) and pressure (<100 MPa). The specific conditions employed should be capable of maintaining a solution phase that provides a labile mass transport path promoting rapid phase transformation kinetics. The combined effect of pressure and temperature also can reduce free energies for various equilibria-stabilizing phases that might not be stable at atmospheric conditions.

Ceramic powders can be prepared via an in situ transformation and/or a dissolution-reprecipitation. The former can be effected in a dispersion when suspended particles undergo a polymorphic or chemical phase transformation. For instance, narrow size distribution spherical submicron titanium hydrous oxide can be chemically transformed into polycrystalline anhydrous anatase spheres (TiO_2) by refluxing the powder for 12 h in a pH 10 ammonium hydroxide–water suspension at atmospheric pressure and a temperature of 100°C [69].

Dissolution-reprecipitation can be effected when suspended particles dissolve into solution, supersaturate the solution phase, and subsequently precipitate particles (Fig. 4). In many cases, the suspended solids are not very soluble in water, and solubilizing components (mineralizers) such as acids, bases, or other complex-forming substances must be added. For instance, in the preparation of zirconia powder from a suspension of hydrous zirconium oxide, a hydrothermal reaction at low pH (<3.5) dissolves the precipitate as soluble species and reprecipitates the oxide [67]. This process allows the rate of supersaturation in solution to be controlled. In many systems, both in situ transformation and dissolution-reprecipitation occur in the hydrothermal medium. For instance, Watson et al. [71] found both mechanisms operating when they prepared $PbTiO_3$ from suspensions of alkoxy-derived hydrous titania in alkaline lead acetate solutions.

Hydrothermal reactions are sometimes referred to as PFHS systems since uniform solutions of aqueous complexes prepared under atmospheric conditions are reacted at high temperature and pressure to form ceramic powders. For instance, 200°C thermal decomposition of calcium EDTA (ethylene diamine tetraacetate) complexes in pH 7.5 phosphate solution has been used to prepare 100 μm hexagonal rods of carbonate-substituted hydroxyl apatite [70].

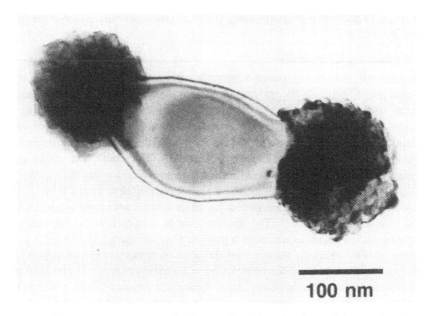

FIG. 4 Barium titanate crystallized on a dissolving titania seed in a hydrothermal solution.

Hydrothermal processing has been useful in producing a large number of multicomponent materials. Early studies utilized dispersed oxides and/or carbonates to make a whole family of lithium aluminosilicates (LAS) [72]. In other cases, uniform (single-phase) or segregated (diphasic) gels have been used to produce a spectrum of compositions [73,74]. At 600°C (100 MPa, 12 h), single-phase hydrous zirconium silicon oxide gels (prepared via sol-gel techniques) hydrothermally convert to vermicular aggregates of zircon ($ZrSiO_4$), while diphasic gels (formed by mixing zirconia and silica sols) result in the formation of equiaxed isolated particles [74]. Gels of coprecipitated oxides can be aged under much less severe hydrothermal conditions (e.g., at temperatures as low as 90°C) to prepare oxide materials such as ferrites [75,76] and zeolites [77] having as many as four cation components. However, due to factors (usually kinetic), the oxides obtained do not have the same stoichiometry as the initial gel, so empirical correlation of initial suspension stoichiometry with that of the precipitate is necessary. In other cases, kinetics and thermodynamics work in unison to produce a precipitate whose cation stoichiometry is the same as that of initial suspension; for example, a mixed hydrous titanium zirconium oxide gel suspended in barium hydroxide solution is fully converted to barium titanate zirconate at 130°C in less than 6 h depending on the Ti:Zr ratio [78].

In hydrothermal synthesis, solution environment and feedstock require attention, since particle size and shape can be modified by the types of mineralizers added to the solution [79]. As with the PFHS systems discussed earlier, these additives, many of which are proprietary, play a complexometric role in the nucleation, growth, and aging of the precipitate. For instance, the use of alkali hydroxides and/or alkali halides in the synthesis of lead titanate zirconate from a hydrous titanium zirconium oxide–lead oxide diphasic gel can affect the morphology (cubelike or rounded) and/or the size (micron or submicron) of the powder [80]. Chemical and phase purity are also affected. Takamori and David [81] found that pretreatments such as heat treatment or mechanical comminution of freeze-dried salts prior to hydrothermal synthesis dramatically altered the morphology, size, and size distribution of terbium-activated yttrium aluminum garnet (YAG). It is believed that the pretreatments control the nuclei density during the dissolution-reprecipitation process.

Hydrothermal oxidation of metals is another interesting route for preparing ceramic powders. Metals such as iron, zirconium, aluminum, or hafnium can be reacted with water at temperatures as low as 500°C and 100 MPa [82,83]:

$$mM + nH_2O = M_mO_n + nH_2 \tag{16}$$

Multicomponent oxides such as alumina-zirconia composite powders can be prepared from a Zr-Al metal alloy [84]. Metals such as titanium or zirconium can be reacted with ammonium chloride in an ammonia medium (rather than water) to form nitrides at temperatures of 500–600°C and 1 MPa [85].

2. Molten Salt Synthesis

Molten (fused) salt synthesis is similar in many ways to a dissolution-reprecipitation hydrothermal synthesis. An inert molten ceramic material in which the ceramic powders are somewhat soluble acts as a solvent medium. The single-component oxides dissolve, react in solution to form multicomponent species, and precipitate from solution. For instance, lead zirconate titanate can be prepared by first mixing together TiO_2, ZrO_2, PbO, and any desired dopants to form an aqueous slurry [86]. A KCl-NaCl molten salt eutectic is mixed into the slurry, causing the system to gel. Upon drying, the mixture is heated to ~980°C for 1 h to form the lead zirconate titanate, which subsequently is rinsed free of the molten salts and any excess unreacted ions. In some systems, the solvent can react with the other components to yield multicomponent powders [87]:

$$TiO_2 + 2MgCl_2 + 2SiO_2 = TiCl_4 + 2MgSiO_3 \tag{17}$$

Since $TiCl_4$ is highly volatile, purification of the precipitated powder is easily accomplished. Materials such as spinel, cordierite, and magnesium titanate have been prepared in this manner.

Particle formation in molten salt synthesis has been observed to depend upon the dissolution characteristics of the individual components. When a slowly dissolving component is in the presence of a rapidly dissolving one, the multicomponent reaction product precipitates on the surface of the less soluble component. Unidirectional diffusion of the more soluble component through the product layer carries the process to completion, resulting in a powder with a morphology reflective of the more slowly dissolving component. Both needlelike barium titanate and platelike zinc ferrite have been prepared in this manner [88–90]. On the other hand, feedstocks with somewhat equivalent dissolution characteristics precipitate in bulk solution, producing either well-defined or lumpy particles. When lumpy particles result, faceting follows, and the final particle shape depends on the growth rate anisotropy. Oxides such as Bi_2WO_6 and $PbNb_2O_6$ are believed to form via this mechanism.

3. Pyrolysis

Pyrolysis processes form a ceramic powder by solid state decomposition of a solid or molten precursor (typically a molecular or polymeric compound) in the presence of an inert (e.g., vacuum, argon) or reactive (e.g., oxygen, nitrogen) atmosphere. This method is distinguishable from a solution precipitation/calcination method because it is the pyrolysis rather than the chemical synthesis that influences the physical characteristics of the powder. For instance, agglomeration and excessive crystallite growth can be caused by excessive exothermic decomposition. Decomposition residue (e.g., carbon) can result from factors such as poor choice and control of pyrolysis atmosphere, improper gas flow rates, and furnace design. Mechanistic studies of the decomposition and its interaction with the processing variables are essential for control of the powder characteristics of the pyrolysis product.

Metal-organic molecular compounds such as carboxylates are useful materials for preparing oxides and nonoxides. These compounds can be derived in-house from starting reagents such as metal alkoxides, nitrates, or chlorides using simple procedures. Neodecanoate salts of various metals form uniform solutions in xylene. These solutions can be pyrolyzed to form submicron powders. For instance, $PbTiO_3$ can be prepared from a solution of dimethoxytitanium dineodecanoate $((MeO)_2Ti(OOCC_9H_{19})_2)$ and lead neodecanoate $(Pb(OOCC_9H_{19})_2)$ by pyrolyzing at temperatures of ~600°C [91]. Molecular metal organic compounds such as $B(N(C_2H_4O)_3)$ derived from condensation products of boric acid $(B(OH)_3)$ and triethanolamine $(N(CH_2CH_2OH)_3)$ are useful for the preparation of boron nitride (when pyrolyzed in nitrogen for 2 h at 1400°C) or boron carbide–boron nitride composite powders (when decomposed in argon for 2 h at 1400°C) [93].

Choice of ligand in the molecular precursor is an important processing variable since it controls pyrolysis behavior. Crystalline solids such as the hydrazinium metal hydrazinecarboxylate hydrates, easily prepared from aqueous metal sulfate solutions, autocatalytically decompose at low temperatures to form oxide powders [92]. For instance, a compound such as $N_2H_5Mn_{1/3}Fe_{2/3}(OOCN_2H_3)_3 \cdot H_2O$ decomposes in a single step in a temperature range of 75–120°C to form agglomerates of nanometer-size $MnFe_2O_4$. Inorganic nitrates (oxidizers) can be combined with organics such as urea (fuel) in stoichiometric compositions using principles of propellant chemistry to produce 4–5 μm agglomerates of submicron particles at temperatures as low as 500°C [95]. A diverse number of oxides, including spinel $(MgAl_2O_4)$, $Y_3Al_5O_{12}$ (YAG), ZrO_2-Al_2O_3 composites, and β'-Al_2O_3, have been prepared with this method.

Molecular chemistry can be used to control the cation stoichiometry of precursors useful for multicomponent ceramic oxides. Heistand and Gregory [94] utilized strongly binding catechol complexes to precipitate crystalline multicomponent pyrolysis precursors useful for dielectric applications. Under strongly basic conditions trisoctahedral anions combine with all transition metals. Nearly any cation can serve as the counterion. Resultantly, it is possible to precipitate stoichiometric $(Ba_{1.002}Ti_{1.000}O (C_6H_4O_2)_2 \cdot$ solvate) as well as nonstoichiometric $([Ba_{0.82}Sr_{0.18}][Ti_{0.90}Zr_{0.10}] O(C_6H_4O_2)_2 \cdot$ solvate) compounds that exothermically pyrolyze to form multicomponent oxides.

Polymeric metal organic precursors prepared from water-soluble metallic salts of polyfunctional acids are useful for the preparation of multicomponent oxides [96,97]. Typically, α-hydroxy acids (e.g., hydroxybutyric acid, citric acid) and/or α,β-unsaturated acids (maleic acid, acrylic acid) are mixed with metal salts to form a uniform solution of metal cation complexes. The solution subsequently is evaporated until a viscous mass is obtained, followed by a drying step at ~120°C. During the dissolution and solvent removal stages, it is essential that none of the components crystallize, so that powder homogeneity is ensured. Finally, the precursor powder is calcined (>600°C). Materials such as $BaFe_{12}O_{19}$, $LaCrO_3$, and $Y_3Fe_{4.9}Al_{0.1}O_{12}$ have been prepared with this method.

A popular metal organic polymer pyrolysis method is the Pechini process [98–100], also known as the liquid mix technique. As before, chemical complexation of conventional metal salts is accomplished with hydroxycarboxylic acids to form a uniform aqueous solution. Subsequently, a polyhydroxyl alcohol (usually ethylene glycol) is introduced, and the solution is heated to ~80°C to achieve homogeneity. As the temperature is raised above this point, alcohol esterifies the complexed and uncomplexed carboxylic acid to generate water. Since both the acid and the alcohol are polyfunctional, cross-linking can occur, and an organic polymer network is established with the chelated cations atomically distributed along the backbone. Upon removal of the water, a resin is formed that yields an oxide when calcined at temperatures ranging from 500 to 900°C. Over 100 oxides, including aluminates ($LaAlO_3$), silicates (Zn_2SiO_4), zirconates ($CaZrO_3$), and titanates ($Pb(Zr,Ti)O_3$), have been prepared with this method. Polymers prepared by the Pechini process can also be pyrolyzed to form nonoxides by varying the firing atmosphere [177]. For instance, nanometer-size agglomerated aluminum nitride can be prepared by firing aluminum-based polymers in nitrogen for 4 h (Fig. 5).

Organometallic polymers are considered desirable for nonoxide materials since the metal center is bonded to carbon rather than to oxygen. However, the synthesis of organometallic compounds requires air-sensitive procedures that can be complex. Many problems must be overcome with this approach, such as obtaining the desired phase assemblage, controlling chemical purity, diminishing the loss of volatile species that belong in the final ceramic product (e.g., Si, C, or N), minimizing the weight loss of organic species, and controlling the physical characteristics of the powder. Various nonoxide powders such as silicon nitride and silicon carbide have been prepared by decomposing polymeric precursors, sometimes producing a powder and other times yielding a porous monolithic form [101,102]. The decomposition of the materials to produce a desired phase requires a combination of skill and art. For instance, silazane polymer obtained by KH treatment of $[CH_3SiHNH]_x$ cyclics in diethyl ether pyrolyzes in nitrogen to form α-silicon nitride at 1420°C. When the synthesis is conducted in tetrahydrofuran, silicon nitride/silicon carbide composite powders form instead.

Alkyl aluminum amides ($\{R_2AlNH_2\}_3$) can be prepared by reaction between trialkyl aluminum and ammonia and subsequently pyrolyzed at ~1000°C to form aluminum nitride powder with an 8 nm crystallite size [103]. Polycarbosilane ($\{[((CH_3)_3Si)_{0.5}(CH_2CHSi(CH_3))_{1.0}(Si(CH_3)_2)_{1.0}]_n\}$)

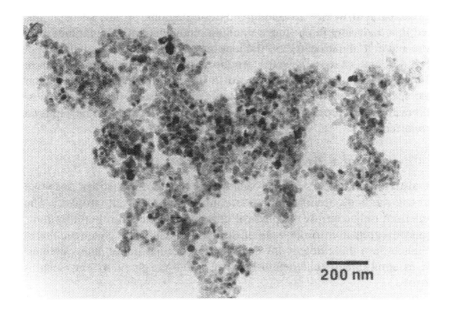

FIG. 5 Aluminum nitride prepared by pyrolyzing aluminum-based Pechini polymers in nitrogen at 1500°C for 4 h.

also can be copyrolyzed with diisobutylaluminum amide to form composite powders of silicon nitride and aluminum nitride [104]. However, the Si:Al ratio does not appear to be uniform throughout the bulk of the powder.

4. Spark Erosion

Another method that utilizes reactions between heterogeneous phases for preparing submicron ceramic powders is spark erosion [105]. In this technique, two electrodes are immersed in a dielectric liquid (aqueous, organic, or cryogenic) with a charge of large granules of the electrode material between the electrodes. A field strength of ~10^5 V/cm is applied in a pulsed mode (230–610 kHz) to get a spark discharge that takes place at the asperities of the electrode and granular charge. Since the material is heated in localized regions well above its melting point via plasma channel transient pressure zones, molten droplets or vapor species are produced. When the species leave the localized region they either freeze or condense to form particles. Subsequent reaction between the dielectric fluid or decomposition products and the metal (either as molten droplets or as vapor) produces a ceramic powder.

Spherical narrow size distribution ~34 µm titanium carbide has been prepared in a dodecane fluid using a titanium metal electrode, a graphite electrode, and a granular charge of titanium metal. Spark erosion of aluminum electrodes and charge in water produces submicron acicular aluminum hydroxide. This product is considered to form from condensed aluminum vapor species instead of molten droplets. There exist numerous other possibilities for spark erosion of ceramic materials that have yet to be explored or reported.

C. Chemistry in a Droplet

Chemistry in a droplet is an approach in which one of the earlier described synthetic methods occurs within the confines of a solvent droplet. The droplet not only controls the physical powder characteristics but also limits chemical segregation to the scale of the droplet diameter. Synthetic methods falling into this category involve the use of either liquid phase systems, such as emulsions or microemulsions, or liquid-gas phase systems such as aerosols.

1. Emulsions

An emulsion is a mixture of two immiscible liquids in which one liquid is dispersed in the other with the aid of a surfactant (Fig. 6). The surfactant stabilizes the emulsion by reducing interfacial tension, reinforcing the interfacial film, and preventing coalescence of the droplets. Mixing conditions and surfactant variables control the size of the droplet. High shear rates and surfactant levels promote smaller droplet sizes, which can range from 1 to 50 µm [106,107]. A single surfactant such as Span 80 (sorbitan monooleate) can be employed for water-in-2-ethyl hexanol emulsions. In some cases, a mixture of surfactants is employed. Span 60 (sorbitan monostearate) and Tween 20 (polyoxyethylene sorbitan monolaurate) mixtures are useful for water-in-toluene emulsions.

The water-in-oil emulsion has been the most widely used type of emulsion for the preparation of single-component oxides such as titania [108], multicomponent oxides as complex as synroc (a mixture of $BaAl_2Ti_6O_{16}$, $CaZrTi_2O_7$, and $CaTiO_3$) [108], and nonoxides such as the carbides of thorium or uranium [109,110]. In this type of emulsion, a stable colloidal sol consisting of one or more metal complexes is prepared and emulsified in an oil medium to form spherical droplets. Gelation of the spheres is induced in one of two ways. First, a liquid removal technique such as dehydration or evaporation can be used. For example, large volumes of an organic phase such as a long-chain alcohol (2-ethyl hexanol) can be used to concentrate the sol by dehydrating the droplets. Upon sufficient dehydration, van der

(a)

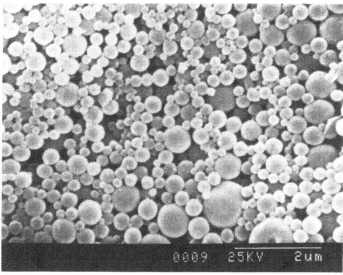

(b)

FIG. 6 (a) Oil-in-water emulsion (courtesy of Professor J. M. Wincek, Rutgers University). (b) Titanium hydrous oxide powder prepared by an alkoxide emulsion method (courtesy of Anne Hardy and the Massachusetts Institute of Technology, Ceramics Processing Research Laboratory).

Waals forces predominate and flocculate the sol within the droplet. It is conceivable that a precipitation reaction is occurring inside the droplet as well. Evaporation methods such as the "hot kerosene method," in which an emulsion is added to a hot oil phase [29,107] or the water is distilled from the emulsion using a Dean Stark moisture trap [111], also have been employed to remove the liquid phase from the droplet and flocculate the sol.

The second technique induces gelation internally by changing the pH inside the sol droplet. This reduces electrostatic attraction between colloidal species and/or initiates a chemical reaction inside the droplet. For example, urea (NH_2CONH_2) added to a stable sol decomposes and releases ammonia in the aqueous phase during heating of the emulsion [112]. For droplets smaller than 50 µm, amines can be added that diffuse into the droplet and effect an internal pH change [113]. Organic primary amines (R_3C-NH_2) added to the sol remain in the organic phase while extracting protons and anions from the droplet, whereas bubbling ammonia diffuses into the droplet to raise the pH. The organic amine route is preferable to the use of bubbling ammonia or dehydration of sol droplets because it minimizes anion contamination in the powder. This is particularly important when the anions present are difficult to eliminate during calcination. In addition, the amine route requires less solvent than the dehydration route, with a 10:1 ratio of gelation solvent to sol for the primary amine route, compared to a 200:1 ratio for dehydration [106].

"Gel sphere" particles also can be prepared by delivering metal organic monomers to the droplet [29]. First, a hydrochloric acid solution is emulsified in an oil phase (mineral oil and heptane). Silicon tetraethoxide-in-heptane solution subsequently is added to the emulsion to facilitate transport of alkoxide monomers to the acidified droplets. This method produces silica gel spheres in the submicron to micron range.

Very little work has examined the use of emulsions of nonpolar nonaqueous phases in polar nonaqueous phases to make ceramic powders. Hardy [29,114] has investigated methods for hydrolyzing droplets of nonpolar metal alkoxides (or alkoxides dissolved in hexane) dispersed in a water-soluble polar continuous phase. Continuous phases such as acetonitrile or propylene carbonate have been employed successfully for insoluble alkoxides such as titanium tetra-n-butoxide or aluminum tri-sec-butoxide. Systems utilizing propylene carbonate as a continuous phase did not require a surfactant to stabilize the emulsion, while those utilizing acetonitrile did. Single- and multicomponent hydrous oxides yield spherical submicron atomically homogeneous oxide powders such as titania, aluminum titanate, or alumina-zirconia. Unlike other emulsion powders synthesized with inorganic salts, these powders are free of anions such as nitrates and chlorides.

2. Microemulsions

Unlike emulsions, which are kinetically stable, microemulsions are thermo-dynamically stable [115]. In addition, the dispersed phase of a microemulsion is on a size scale less than 0.15 μm and therefore is transparent, while emulsions, having a droplet size greater than 1 μm, appear turbid. The reduction in size provides a means of forming finer particles than are possible with emulsions. Since only a handful of nonoxides and oxides have been prepared from microemulsions, this method is relatively unexplored. Water-in-oil microemulsions utilizing lyotropic liquid crystals or normal or inverse micelles have been investigated.

Gobe et al. [116] have prepared ultrafine (~4 nm) narrow size distribution magnetite (Fe_2O_3) using inverse micelles. The first step is to mix the following two microemulsions: (1) $FeCl_3$ aqueous droplets stabilized with an Aerosol OT surfactant (dioctylester of sodium sulfosuccinic acid) in a continuous phase of cyclohexane and (2) ammonium hydroxide droplets (microemulsified similarly). Upon the addition of aqueous $FeCl_2$ to the vigorously stirred microemulsion, Fe_2O_3 is formed in the aqueous droplet phase. Precipitation is effected both by diffusion of $FeCl_2$ species and by communication between microemulsion droplets, a process in which two droplets coalesce, exchange contents, and separate, reforming two droplets with a new composition. Communication allows chemical species (e.g., NH_4OH, $FeCl_2$, $FeCl_3$) to be exchanged, activating chemical and precipitation reactions in a manner analogous to a titration process.

Ceramic powders such as 50–120 nm narrow size distribution carbonates of barium or calcium can be formed by bubbling carbon dioxide into a microemulsion of aqueous alkaline earth hydroxide in cyclohexane, using CaOT (a derivative of Aerosol OT) as a surfactant [117]. Hydrous silica spheres of 10 nm can be obtained by introducing silicon tetraethoxide into an aqueous ammonium hydroxide–Aerosol OT–isopentane microemulsion [118]. Nonoxides such as Ni_2B or Co_2B catalysts in the 2–7 nm range can be prepared from aqueous metal complex solution droplets stabilized by cetyltrimethylammonium bromide (CTAB) in a continuous phase of hexan-1-ol [119]. The pools of aqueous nickel or cobalt chloride are reduced by $NaBH_4$ to form the boride by diffusion of the $NaBH_4$ species into the aqueous pools. Reverse microemulsions also can be prepared using supercritical fluids [120]. Submicron aluminum hydroxide particles have been formed by introducing ammonia into a microemulsion of supercritical aqueous aluminum nitrate in propane stabilized by sodium bis(2-ethyl hexyl)sulfosuccinate (AOT) at 100°C and ~20 MPa.

Research has been directed to the preparation of ceramic powders using lyotropic liquid crystals [115,170]. Monodisperse basic copper sulfate

$(Cu_4(OH)_6SO_4)$ has been prepared from a hexagonal liquid crystal formed by using a mixture of Tween 80 (sorbitan monooleate with 20 ethylene oxide groups) and aqueous copper sulfate $(CuSO_4)$ solution. In this system, precipitation occurs in the interstitial zones between the columns of surfactant molecules arranged in hexagonal arrays. This compartmentalization approach also has been utilized with routes that mimic the way nature precipitates ceramic materials such as bone or teeth [171-175]. For instance, aluminum oxide in the nanometer size range (10-30 nm) has been prepared using single-compartment unilamellar vesicles [173]. A unilamellar vesicle is a hollow sphere with an internal diameter in the range of tens of nanometers. The thin walls facilitate small anion transport but minimize cation diffusion. Vesicles containing aluminum ions are formed by ultrasonically mixing egg yolk phosphatidylcholine with an aqueous solution of aluminum nitrate. After removing all aluminum ions outside the vesicle wall by ion exchange, an aqueous sodium hydroxide solution is mixed with the vesicle suspension. Hydroxyl ions preferentially permeate the vesicle wall to effect a precipitation reaction. A synthetic polymer (e.g., polydimethyl siloxane, polymethyl methacrylate) also can be utilized as a porous network. Using a two-stage infiltration process, a hydrolytically sensitive precursor such as a metal alkoxide is introduced into the network followed by the addition of a hydrolysis solution. Materials such as silica, titania, and barium titanate with sizes in the 1-10 nm range have been prepared using this approach [174,175].

3. Aerosols

Aerosols or sprays are liquids dispersed in a gas phase. As with micelle-based systems, particle size control is directly related to the droplet size, which can be controlled by selection of both the precursor system and the method of aerosol generation.

Submicron aerosols can be generated by vapor condensation. This approach has been used primarily in low-temperature reactions in which particles form within the precursor droplets as they react with the species in the vapor phase. For instance, narrow size distribution 0.2 μm powders can be precipitated from monodisperse 0.4 μm aerosols of alkoxides ($Ti(OR)_4$, $Al(O\text{-}s\text{-}Bu)_3$) or metal tetrachlorides ($SiCl_4$, $TiCl_4$) by seeding the vapor phase with silver chloride and subsequently hydrolyzing the droplets with water vapor in one or more successive stages at temperatures $\geq 0°C$ [121-123]. The hydrolysis is completed by elevating the temperature of the aerosol to 100°C or more.

Multicomponent systems such as titania-silica powders have been prepared by diffusing $SiCl_4$ vapor into monodisperse $Ti(OEt)_4$ droplets and subsequently hydrolyzing the multicomponent droplet. However, silver chloride contamination and chlorine contamination from the precursor may present problems for ceramics processing applications. Purity problems can be avoided by using a two-stage reaction. First, alkoxide vapor is hydrolytically decomposed to form hydrous oxide seed particles to nucleate an aerosol. This aerosol subsequently can be hydrolyzed to form spherical hydrous oxide particles [124,125].

Using mechanical rather than vapor condensation methods for generating droplets, sols have been sprayed into an immiscible liquid to form "gel microspheres" by a subsequent dehydration or pH change step as described for emulsions [112,126]. This technique usually produces particles with sizes on the order of tens of microns. Mechanically generated aerosols also have been used with other types of precipitation chemistries. For instance, an ultrasonic spray atomizer has been used to react aqueous zirconium nitrate with ammonia gas to precipitate a hydrous oxide [126]. Electrostatic aerosols of aqueous solutions of zirconium acetate and polyethylene oxide yielded 0.1–5 μm diameter acetate particles, which were subsequently calcined at 1100°C to form zirconia [128]. In this system, varying concentrations of polyethylene oxide (<0.06 wt %) served as means to control the particle size. Salts can be spray dried as well; however, significantly larger particle sizes result. Iron oxides spray dried as various salts typically have particle sizes in the 3–20 μm range [129,130].

Because of their unique solvating properties, supercritical fluids have been used to generate aerosols [120,131]. With this method, a solute soluble in a supercritical fluid produces supersaturated species by an abrupt loss of solvency as the fluid expands into a low-pressure environment. Selection of the proper pressure and temperature of the preexpansion and postexpansion conditions produces either one-phase (vapor) or two-phase (liquid-vapor) systems. Two-phase aqueous systems have been used to precipitate silica, germanium oxide, and uniform mixtures of the two components, while nonaqueous systems have been used to prepare zirconium oxynitrate and polycarbosilane, useful precursors for zirconium oxide and silicon carbide, respectively.

Ceramic powders have been prepared by freeze drying, in which multicomponent liquid solutions are sprayed into a cold liquid to freeze the droplets, then isolated from the liquid and subsequently sublimed. Spherical alumina, magnesium oxide, and spinel powders with submillimeter particle sizes have been prepared from calcined sulfate salts by pneumatically spraying aluminum sulfate solutions into hexane solutions chilled with a dry ice–acetone bath [132]. Acidic barium–yttrium–copper formate solutions have been ultrasonically sprayed into liquid nitrogen (−196°C) to prepare

10–90 μm spheres that calcine as $Ba_2YCu_3O_7$ [133]. The results of these two examples demonstrate the importance of the method of aerosol generation to controlling the particle size. Other variables affecting size and the related chemical and physical particle homogeneity, such as spray height distance above bath and cooling bath temperature, have been examined [133]. In general, it is desirable to spray at a distance that minimizes droplet evaporation and at a temperature that minimizes rejection of solute as the ice front propagates toward the center of the droplet. In a homogeneous powder, the freezing process imparts a porous high surface area microstructure to the droplet [134,135].

Powders also can be precipitated in spray droplets by solvent replacement. This technique involves the addition of an aerosol of an aqueous solution of metal salts to a "nonsolvent" that solubilizes water but not the dissolved salts. O'Toole and Card [136] reported how various factors such as the choice of aerosol generator and "nonsolvent" control the size distribution and morphology of yttria-stabilized zirconia powder derived from sulfate salts. For instance, acetone produced fused masses, while isopropanol formed spheres almost exclusively. In ethanol, dropwise addition with a pipet yielded a precipitate with a diameter (214 nm) five times that produced by atomization (42 nm).

In spray pyrolysis, liquid droplets are sprayed into a reactor that has a low-temperature chamber for precipitating salts and a high-temperature chamber for subsequent pyrolysis [137]. Powders range from micrometer-size shell-like aggregates to narrow size distribution submicron solid spheres. Efforts currently are underway to understand how dense equiaxed ceramic powders can be formed, such ZnO and MgO derived from acetate precursors. The effect of cation source (e.g., nitrates, metal alkoxides, prehydrolyzed metal alkoxides), solvent (e.g., ethanol, water), and chemical additives (e.g., polyvinyl alcohol (PVA), glycerin) on the powder characteristics are being examined [137–139].

The problems inherent in spray pyrolysis are magnified because reactor and precursor systems must incorporate the ideal attributes of both spray drying and calcination processes. During drying, the kinetics of both solvent evaporation and salt precipitation propagating inward from the surface of the droplet should be optimized while maintaining the spherical morphology. During pyrolysis, attention must be paid to the large amounts of volatile products that are associated with rate-limiting processes such as heat transfer, diffusion through a product layer, and crust formation. In addition, in each case it must be determined whether or not spray pyrolysis is a liquid-gas phase reaction or a vapor phase process utilizing a liquid precursor, as described below. If the latter process is occurring, particle size will not be controlled by droplet size.

D. Chemistry in the Vapor Phase

Chemistry in the vapor phase is an approach in which the physical characteristics of ceramic powders are controlled by precipitation processes in the gas phase. In a vapor phase reaction, molecular or atomic species undergo reactive chemistry to form monomeric vapor species that supersaturate in the gas phase to form a precipitate. Vapor phase reactions have been used to prepare a variety of ceramic oxides and nonoxides. Depending on the material and reaction conditions, powders can be prepared as amorphous or crystalline materials exhibiting low amounts of volatile material and extremely high purities. Vapor phase reactions typically form 0.1 μm agglomerates of particles in the 5–10 nm range (Fig. 7). However, various studies [140–148] have demonstrated how control of the physical powder characteristics can be accomplished by utilizing computational modeling and systematic variation of the processing variables, as guided by process fundamentals.

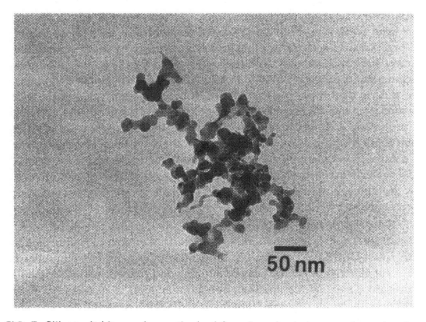

FIG. 7 Silicon nitride powder synthesized from laser-heated vapor phase chemistry (courtesy of S. C. Danforth, Rutgers University).

In general, efforts in vapor phase synthesis have focused on design of the reactor and type of heating zone rather than on molecular design and synthesis of precursors. Vapor phase reactions can employ plasma, furnace, laser, or other novel heating methods. In plasma-heated reactions, thermal or cold plasmas have been used to volatilize a variety of precursors [149,150]. In thermal plasmas, ions, electrons, and neutral atoms are all at the same temperature (~10,000 K). Direct current (DC) and radio frequency (RF) induced plasmas are common thermal plasmas capable of volatilizing solid, liquid, and gaseous precursors. As volatilized species leave the plasma zone, the vapor is quenched (at ~10,000 K/s) to achieve a very high degree of supersaturation. Microwave plasmas have been used to perform seeded growth of diamond powder [151].

With cold plasmas, only the electrons are hot. Consequently, reaction temperatures are not high. The novel chemistries provided in this new powder synthesis technology by both the precursor and the reactive intermediates produced by energetic electrons invite exploration. Silicon nitride has been prepared using a radio frequency glow discharge cold plasma. The stoichiometry (Si:Ni ratio) of the powder can be controlled; however, unlike most plasma reactions, subsequent calcination of the powder is necessary to convert silicon diimide $[Si(NH)_2]_n$ species to silicon nitride.

As furnace heating sources, resistance, electric arc, and radio frequency have been used to conduct reactions at temperatures as high as 2800°C [147]. Combustion mixtures of oxygen and methane or other oxidant-fuel mixtures have been employed to react with precursors (e.g., $SiCl_4$) at room temperature to produce ~2000°C flames [152,153]. Laser beams also have been used to heat vapor phase reactants by coupling absorption of the laser radiation with one of the gaseous species in the vapor stream [144]. Unlike the previously discussed methods, laser beam heating provides a reaction zone (defined by the cross-sectional area of the laser beam) that facilitates the controlled nucleation and growth of powders. Since reactions proceed in a cold-walled cell, both heterogeneous nucleation and corrosion problems are eliminated and access to in situ particle growth and temperature measurements is facilitated. In laser heating, as with thermal plasma methods, vapor species are both heated (10^6 °C/s) and cooled (10^5 °C/s) rapidly to supersaturate the gas phase. Although reaction temperatures are significantly lower (~2000°C), volatile-free oxides and nonoxides can be prepared [154]. An electron beam also has been utilized to prepare nonoxides [176]. In this case, a solid such as titanium or aluminum metal is irradiated with an electron beam in an ammonia atmosphere to form a nitride phase. The electron beam subsequently volatilizes the solid phase as a nitride vapor to supersaturate in the gas phase, forming nitride powder. The problem with the above beam-oriented methods is the small cross-sectional area of the beam, which limits the extent to which the process can be scaled up.

Vapor phase reactions have been utilized with a variety of precursors for both single- and multicomponent ceramic materials. Halide precursors such as chlorides or bromides typically are used, since they are readily available in the gaseous, highly volatile liquid, or sublimable solid state. The reactive atmosphere utilized determines whether an oxide or a nonoxide forms. For instance, titania can be prepared from a volatile liquid such as titanium tetrachloride ($TiCl_4$) by oxidation (1050°C) [155]:

$$TiCl_4 \text{ (g)} + O_2 \text{ (g)} = TiO_2 + 2Cl_2 \text{ (g)} \tag{18}$$

Alternatively, oxides can be formed by using water instead of oxygen:

$$TiCl_4 \text{ (g)} + 2H_2O \text{ (g)} = TiO_2 + 4HCl \text{ (g)} \tag{19}$$

To form a nonoxide such as titanium nitride, $TiCl_4$ can be reacted in an ammonia/hydrogen atmosphere using a plasma jet [146,156,157]. The reaction chemistry can take any of several paths, depending on the temperature. When the gases are mixed at temperatures below 250°C, $TiCl_4 \cdot nNH_3$ adducts precipitate in the vapor phase, decomposing to TiN when heated to temperatures above 500°C. When mixing proceeds at temperatures above 600°C, TiN precipitates directly from the vapor phase. Such an example demonstrates the importance of acquiring a detailed understanding of reaction thermodynamics and kinetics.

Volatile $TiCl_4$ vapors can be mixed with subliming $ZrCl_4$ vapors to make a multicomponent mixture of TiO_2 and $ZrTiO_4$ powders [155]. Differing volatilities, oxidation kinetics, and precipitation kinetics (e.g., homogeneous versus heterogeneous nucleation) of the various phases create difficulties for the preparation of phase homogeneous powders. However, phase nonuniformities sometimes are desirable, e.g., in the preparation of (α or β)-alumina-titania composite powders from chloride flame oxidation reactions to be employed in low-expansion ceramics [158].

The contamination problems caused by chloride precursors (e.g., NH_4Cl in plasma-derived titanium nitride powder [156]) have encouraged investigators to turn to other volatile inorganic precursors. Rather than silicon tetrachloride ($SiCl_4$), a reactive gas such as silane (SiH_4) is coreacted with ammonia to form silicon nitride. Silane also can be reacted with methane to form silicon carbide [141]. In laser-heated reactions in which the precursor does not absorb carbon dioxide radiation, passive reactants are necessary. For example, SF_6 has been used to assist in the decomposition of $Zr(BH_4)_4$ to form ZrB_2 [159].

Both metal organic and organometallic compounds have been utilized to prepare powders free of chemical contamination. A tertiary tetrabutoxide of zirconium ($Zr(O\text{-}t\text{-}Bu)_4$) has been used to prepare zirconia powder via pyrolytic decomposition at relatively low temperatures (350–500°C) [160]. Aluminum nitride can be prepared by reaction between trimethyl aluminum and ammonia at temperatures as low as 400°C [161]. Silicon and carbon can be incorporated within the same molecule to prepare silicon carbide using precursors such as methylsilane ($Si(Me)_4$) at temperatures below 1500°C [146]. While the above precursors appear promising, work is being directed to the development of less expensive precursors. Silicon nitride is being prepared from a trisdimethylaminosilane ($HSi(NMe_2)_3$) precursor that is expected to cost less than preparations from SiH_4-NH_3 mixtures [162].

Molten metals also provide a high-purity route for preparing oxides and nonoxides. MgO is prepared by reacting a saturated vapor of magnesium with oxygen to form a diffusion-limited flame [147]. Titanium and aluminum nitride have been prepared by evaporating metals in ammonia [163].

Solutions can be used as vapor phase reaction precursors as long as the temperatures are high enough to create atomic species. Metal organic compounds such as acetates and alkoxides can be dissolved in alcoholic media and ignited in a flame to form complex oxides such as $BaTiO_3$ [164,165] or burned in a plasma to prepare doped alumina [153]. Multicomponent aqueous metal nitrate solutions have been introduced into plasma flames to make materials such as a homogeneous $MgO \cdot Al_2O_3$ spinel [166]. However, as in other vapor phase systems, segregation processes can persist in multicomponent systems such as the alumina-zirconia system [167]. As mentioned earlier, rapid expansion of supercritical solutions is another useful way to supersaturate in the vapor phase [131].

Solids can be introduced directly into the heated zone as a source of volatile reactive species. Aluminum nitride has been prepared via direct reaction between aluminum powder and a nitrogen thermal plasma spray [168]. SiC can be made by fabricating a rod comprised of silica and carbon, creating a high-temperature arc between the rod and an electrode to form vapor species [169].

IV. SUMMARY

Ideal physicochemical characteristics of a ceramic powder include submicron particle size, controlled particle size distribution, uniform and equiaxed shape, minimum degree of agglomeration, high degree of chemical purity, controlled chemical and phase homogeneity, maximized bulk

TABLE 1 Synthetic Technologies for Ceramic Powders

Processing approach	Synthesis technology
Chemistry in the liquid phase	Direct strike
	Nonsolvent addition
	Solvent removal
	Gel drying
	PFHS
Chemistry between heterogeneous phases	Hydrothermal synthesis
	Molten salt synthesis
	Pyrolysis
	Spark erosion
Chemistry in a droplet	Emulsions
	Microemulsions
	Aerosols
Chemistry in the vapor phase	Heating method
	Vapor precursors
	Liquid precursors
	Solid precursors

particle density, minimum weight loss upon heating, and controlled surface chemistry. The benefits of processing with an ideal (chemically derived) powder versus a conventionally processed multicomponent mixture and the various synthetic technologies for preparing a spectrum of submicron ceramic oxides and nonoxides have been discussed. Present technology offers the investigator many alternative approaches for preparing powders with ideal or near ideal characteristics. Current technologies are summarized in Table 1.

ACKNOWLEDGMENTS

The author would like to acknowledge the support of the Center for Ceramic Research and the New Jersey State Commission on Science and Technology in preparing this manuscript. The author would also like to thank Ms. D. Netinho and Mrs. L. Chirichillo for their assistance in the preparation of this manuscript and Ms. K. Griffin for editorial assistance. Finally, the assistance of Dr. H. Dess and the staff of the Rutgers Library of Science and Medicine for the literature search is appreciated.

REFERENCES

1. P. E. Debely, E. A. Barringer, and H. K. Bowen, *J. Am. Ceram. Soc. 68*(3): C-76–C-78 (1985).
2. R. J. Higgins, H. K. Bowen, and E. A. Giess, in *Advances in Ceramics, Volume 21, Ceramic Powder Science* (G. L. Messing, K. S. Mazdiyasni, J. W. McCauley, and R. A. Haber, eds.). American Ceramic Society, Westerville, Ohio, 1987, pp. 691–698.
3. G. R. Wiese and T. W. Healy, *J. Coll. Inter. Sci. 52*(3): 458–467 (1975).
4. R. Jin, W. Hu, and X. Hou, *Colloids and Surfaces 26*: 317–331 (1987).
5. E. M. De Liso, W. R. Cannon, and A. S. Rao, in *Materials Research Society Symposia Proceedings* (Y. Chen, W. D. Kingery, and R. J. Stokes, eds.), Vol. 60. Materials Research Society, Pittsburgh, Pa., 1986, pp. 43–50.
6. R. Hogg, *Ceram. Bull. 60*(2): 206–211 (1981).
7. W. D. Kingery, H. K. Bowen, D. R. Uhlmann, in *Introduction to Ceramics*, 2d ed., John Wiley, New York, 1976, pp. 381–447.
8. T. Ishii, R. Furuichi, T. Nagasawa, and K. Yokoyama, *J. Thermal Anal. 19*: 467–474 (1980).
9. M. Murata, K. Wakino, K. Tanaka, and Y. Hamakawa, *Mat. Res. Bull. 2*: 323–328 (1976).
10. A. Ochi, H. K. Bowen, and W. E. Rhine, in *Materials Research Society Symposia Proceedings* (C. J. Brinker, D. E. Clark, D. R. Ulrich, eds.), Vol. 121. Materials Research Society, Pittsburgh, Pa., 1988, pp. 663–666.
11. G. M. Crosbie, R. L. Predmesky, J. M. Nicholson, and E. D. Stiles, *Ceram. Bull. 68*(5): 1010–1014 (1989).
12. T. Iwai, T. Kawahito, and T. Yamada, U.S. Patent 2,020,264 A to UBE Industries Ltd., Japan, 12 (1979).
13. T. Iwai, T. Kawahito, and T. Yamada, U.S. Patent 4,405,589 to UBE Industries Ltd., Japan (1983).
14. R. S. Kalyoncu, Bureau of Mines Report of Investigations, RI9012, Pittsburgh, Pa., 1986.
15. R. S. Kalyoncu, in *Ceramic Engineering and Science Proceedings*, Vol. 6. American Ceramic Society, Westerville, Ohio, 1985, pp. 1356–1364.
16. B. Yu and A. Goldman, in *Ferrites, Proc. ICF 3d 1980* (H. Watanabe, S. Iida, and M. Sugimoto, eds.). Cent. Acad. Publ. Jpn., Tokyo, 1982, pp. 68–73.
17. N. D. Spencer, S. D. Murphy, G. Shaw, A. Gould, E. M. Jackson, and S. M. Bhagat, *Japan. J. Appl. Phys. 28*(9): L1564–1567 (1989).
18. D. Hennings and W. Mayr, *J. Sol. State Chem. 26*: 329–338 (1978).
19. P. K. Gallagher and F. Schrey, *J. Am. Ceram. Soc. 46*(12): 567–573 (1963).
20. K. Saegusa, W. E. Rhine, and H. K. Bowen, personal communication, MIT, July, 1989.
21. I. M. Thomas, U.S. Patent 3,761,500 to Owens-Illinois, Inc. (1972).
22. M. Sugiura and O. Kamigaito, *Yogyo-Kyokai-Shi 92*(11): 605–611 (1984).
23. D. C. Bradley, R. C. Mehrotra, and D. P. Gaur, in *Metal Alkoxides*. Academic Press, New York, 1978.

24. R. C. Mehrotra, S. Goel, A. B. Goel, R. B. King, and K. C. Nainan, *Inorganica Chimica Acta 29*: 131–136 (1978).
25. K. S. Mazdiyasni and L. M. Brown, *J. Am. Ceram. Soc. 55*(11): 548–552 (1972).
26. K. S. Mazdiyasni, C. T. Lynch, and J. S. Smith II, *J. Am. Ceram. Soc. 50*(10): 532–537 (1967).
27. G. F. Hauck, K. W. Hass, and A. Lenz, U.S. Patent 3,458,552 to Dynamit Nobel Aktiengesellschaft, Troisdorf, Bezirk Cologne, Germany (1969).
28. B. E. Yoldas, *J. Mat. Sci. 14*: 1843–1849 (1979).
29. A. R. Hardy, G. Gowda, T. J. McMahon, R. E. Riman, W. E. Rhine, and H. K. Bowen, in *Ultrastructure Processing of Advanced Ceramics* (J. D. Mackenzie and D. R. Ulrich, eds.). John Wiley, New York, 1988, pp. 407–428.
30. I. M. Thomas, in *Sol-Gel Technology For Thin Films, Fibers, Preforms, Electronics, and Speciality Shapes* (L. C. Klein, ed.). Noyes, Park Ridge, N.J., 1988, pp. 2–15.
31. R. Dosch, in *Materials Research Society Symposia Proceedings* (C. J. Brinker, D. E. Clark, and D. R. Ulrich, eds.), Vol. 32. North Holland, New York, 1984, pp. 157–162.
32. S. R. Gurkovitch and J. B. Blum, in *Ultrastructure Processing of Ceramics, Glasses and Composites* (L. L. Hench and D. R. Ulrich, eds.). John Wiley, New York, 1984, pp. 152–160.
33. H. Perthuis and P. Colomban, *Mat. Res. Bull. 19*: 621–631 (1984).
34. D. L. Monroe, J. B. Blum, and A. Safari, *Ferroelectric Letters 5*: 39–46 (1986).
35. C. P. Cheng, P. A. Iacobucci, and E. N. Walsh, U.S. Patent 4,619,908 to Stauffer Chemical Company (1986).
36. T. Gallo, personal communication, Akzo Chemicals, Inc., Jan., 1990.
37. B. J. Mulder, *Am. Ceram. Soc. Bull. 49*(11): 990–993 (1970).
38. D. M. Ibrahim and H. W. Hennicke, *Trans. J. Br. Ceram. Soc. 80*: 8–12 (1981).
39. R. E. Jaeger and T. J. Miller, *Am. Ceram. Soc. Bull. 53*(12): 855–859 (1974).
40. K. Kiss, J. Magder, M. S. Vulkasovich, and R. J. Lockhart, *J. Am. Ceram. Soc. 49*(6): 291–295 (1966).
41. K. S. Mazdiyasni, R. T. Dolloff, and J. S. Smith, *J. Am. Ceram. Soc. 52*(10): 523–526 (1969).
42. F. A. Cotton and G. Wilkinson, in *Advanced Inorganic Chemistry, A Comprehensive Text*, 4th ed. John Wiley, New York, 1980, pp. 733, 857.
43. C. Duval, in *Inorganic Thermogravimetric Analysis*. Elsevier, New York, 1953, pp. 397–398.
44. J. Voigt, An integrated study of the ceramic processing of yttria, chemical engineering, Ph.D. diss., Iowa State University, Ames, Iowa, 1986.
45. H. Anzai, *Chem. Abs. 107*(18): 160185v (1987).
46. L. Gordon, M. L. Salutsky, and H. H. Willard, in *Precipitation from Homogeneous Solution*. John Wiley, New York, 1959.
47. M. Munson and R. E. Riman, in *Ceramic Transactions, Vol. 8, Ceramic Dielectrics: Composition, Processing and Properties* (H. C. Ling and M. F. Yan, eds.). American Ceramic Society, Westerville, Ohio, 1990, pp. 213-220.

48. E. Matijevic, in *Ultrastructure Processing of Advanced Ceramics* (J. D. Mackenzie and D. R. Ulrich, eds.). John Wiley, New York, 1988, pp. 429–442.
49. E. Matijevic, *J. Coll. Inter. Sci. 58*: 374 (1977).
50. E. Matijevic, *Acc. Chem. Res. 14*: 22–29 (1981).
51. P. Gherardi and E. Matijevic, *Colloids and Surfaces 32*(3-4): 257–274 (1988).
52. M. Takenaka, T. Takahashi, and T. Momobayashi, *Chem. Abs. 97*(10): 72982x (1982).
53. P. Gherardi and E. Matijevic, *J. Coll. Inter. Sci. 109*(1): 57–68 (1986).
54. P. M. Kayima and S. Qutubuddin, *J. Mat. Sci. Lett. 8*: 171–172 (1989).
55. J. H. Blok and J. W. Geus, *Chem. Abs. 73*(24): 125072p (1970).
56. E. A. Barringer and H. K. Bowen, *Langmuir 1*(4): 414–420 (1985).
57. W. Stober, A. Fink, and E. Bohn, *J. Coll. Inter. Sci. 26*: 62–69 (1968).
58. R. H. Heistand II, Y. Oguri, H. Okamura, W. C. Moffatt, B. Novich, E. A. Barringer, and H. K. Bowen, in *Science of Ceramic Chemical Processing* (L. L. Hench and D. R. Ulrich, eds.). John Wiley, New York, 1986, pp. 482–496.
59. E. Barringer, N. Jubb, B. Fegley, R. L. Pober, and H. K. Bowen, in *Ultrastructure Processing of Ceramics, Glasses and Composites* (L. L. Hench and D. R. Ulrich, eds.). John Wiley, New York, 1984, pp. 315–333.
60. B. Fegley, Jr., P. White, and H. K. Bowen, *Am. Ceram. Soc. Bull. 64*(8): 1115–1120 (1985).
61. H. Okamura, E. A. Barringer, and H. K. Bowen, *J. Am. Ceram. Soc. 69*(2): C-22-C-24 (1986).
62. J. H. Jean and T. A. Ring, *Colloids and Surfaces 29*: 273–291 (1988).
63. F. G. Sherif and F. A. Via, U.S. Patent 4,764,357 to Akzo America Inc. (1988).
64. R. E. Riman, The role of the chemical processing variables for the synthesis of ideal alkoxy-derived $SrTiO_3$ powder, Ph.D. diss., MIT, Cambridge, Mass., 1987.
65. M. K. Gallagher, W. E. Rhine, and H. K. Bowen, in *Ultrastructure Processing of Advanced Ceramics* (J. D. Mackenzie and D. R. Ulrich, eds.). John Wiley, New York, 1988, pp. 901–906.
66. J. J. Ritter, in *Materials Research Society Symposia Proceedings* (C. J. Brinker, D. E. Clark, and D. R. Ulrich, eds.), Vol. 73. Materials Research Society, Pittsburgh, Pa., 1986, pp. 367–372.
67. J. H. Adair, R. P. Denkewicz, F. J. Arriagada, and K. Osseo-Asare, in *Ceramic Transactions, Ceramic Powder Science II* (G. L. Messing, E. R. Fuller, and H. Hausner, eds.). American Ceramic Society, Westerville, Ohio, 1987, pp. 135–145.
68. A. Rabenau, *Journal of Materials Education 10*(5): 545–591 (1988).
69. Y. Oguri, R. E. Riman, and H. K. Bowen, *J. Mat. Sci. 23*: 2897–2904 (1988).
70. N. Christiansen and R. E. Riman, in *Proceedings from the 5th Scandinavian Symposium on Materials Science* (I. L. H. Hansson and H. Liholt, eds.). SBI, Horsholm, Denmark, 1989, pp. 209–220.
71. D. J. Watson, C. A. Randall, R. E. Newnham, and J. H. Adair, in *Ceramic Transactions, Ceramic Powder Science II* (G. L. Messing, E. R. Fuller, and H, Hausner, eds.). American Ceramic Society, Westerville, Ohio, 1987, pp. 154–162.

72. R. Roy, D. M. Roy, and E. F. Osborn, *J. Am. Ceram. Soc. 33*(5): 152-159 (1950).
73. V. G. Hill, R. Roy, and E. F. Osborn, *J. Am. Ceram. Soc. 35*(6): 135-142 (1952).
74. S. Komarneni, R. Roy, E. Breval, M. Ollinen, and Y. Suwa, *Advanced Ceramic Materials 1*: 87-92 (1986).
75. E. Matijevic, C. M. Simpson, N. Amin, and S. Arajs, *Colloids and Surfaces 21*: 101-108 (1986).
76. A. E. Regazzoni and E. Matijevic, *Colloids and Surfaces 6*: 189-201 (1983).
77. E. Narita, *J. Crys. Growth 78*: 1-8 (1986).
78. R. Vivekanandan, S. Philip, and T. R. N. Kutty, *Mat. Res. Bull. 22*: 99-108 (1986).
79. E. P. Stambaugh and J. F. Miller, in *Proceedings of the First International Symposium on Hydrothermal Reactions* (S. Somiya, ed.). Gakujutsu Bunken Fukyu-Kai, Tokyo, Japan, 1982, pp. 858-871.
80. K. C. Beal, in *Advances in Ceramics, Volume 21, Ceramic Powder Science* (G. L. Messing, K. S. Mazdiyasni, J. W. McCauley, and R. A. Haber, eds.). American Ceramic Society, Westerville, Ohio, 1987, pp. 33-42.
81. T. Takamori and L. D. David, *Ceram. Bull. 65*(9): 1282-1286 (1986).
82. S. Hirano, *Ceram. Bull. 66*(9): 1342-1344 (1987).
83. H. Toraya, M. Yoshimura, and S. Somiya, *J. Am. Ceram. Soc. 66*(2): 149-150 (1983).
84. S. Somiya, in *Materials Research Society Symposia Proceedings* (J. H. Crawford, Jr., Y. Chen, and W. A. Sibley, eds.), Vol. 24. Elsevier, New York, 1984, pp. 255-271.
85. S. Somiya, K. Suzuki, and M. Yoshimura, in *Advances in Ceramics, Volume 21, Ceramic Powder Science* (G. L. Messing, K. S. Mazdiyasni, J. W. McCauley, and R. A. Haber, eds.). American Ceramic Society, Westerville, Ohio, 1987, pp. 279-288.
86. R. H. Arendt, J. H. Rosolowski, and J. W. Szymaszek, *Mat. Res. Bull. 14*: 703-709 (1979).
87. E. I. Cooper and D. H. Kohn, *Ceramics International 9*(2): 68-72 (1983).
88. T. Kimura, T. Takahashi, and T. Yamaguchi, in *Ferrites, Proc. ICF 3d 1980* (H. Watanabe, S. Iida, and M. Sugimoto, eds.). Cent. Acad. Publ. Jpn., Tokyo, 1982, pp. 27-29.
89. T. Kimura and T. Yamaguchi, *Advances in Ceramics, Volume 21, Ceramic Powder Science* (G. L. Messing, K. S. Mazdiyasni, J. W. McCauley, and R. A. Haber, eds.). American Ceramic Society, Westerville, Ohio, 1987, pp. 169-177.
90. Y. Hayashi, T. Kimura, and T. Yamaguchi, *J. Mat. Sci. 21*: 757-762 (1986).
91. A. S. Shaikh and G. M. Vest, *J. Am. Ceram. Soc. 69*(9): 682-688 (1986).
92. P. Ravindranathan and K. C. Patil, *Am. Ceram. Soc. Bull. 66*(4): 688-692 (1987).
93. H. Wada, K. Nojima, K. Kuroda, and C. Kato, *Yogyo-Kyokai-Shi 95*(1): 140-144 (1987).

94. R. H. Heistand II and T. D. Gregory, in *Ceramic Transactions, Ceramic Powder Science II* (G. L. Messing, E. R. Fuller, and H. Hausner, eds.). American Ceramic Society, Westerville, Ohio, 1988, pp. 94–110.
95. J. J. Kingsley and K. C. Patil, *Mat. Lett.* 6(11–12): 427–432 (1988).
96. L. Koppens, *Science of Ceramics*, 8th ed. British Ceramic Society, 1976, pp. 101–109.
97. C. Marcilly, P. Courty, and B. Delmon, *J. Am. Ceram. Soc.* 53(1): 56–57 (1970).
98. M. P. Pechini, U.S. Patent 3,231,328 to Sprague Electric Co. (1967).
99. P. A. Lessing, *Ceram. Bull.* 68(5): 1002–1007 (1989).
100. N. G. Eror and H. U. Anderson, in *Materials Research Society Symposia Proceedings* (C. J. Brinker, D. E. Clark, and D. R. Ulrich, eds.), Vol. 73. Materials Research Society, Pittsburgh, Pa., 1986, pp. 571–575.
101. D. Seyferth and G. H. Wiseman, *Science of Ceramic Chemical Processing* (L. L. Hench and D. R. Ulrich, eds.). John Wiley, New York, 1986, pp. 354–362.
102. R. H. Baney, *Ultrastructure Processing of Ceramics, Glasses and Composites* (L. L. Hench and D. R. Ulrich, eds.). John Wiley, New York, 1984, pp. 245–255.
103. L. V. Interrante, L. E. Carpenter II, C. Whitmarsh, W. Lee, M. Garbauskas, and G. A. Slack, in *Materials Research Society Symposia Proceedings* (C. J. Brinker, D. E. Clark, and D. R. Ulrich, eds.), Vol. 73. Materials Research Society, Pittsburgh, Pa., 1986, pp. 359–366.
104. M. L. J. Hackney, L. V. Interrante, G. A. Slack, and P. J. Shields, in *Ultrastructure Processing of Advanced Ceramics* (J. D. Mackenzie and D. R. Ulrich, eds.). John Wiley, New York, 1988, pp. 99–111.
105. A. E. Berkowitz and J. L. Walter, *J. Mater. Res.* 2(2): 277–288 (1987).
106. J. L. Woodhead, *JME* 6(6): 887–925 (1984).
107. M. Akinc and K. Richardson, in *Materials Research Society Symposia Proceedings* (C. J. Brinker, D. E. Clark, and D. R. Ulrich, eds.), Vol. 73. Materials Research Society, Pittsburgh, Pa., 1986, pp. 99–109.
108. J. L. Woodhead, K. Cole, J. T. Dalton, J. P. Evans, and E. L. Paige, *Sci. Ceram.* 12: 179–185 (1980).
109. J. L. Kelly, A. T. Kleinsteuber, S. D. Clinton, and O. C. Dean, *I & EC Process Design and Development* 4(2): 212–216 (1965).
110. J. L. Woodhead, *Sci. Ceram.* 4: 105–111 (1967).
111. T. Kanai, W. E. Rhine, and H. K. Bowen, in *Ceramic Transactions, Ceramic Powder Science II* (G. L. Messing, E. R. Fuller, and H. Hausner, eds.). American Ceramic Society, Westerville, Ohio, 1988, pp. 119–126.
112. M. E. A. Hermans, *Powd. Metal. Inter.* 5(3): 137–140 (1973).
113. M. E. A. Hermans, in *Science of Ceramics* (C. Brosset and E. Knopp, eds.), Vol. 5. Swedish Institute for Silicate Research, 1970, pp. 523–538.
114. A. R. Hardy, Preparation of submicrometer, unagglomerated oxide particles by reaction of emulsion droplets, Ph.D. diss., MIT, Cambridge, Mass., 1988.
115. A. J. I. Ward and S. E. Friberg, *MRS Bull. December:* 41–46 (1989).
116. M. Gobe, K. K. No, K. Kanori, and A. Kitahara, *J. Coll. Inter. Sci.* 93(1): 293–295 (1983).

117. K. Kandori, K. N. Kijiro, and A. Kitahara, *J. Coll. Inter. Sci. 122*(1): 78–82 (1988).
118. H. Yamauchi, T. Ishikawa, and S. Kondo, *Colloids and Surfaces 37*: 71–80 (1989).
119. J. B. Nagy, *Colloids and Surfaces 35*: 201–220 (1989).
120. D. W. Matson and R. D. Smith, *J. Am. Ceram. Soc. 72*(6): 871–881 (1989).
121. M. Visca and E. Matijevic, *J. Coll. Inter. Sci. 68*(2): 308–319 (1979).
122. B. J. Ingebrethsen and E. Matijevic, *J. Aerosol Sci. 11*: 271–280 (1980).
123. A. Balboa, R. E. Partch, and E. Matijevic, *Colloids and Surfaces 27*: 123–131 (1987).
124. A. Sood and R. A. Marra, U.S. Patent 4,678,657 to Aluminum Company of America (1987).
125. T. T. Kodas, A. Sood, and S. E. Pratsinis, *Powder Tech. 50*: 47–53 (1987).
126. P. A. Haas, *Nuclear Tech. 10*: 283–292 (1971).
127. H. Kirchhoefer, *Chem. Abs. 108*(24): 207167d (1988).
128. E. B. Slamovich and F. F. Lange, in *Materials Research Society Symposia Proceedings* (C. J. Brinker, D. E. Clark, and D. R. Ulrich, eds.), Vol. 121. Materials Research Society, Pittsburgh, Pa., 1988, pp. 257–262.
129. J. G. M. de Lau, *Am. Ceram. Soc. Bull. 49*(6): 572–574 (1970).
130. P. K. Gallagher, D. W. Johnson, Jr., and F. Schrey, *Am. Ceram. Soc. Bull. 55*(6): 589–93; *Chem. Abs. 85*(14): 103113y (1976).
131. D. W. Matson, R. C. Petersen, and R. D. Smith, in *Advances in Ceramics, Volume 21, Ceramic Powder Science* (G. L. Messing, K. S. Mazdiyasni, J. W. McCauley, and R. A. Haber, eds.). American Ceramic Society, Westerville, Ohio, 1987, pp. 109–120.
132. F. J. Schnettler, F. R. Monforte, and W. W. Rhodes, *Sci. Ceram. 4*: 79–90 (1968).
133. P. J. McGrath, Freeze drying spray frozen aerosols: Applications in synthesizing high T_c ceramic superconductors, M. S. thesis, University of Washington, Seattle, Wash., 1989.
134. D. W. Johnson and F. J. Schnettler, *J. Am. Ceram. Soc. 53*(8): 440–444; *Chem. Abs. 73*(16): 81510y (1970).
135. I. K. Lloyd and R. J. Kovel, *J. Mat. Sci. 23*: 185–188 (1988).
136. M. P. O'Toole and R. J. Card, *Am. Ceram. Soc. Bull. 66*(10): 1486–1489 (1987).
137. D. W. Sproson and G. L. Messing, in *Advances in Ceramics, Volume 21, Ceramic Powder Science* (G. L. Messing, K. S. Mazdiyasni, J. W. McCauley, and R. A. Haber, eds.). American Ceramic Society, Westerville, Ohio, 1987, pp. 99–108.
138. O. Sakurai, N. Mizutani, and M. Kato, *Yogyo-Kyokai-Shi 94*(8): 117–121 (1986).
139. Y. Kanno and T. Suzuki, *J. Mat. Sci. Lett. 7*: 386–388 (1988).
140. G. D. Ulrich, *C&EN August 6*: 22–29 (1984).
141. J. H. Flint and J. S. Haggerty, in *Ceramic Transactions, Ceramic Powder Science II* (G. L. Messing, E. R. Fuller, and H. Hausner, eds.). American Ceramic Society, Westerville, Ohio, 1988, pp. 244–252.

142. M. Aoki, J. H. Flint, and J. S. Haggerty, in *Ceramic Transactions, Ceramic Powder Science II* (G. L. Messing, E. R. Fuller, and H. Hausner, eds.). American Ceramic Society, Westerville, Ohio, 1988, pp. 253–260.
143. R. A. Marra, Homogeneous nucleation and growth of silicon powder from laser heated gas phase reactants, Ph.D. diss., MIT, Cambridge, Mass., 1983.
144. W. R. Cannon, S. C. Danforth, J. H. Flint, J. S. Haggerty, and R. A. Marra, *J. Am. Ceram. Soc. 65*(7): 324–330 (1982).
145. Y. Suyama, R. M. Marra, J. S. Haggerty, and H. K. Bowen, *Am. Ceram. Soc. Bull. 64*(10): 1356–1359 (1985).
146. A. Kato, in *Advances in Ceramics, Volume 21, Ceramic Powder Science* (G. L. Messing, K. S. Mazdiyasni, J. W. McCauley, and R. A. Haber, eds.). American Ceramic Society, Westerville, Ohio, 1987, pp. 181–192.
147. A. Nishida, A. Ueki, and K. Yoshida, in *Advances in Ceramics, Volume 21, Ceramic Powder Science* (G. L. Messing, K. S. Mazdiyasni, J. W. McCauley, and R. A. Haber, eds.). American Ceramic Society, Westerville, Ohio, 1987, pp. 265–269.
148. F. Kirkbir and H. Komiyama, *Canadian J. Chem. Eng. 65*: 759–766 (1987).
149. P. Ho, R. J. Buss, and R. E. Loehman, *J. Mater. Res. 4*(4): 873–881 (1989).
150. H. Anderson, T. T. Kodas, and D. M. Smith, *Ceram. Bull. 685*: 996–1000 (1989).
151. M. Kamo, Y. Sato, and N. Sedaka, *Chem. Abs. 109*(18): 154993g (1988).
152. G. D. Ulrich and J. W. Riehl, *J. Coll. Inter. Sci. 87*(1): 257–265 (1982).
153. R. K. Buchanan, P. J. Grose, and M. S. J. Gani, *Proc. Aust. Ceram. Conf. 9*: 102–104, *Chem. Abs. 95*(4): 29022f (1980).
154. J. S. Haggerty, in *Ultrastructure Processing of Ceramics, Glasses and Composites* (L. L. Hench and D. R. Ulrich, eds.). John Wiley, New York, 1984, pp. 353–366.
155. Y. Suyama, M. Tanaka, and A. Kato, *Ceramurgia Int'l. 5*(2): 84–88 (1979).
156. M. Fukuhara, H. Nagai, and H. Mitani, *J. Japan Inst. Metals 42*(6): 558–592 (1978).
157. A. Kato, J. I. Hojo, and T. Watar, *Mat. Sci. Res. 17*(1): 123–135 (1984).
158. S. Hori and Y. Ishii, *Chem Abs. 106*(24): 200690u (1987).
159. M. Cauchetier, O. Croix, M. Luce, M. Michon, J. Paris, and S. Tistchenko, *Ceramics Int'l. 13*: 13–17 (1987).
160. K. S. Mazdiyasni, C. T. Lynch, and J. S. Smith, *J. Am. Ceram. Soc. 48*(7): 372–375 (1965).
161. M. Koizumi, J. Oosawa, I. Suzuki, and K. Kawada, *Chem. Abs. 109*(22): 195992z (1988).
162. W. M. Shen and C. F. Chang, in *Advances in Ceramics, Volume 21, Ceramic Powder Science* (G. L. Messing, K. S. Mazdiyasni, J. W. McCauley, and R. A. Haber, eds.). American Ceramic Society, Westerville, Ohio, 1987, pp. 193–202.
163. S. Iwama, K. Hayakawa, and T. Arizumi, *J. Cryst. Growth 56*(2): 265–269 (1982).
164. S. DiVita and R. J. Fischer, U.S. Patent 2,985,506 to U.S. Department of the Army (1961).

165. R. J. Walsh, U.S. Patent 2,988,422 to Monsanto Chemical Co. (1961).
166. J. P. Pollinger and G. L. Messing, in *Advances in Ceramics, Volume 21, Ceramic Powder Science* (G. L. Messing, K. S. Mazdiyasni, J. W. McCauley, and R. A. Haber, eds.). American Ceramic Society, Westerville, Ohio, 1987, pp. 217-228.
167. M. Kagawa, M. Kikuchi, and Y. Syono, *J. Am. Ceram. Soc. 66*(1): 751-754 (1983).
168. R. Watanabe and I. Yazaki, in *Horiz. Powder Metall., Proc. Int. Powder Metall. Conf. Exhib.* (W. A. Kaysser and W. J. Huppmann, eds.), Vol. 1, 1986, pp. 105-108; *Chem. Abs. 108*(26): 225235d.
169. W. E. Kuhn, U.S. Patent 3,166,380 to Carborundum Co. (1965).
170. J. F. Wang, Small particle formation from lyotropic liquid crystals, Ph.D. diss., Clarkson University, Potsdam, N.Y., 1989.
171. J. H. Fendler, *Chem. Rev. 87*: 877-899 (1987).
172. G. H. Nancollas, *Adv. Colloid Interface Sci. 10*: 215-252 (1979).
173. S. Bhandarkar and A. Bose, *J. Coll. Inter. Sci. 135*(2): 531-538 (1990).
174. J. E. Mark, in *Ultrastructure Processing of Advanced Ceramics* (J. D. Mackenzie and D. R. Ulrich, eds.). John Wiley, 1988, New York, pp. 623-633.
175. Calvert, P. D., in Proceedings of the 4th International Conference on Ultrastructure Processing of Ceramics, Glasses and Composites, February 19-24, 1989, Tucson, Ariz.
176. S. Iwama, K. Hayakawa, and T. Arizumi, *J. Cryst. Growth 56*(2): 265-269 (1982).
177. L. Dupuy, Synthesis of aluminum nitride via molecular carbothermal reduction, Master's thesis, Rutgers University, Piscataway, N.J., 1991.

3

Surface Chemical Characterization of Ceramic Powders

LENNART BERGSTRÖM Institute for Surface Chemistry, Stockholm, Sweden

I. INTRODUCTION

The forming of ceramic bodies from colloidal suspensions of ceramic powders includes several steps, such as powder mixing, suspension preparation, and consolidation. Structural heterogeneities such as agglomerates or density gradients can arise in any of these forming steps. Many of these heterogeneities can be eliminated if the interparticle forces can be controlled. The chemical nature of the ceramic powder surface is generally of crucial importance in controlling the interparticle forces. The adsorption of dispersants, surface charge properties, and even solubility are often related to the surface chemical properties of the ceramic powders. In the production of advanced and high-performance ceramic materials, the characterization and control of the surface chemical properties of the ceramic powders are therefore of major importance.

The rapid development of new surface-sensitive analytical techniques and the improvement of already existing techniques have provided new insight into the chemical identity and molecular interactions at the surface film. Several methods, like secondary ion mass spectroscopy, can identify atomic monolayers on surfaces, and other methods, like infrared spectroscopy, can provide information about the functional groups on the powder surface. Presently, there are several different techniques that have the potential to improve our knowledge of the surface chemistry of different ceramic powders.

The structure of this chapter is based on important questions, vital for optimal processing, to which a thorough surface analysis should provide answers.

The main questions are

1. What is the chemical composition of the surface film (stoichiometry, extent of oxidation, surface impurities, etc.)?
2. What is the chemistry of the surface functional groups? How is their stability and reactivity to thermal and chemical treatments?
3. What is the behavior at the solid-liquid interface? What is the dissolution, surface charge, and electrokinetic behavior in aqueous solutions? What types of interaction govern the adsorption behavior in nonaqueous media?

The emphasis is put on the interactions at the solid-liquid interface, as this is so important in ceramic processing. However, questions 1 and 2 are important because they seek fundamental information regarding the chemistry of the powder surface. This chapter presents a summary of the main analysis techniques used for the chemical characterization of powder surfaces. Examples in which the techniques have been used to characterize ceramic powders are given when such information is readily available. In a few cases, more general examples relating to silica have been used.

Hopefully, this chapter will provide an insight and stimulate an understanding of the need for more thorough surface characterization and control of ceramic powders.

II. ELEMENTAL COMPOSITION OF POWDER SURFACES

Information about the elemental composition of ceramic powder surfaces can be obtained by a wide range of analytical techniques. All the surface analysis techniques used involve bombarding the surface with one species (photons, electrons, ions, etc.) and detecting and analyzing the species emitted from the surface as a result of the bombardment. A high surface sensitivity is achieved either by a limited depth of penetration of the ingoing species or by a limited depth of escape of the outgoing species.

The three surface analysis techniques that currently have the greatest potential for, or are already used for, inorganic powder characterization are (a) Auger electron spectroscopy (AES), (b) secondary ion mass spectroscopy (SIMS), and (c) electron spectroscopy for chemical analysis (ESCA), also called x-ray photoelectron spectroscopy (XPS). The ingoing and outgoing species used in these and other chemical techniques are schematically shown in Fig. 1, and an overview of the capabilities of XPS,

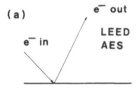

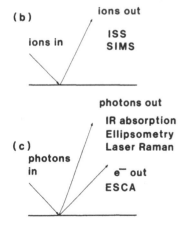

FIG. 1 Pictorial representation of surface analysis techniques. Various combinations of species going in and out determine the various surface analysis techniques.

AES, and SIMS is given in Table 1 [1]. The spatial resolution and detection limits of these methods are developing rapidly.

A. Auger Electron Spectroscopy

When electrons or x-rays with energies in the range of 1–10 keV strike the atoms in a material, electrons with binding energies smaller than the energy of the incident beam may be ejected from the core levels of the atom, resulting in ionization. The electron vacancy created by ionization is filled by an electron from a higher energy level in the atom. This releases energy, which may be transferred to a third electron in the same or another atom. If the binding energy of this electron is less than the transferred energy, it is ejected from the atom. Electrons ejected by this process are called Auger electrons, and the analysis of their energies forms the basis of Auger electron spectroscopy (AES). The energies of Auger electrons are characteristic of each atom and can be used to identify the atoms present. Information about the chemical binding of each atom can be extracted from the Auger spectra because the kinetic energy of the Auger electron depends on the electron density surrounding the atom. The spectra are usually differentiated in order to identify the peak positions more easily.

An excellent review of the measurement technique, tabulated energies for the different elements, and examples of applications of AES are given in Ref. 2. Although AES shows high surface sensitivity and good lateral resolution, the technique has not found wide application in the field of ceramic powders. Because these powders are poor conductors of electricity, the emission of electrons leads to considerable charging of the samples [3,4], and this leads to large shifts in the energies of AES electrons, making a reliable analysis of the spectra very difficult.

However, by recording spectra in defocusing conditions with small acceleration voltages and small currents up to 0.01 µA, hot-pressed Si_3N_4 (HPSN) samples could be analyzed [4]. Air-fractured HPSN fluxed with 10 wt % MgO was studied (Fig. 2). The presence of calcium and fluorine was attributed to the Si_3N_4 powder and the presence of sulphur to contamination during processing.

AES has also been performed on Si powder [5]. In this case an unfired (green) compacted specimen was fractured and analyzed. Oxygen and carbon were the only impurities detected. By measuring the variation in the oxygen and carbon concentrations with sputtering time it was found that the concentrations were rather high in the as-broken condition and decreased rapidly to bulk values. Thus the main sources of contamination on the as-broken surfaces are atmospheric species.

TABLE 1 Capabilities of Techniques [a]

	Depth of analysis	Spatial resolution	Elemental range	Detection limits	Chemical information	Quantification
XPS	~1–5 nm	>1 mm (100 μm)	All > He	0.1–1 at %	Good	Semi-quantitative without standards, quantitative with
AES	~1–5 nm	~1 μm (25 nm)	All > He	0.1–1 at %	Some	As above
SIMS (dynamic)	~5 nm	~150 μm	All elements	>ppb	Poor	Possible by careful use of standards
SIMS (static)	Monolayer	>1 mm	As above	~10^{-6} Monolayer	V good	Very difficult
SIMS (imaging)	~5 nm	1 μm (50 nm)	As above	Complex	Poor	Very difficult

[a]Values quoted are intended as a guide to the performance of typical equipment at the present time; values in parentheses indicate current state-of-the-art performance.
Source: Ref. 1.

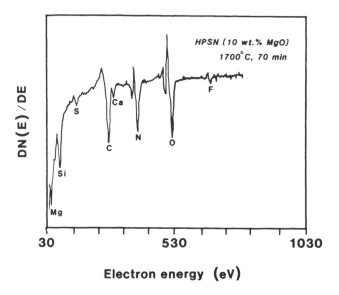

FIG. 2 Auger spectrum of a hot-pressed Si_3N_4 (HPSN) sample containing 10 wt % MgO. Note the presence of Ca, F, and S. The large carbon peak is probably due to atmospheric contamination. (From Ref. 4.)

Recently, practical methods have been developed to reduce surface charging during AES analysis of insulating materials [6]. In the analysis of Si_3N_4, Al_2O_3, and BN, it was found that irradiating the sample with a second electron beam or supplying a positive charge by a low-energy ion beam reduced the surface charge to acceptable levels.

B. Secondary Ion Mass Spectroscopy

Secondary ion mass spectroscopy (SIMS) is based on an analysis of the ions (single atoms or clusters of atoms) ejected when a surface is bombarded by a beam of primary ions or atoms. The ejected ions are analyzed by a mass spectrometer. The sensitivity of the technique is determined by the ratio of ionized to nonionized species. With a high flux of primary ions, elements can be detected in the part per billion region (so-called dynamic SIMS). However, this high flux of primary ions results in rapid erosion of the surface film, and the depth of resolution actually deteriorates with increasing ion energy. Therefore so-called static SIMS, which uses a low flux of primary ions, is important. This increases the lifetime of the surface layer, although the spatial resolution is reduced (Table 1). Basic concepts together with the theoretical background and applications have been extensively discussed in recent reviews [7-9].

An important problem in the use of SIMS is quantification. Secondary ion emission is very sensitive to the electron state (chemical environment) of the atom or molecule to be ionized and to the substrate from which it is emitted. For example, in an analysis of the oxide coating on silicon wafers [10], the intensity of the Si-containing peaks *decreased* when the pure silicon substrate was reached, although the silicon content increased. This was due to the enhancement of the $^{28}Si^+$ signal by the oxygen atoms in the oxide film.

Recently, a new method with much higher spatial resolution has been developed. This method, called imaging SIMS, uses well-focused ion beams to produce submicron spot sizes at the target [9]. Imaging SIMS has been used to examine sintered Si_3N_4 bodies with Y_2O_3 or MgO as sintering aids [11]. The investigators obtained 0.1 µm resolution and could distinguish between Si_3N_4 grains and the intergranular phase.

Very few measurements have been performed on powders. Figure 3 shows a spectrum from a pressed pellet of Si_3N_4 powder [3]. The interpretation is that the Si_3N_4 surface is mainly contaminated by oxygen but that small amounts of Cl, F, and Na and traces of Li, K, Cu, and Fe are present. It is very difficult to quantify the amounts of these impurities. One way is to compare the spectrum obtained with standards, but to prepare standards with high accuracy is a difficult and tedious procedure. Another problem with powders is that it is not possible to depth-profile these samples, as the surface of a pressed sample is too rough. Spectra of clean Si, SiC, and Si_3N_4 have been obtained from thin films [12].

C. Electron Spectroscopy for Chemical Analysis

It was mentioned in the description of AES that irradiation of atoms with x-rays in the energy range 1-10 keV may lead to the ejection of electrons from the core level of the atom (the well-known photoelectric effect). This is the principle of electron spectroscopy for chemical analysis (ESCA, or x-ray photoelectron spectroscopy, XPS), which is a very widely used method for analysis of the surface composition of ceramic powders.

The kinetic energy of the ejected electrons is given by

$$E_k = h\nu - E_b - \Phi \tag{1}$$

where E_k is the kinetic energy of the photoelectron, $h\nu$ is the energy of the x-ray photon, E_b is the binding energy of the photoelectron, and Φ is the spectrometer work function.

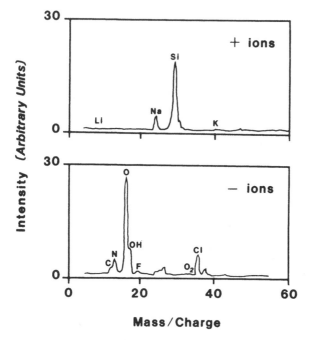

FIG. 3 Positive and negative secondary ion mass (SIMS) spectra of an Si_3N_4 powder. Several impurities are present at the powder surfaces. (From Ref. 3. Reprinted by permission of the American Ceramic Society.)

The binding energies for electrons ejected from different elements are generally well separated, and ESCA is therefore very suitable for elemental analysis. The binding energy is determined mainly by the electrostatic attraction between the electron and the other electrons surrounding the atom, i.e., the total electron density around the atom. The binding energy is therefore dependent on the chemical binding of the ejecting atom, which can thus be identified. For example, in the analysis of ceramic powders it is possible to determine not only the presence of different elements but also their state of oxidation.

The surface sensitivity of ESCA stems from the limited penetration range of the emitted photoelectrons in solid materials. The average distance a photoelectron travels is a function mainly of its kinetic energy and of the density of the solid. This exponential decay of the electron intensity is given by,

$$dI \propto \exp\left(\frac{-x}{\lambda}\right)dx \tag{2}$$

where I is the intensity of photoelectrons, x is the distance from the solid surface, and λ is the inelastic mean free path.

This means that 63% of the signal originates from a surface layer of thickness λ, often called the escape depth. Andrade [13] has collected λ-values calculated by various relations, and the results show mean free paths of about 1–2 nm for metals and 1.5–3.5 nm for oxides when the kinetic energy of the electrons is about 1000 eV.

The irradiation used for ESCA analysis also always gives rise to Auger electrons, which can be observed simultaneously with the x-ray photoelectrons and gives complementary information on the elements present and their binding state. Because they are emitted from core levels, the x-ray photoelectrons are much less sensitive to charging effects than Auger electrons, which is why ESCA is better suited for the analysis of insulating surfaces than AES. Several excellent reviews of the ESCA technique have been published (e.g., [2,13]), and the reader is referred to these for more comprehensive information.

Taylor [14] used ESCA to investigate the surface composition of silicon carbide powders and whiskers. After correcting for charging effects, it was possible to relate the binding energies of the Si(2p) and C(1s) peaks to the degree of surface oxidation and the amount of graphitic carbon on the surface. It was found that ultrafine SiC powders, grown by a plasma process, exhibit graphitic carbon and a thin suboxide coating. The SiC whiskers studied were generally covered by a thicker and more silicalike oxide. In a recent study of SiC whiskers [15] from different manufacturers, the surface chemistry characteristics, analyzed with ESCA, were divided into four general categories according to differences in oxygen content, carbon content, and chemical composition of the surface film.

ESCA can also be used for quantitative measurements. However, only relative values of surface chemical compositions can usually be obtained, and the accuracy is normally not better than 10%. The quantification of ESCA data has been thoroughly examined in recent reviews [2,16].

A simplified expression for the photoelectron intensity I_A emitted from a given electron shell of a given atom is

$$I_A = K_S K_{sp} \sigma_A N_A \lambda_A L_A \tag{3}$$

where K_S is proportional to the incident radiation flow, K_{sp} depends on the analyzer and detector sensitivity, L_A corrects for the angular asymmetry of the photoelectron emission, σ_A is the photoelectron scattering cross section, N_A is the atom density (number of atoms per unit area), and λ_A is the inelastic mean free path of the emitted electrons. For many instruments, K_{sp} varies as $(E_{kin})^{-1}$, where E_{kin} is the kinetic energy of the emitted electrons. On the other hand, $\lambda_A \approx \text{const} \cdot (E_{kin})^a$, where a varies between 0.7 and 1. The product $K_{sp}\lambda_A$ therefore varies rather slowly with E_{kin}. Hence, Eq. (3) can be utilized for a semiquantitative determination of the composition of the surface, provided the scattering cross-sections are known. Scofield [17] has calculated theoretical values of σ_A, and Seah [18] has compiled data for the different parameters.

A detailed evaluation of all the factors in Eq. (3) is somewhat complex, and a more direct approach is often used by writing

$$I_A = N_A S_A \tag{4}$$

where S_A is a sensitivity factor that has to be determined specifically for each element and each spectrometer. Wagner et al. [19] have published a list of sensitivity factors valid for a few spectrometers. The applicability of these sensitivity factors has been extended by calibration of different spectrometer intensity-energy response functions [20,21].

Quantitative ESCA measurements have been used to study the oxidation behavior of ultrafine α-SiC powder in air at temperatures up to 1100°C [22]. The effect of temperature on the oxidation of the surface layer of the powders was followed by analyzing the Si(2p) spectra (Fig. 4). The thickness of the oxide coating and the oxidation rate were determined from the relative intensity of the Si(2p) peak of SiO_2 to the Si(2p) peak of SiC using a uniform overlayer model [13]. A similar treatment has been used to evaluate the thickness and mechanism of deposition of silica on α-Al_2O_3 particles [23] and the thickness of oxide coating on Si_3N_4 [24]. The assumptions inherent in the uniform layer model are, however, probably oversimplifications. For example, the assumption that there is a distinct borderline between the overlayer and the bulk is not fulfilled in the case of Si_3N_4. Several studies have shown that the surface layer on most Si_3N_4 powders corresponds to an intermediate state between silica and silicon oxynitride [3,24,25]. Thus the oxygen-rich overlayer is not pure silica but probably consists of an atomic mixture of nitrogen and oxygen compounds. Another problem is the surface roughness, as this results in an angular dependence of the electron emission. Fulghum [26] has reviewed the important methods for quantification of the surface coverages of nonplanar samples with overlayers, e.g., powders of high surface area. Recently, Johansson [150] showed how different ESCA analysis methods may be used in the characterization of coated powder. Results from elastic peak measurements, depth profiling by sputtering or by varying the analysis angle, and an analysis of both elastic and inelastic parts of an ESCA peak were correlated and discussed. The different ESCA methods were also applied to the surface characterization of coated TiO_2 pigments.

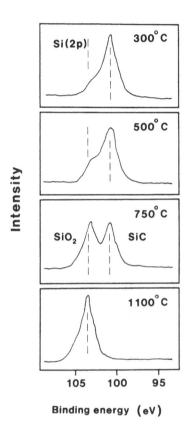

FIG. 4 ESCA spectra of Si(2p) peaks for α-SiC powder oxidized in air at different temperatures. The Si(2p) peak corresponding to SiO$_2$ increases in intensity with increasing temperature. (From Ref. 22.)

The surface layers of α-BN and β-BN have also been investigated [27]. The results show a fairly high oxygen content, and it is concluded from an analysis of the B(1s) and N(1s) peaks that the oxygen is chemically bound both to boron and to nitrogen atoms.

ESCA can also be used to obtain information about the chemical characteristics and relative amount of different oxides in mixtures. Paparazzo [28] has used the O(1s) spectra to analyze binary mixtures of Al$_2$O$_3$, Cr$_2$O$_3$, Fe$_2$O$_3$, and SiO$_2$. From measurements of the intensity ratios of the single O(1s) components, it was found that the single oxides retained their chemical identities in the mixtures.

Dang and coworkers [158–160] have in a number of papers shown how chemical derivatization and ESCA can be used to determine the functional groups on the surfaces of several inorganic materials such as Si_3N_4 and Al_2O_3. In this method, the functional group of interest is specifically reacted with a molecule containing a "tag" atom that is not part of the surface studied. The concentration of the tag atom is then quantitatively analyzed using ESCA. They used this method to determine the amine and silanol functional groups on glass [158] and to quantify the silanol group concentration on various silicon nitride powders [159] and on a wide variety of inorganic materials [160].

III. CHEMISTRY OF FUNCTIONAL SURFACE GROUPS

A. Infrared Spectroscopy

The vibrational spectrum of inorganic powders is the traditional, and still mainly used, source of information about the chemistry of functional surface groups. The development of Fourier transform instruments made it much easier to obtain IR spectra and to apply IR spectroscopy to almost any form of sample. The basic concepts of FTIR spectroscopy have been described in several books [29–32] and review articles [33,34].

The major advantages of FTIR over dispersive instruments are that

1. All frequencies are measured at the same time, which facilitates kinetic studies and makes the time for analysis very short (multiplex advantage).
2. The use of mirrors in the interferometer instead of slits increases the throughput energy 80–200 times.
3. Several scans can be added, and thus the signal-to-noise (S/N) ratio can be increased.
4. The spectrometer is automatically frequency calibrated by the use of an HeNe laser as the source of the modulating beam. No calibration standard for determination of absolute wavenumbers is required.

Accessories for transmission and internal reflection techniques as well as microsampling techniques have been developed and successfully used to monitor powder surfaces.

The most frequently used technique to examine powder surfaces and identify the structure of the surface species is transmission spectroscopy. A thin disc formed by pressing the powder is inserted in the path of the infrared beam. Strictly speaking, transmission spectroscopy is not a surface IR method, as there is no surface selectivity in this method. However, if powders with high surface area are used, the relative concentration of surface species is sufficiently high to produce a spectrum in which these species can be identified and in some cases quantified. It is important to bear in mind that transmission spectroscopy only can be used when the surface groups absorb IR radiation in a region where the bulk is transmitting.

When this is not the case, or when a more surface-sensitive technique is required, diffuse reflectance infrared spectroscopy (DRIFT) may be used. In this technique, the IR beam is focused on a powder bed, and the reflected radiation is collected using ellipsoidal mirrors. Fuller and Griffiths [35] were among the first to present an optical setup for DRIFT. Usually the powder is mixed with 90–99% KBr to enhance the intensity of the reflected radiation. As the sample preparation is simple and the surface signal is enhanced, DRIFT obviously has advantages over transmission measurements. However, one drawback is that the bandwidths and relative intensities of a DRIFT spectrum change with particle size [35]. The scattering pattern also changes with the arrangement of the particles. In a recent development called diffuse transmittance (DT) spectroscopy, which is suitable for fibers, it appears that the orientation effects can be avoided [36]. Probably DT spectroscopy can be useful for IR spectroscopy of ceramic fibers.

In order to obtain reproducible results, it is very important to control the amount of surface impurities and especially the water content on hydrophilic, high surface energy powders such as ceramic powders. Thus it is necessary to pretreat the powders by degassing, and in some cases heating before measuring, and this should preferably be done in situ. Today, IR cells have been developed that can be evacuated down to 10^{-6} mbar and heated up to 1000°C [37]. Also, different types of cells have been constructed to admit the flow of a controlled amount of gas or liquid in contact with the powder.

The types of ceramic powders mainly considered in this chapter are either oxide powders (ZrO_2, Al_2O_3, MgO, Y_2O_3) or thermodynamically unstable nitrides or carbides (SiC, Si_3N_4, B_4C, BN, etc.), which are probably partly oxidized. Thus we expect the powder surfaces to be rich in -OH groups with the addition of -NH and maybe -CH groups. Also, it is important to characterize surface groups stemming from impurities. In most cases, the spectral region 400–4000 cm^{-1} is investigated. The region between 3000 and 4000 cm^{-1} is the most important one, as the surface hydroxyl groups (-OH) adsorb at 3000–3750 cm^{-1} (OH stretch) and the surface amine groups at 3300–3500 cm^{-1} (NH or NH_2 stretch [38]). Fortunately, the bulk powders usually have no inherent infrared absorption in this region.

As an example, Fig. 5 shows a DRIFT spectrum of laser-derived Si_3N_4 before and after exposure to humid air [39]. In the unexposed powder spectrum (Fig. 5a), two broad overlapping peaks occur at 3490 cm^{-1} and 3395 cm^{-1}, due to asymmetric and symmetric N-H stretching absorptions of amines. The peak at 1550 cm^{-1} is assumed to be due to an amino deformation mode. When the powder is exposed to humidity (Fig. 5b) a sharp peak appears at 3760 cm^{-1} that is caused by the stretch of a free Si-OH group. Some of the amino groups remain as a broad unresolved peak at 3350 cm^{-1} due to hydrogen bonding. The silanol groups (Si-OH) on silica have been extensively studied over several decades. Early work on, for example, silica

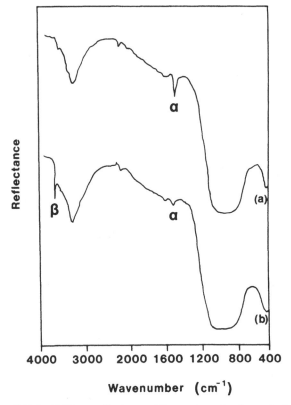

FIG. 5 Diffuse reflectance IR spectra of a laser-derived Si_3N_4 powder (a) unexposed and (b) exposed to the humid air. Note the appearance of a peak at 3760 cm^{-1} (SiOH) after exposure. (From Ref. 39. Reprinted by permission of the American Ceramic Society.)

has been reviewed in three excellent books [40–42]. More recent reviews [43,44] show a similar assignment of the different IR absorption peaks to different types of silanol groups. Silica surface chemistry will not be discussed here; the reader is referred to these reviews for further information.

The chemistry of the functional surface groups can be strongly affected by heat treatment. The thermal stability of Si-OH, Si-H, and Si-NH groups on silicon-oxynitride films deposited by plasma-enhanced CVD has been investigated [45]. The -NH group at 3330 cm^{-1} was found to be stable up to 400°C in H_2/N_2 gas mixtures, whereas the Si-H group at 2180 cm^{-1} was stable up to 500°C.

Hydrogenated silicon powder made by laser-induced decomposition of silane was investigated by IR spectroscopy [46], and two bands, one at 2094 cm^{-1} and one at 2080 cm^{-1}, were assigned to Si-OH and Si-H, respectively. Controlled oxidation and the detection of a shift of these bands to higher frequencies showed that the hydride groups existed on the powder surface.

IR spectroscopy can also be used to follow changes in the formation of ceramic powders. Yttrium hydroxide formed by precipitation from emulsion was heated to 1200°C to form yttria [47]. The disappearance of organic material at low temperatures and of hydroxide and carbonate groups at high temperatures was followed by IR spectroscopy.

It is very important to study the water adsorption on the powder surfaces when dehydration and rehydration of the surface groups are investigated. One example of an IR study of water sorption is an investigation of monoclinic and tetragonal zirconia heated to elevated temperatures [48]. Two types of hydroxyl groups bound to one or two cations, respectively, were found at 3760 cm^{-1} and 3660 cm^{-1} on monoclinic zirconia. On tetragonal zirconia, the signals from each of these groups were split into two, and absorption peaks was observed at 3740, 3720 and 3662, 3645 cm^{-1}, respectively. Water was shown to sorb strongly onto the powder surfaces with a broad band at 2800–3600 cm^{-1}. Desorption at elevated temperatures indicated that the absorption bands between 3440 and 3595 cm^{-1} should be attributed to water molecules strongly bound to the surface.

A further aid in characterizing and distinguishing surface hydroxyl groups is isotopic exchange. Hair [43] has reviewed the use of D_2O exchange to characterize the silanol group. By substitution of H by D in surface silanol groups, the stretching frequency is moved from 3747 to 2750 cm^{-1}. Si_3N_4 powders have been subjected to D_2 exchange [49]. The silanol group at 3742 cm^{-1} and the Si_2-NH group at 3355 cm^{-1} were both shifted to lower absorption frequencies after D_2 exchange (Fig. 6). Almost all silanol groups are available for deuteration, as shown by the very small peak at

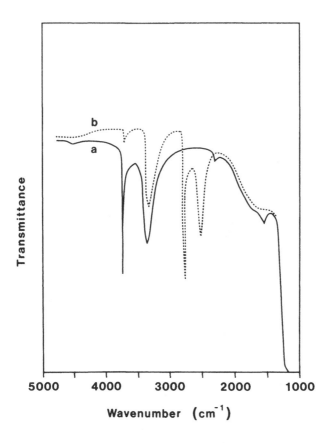

FIG. 6 FT-IR transmission spectra of a Si_3N_4 powder (a) activated at 400°C and (b) after D_2 exchange. Both the silanol and the amino group were shifted to lower absorption frequencies after D_2 exchange. (From Ref. 49.)

3742 cm^{-1} in Fig. 6b. Only some of the amino groups are, however, perturbed to lower absorption frequencies (~ 2500 cm^{-1}) and thus available for deuteration. This can be due to the existence of amino groups inside the powder.

It is also possible to characterize surface groups by reaction with a strong reagent that forms different species on the powder surface. By analyzing these species with IR spectroscopy, information about the surface groups can be deduced. For example, γ-alumina was reacted with trimethylgallium (TMG) [50], and two different types of surface species were formed depending on the nature of the surface group.

$$AlOH + Ga(CH_3)_3 \rightarrow \underset{I}{AlOGa(CH_3)_2} + CH_4 \tag{5}$$

$$Al^+ AlO^- + Ga(CH_3)_3 \rightarrow \underset{I}{AlOGa(CH_3)_2} + \underset{II}{Al\text{-}CH_3} \tag{6}$$

When the alumina powder was treated at different temperatures, it was found that the number of Lewis acid-base sites (Al^+/AlO^-) increased with the temperature of activation, and the ratio of the number of molecules that reacted with Lewis acid-base sites to those that reacted with AlOH sites was determined. The interaction between sorbate molecules and surface groups is discussed more extensively in Section IV.

B. Miscellaneous Techniques

Electron spin resonance (ESR) spectroscopy, which involves the study of paramagnetic ions and free radicals [51], has been used to investigate surface species on solids. ESR spectroscopy has been used to study electron transfer processes involved in the formation of stable radicals upon adsorption on oxides. In one study of MgO, γ-Al_2O_3, CaO, and SiO_2-Al_2O_3 [52] powders heated to 700–900°C, the concentrations of anionic and cationic radicals upon adsorption were measured. It was found that the concentration of electron donor (anion-forming) sites correlated with an increasing Lewis base strength (SiAl < Al_2O_3 < MgO < CaO). The concentration of electron acceptor sites was found to increase with increasing Lewis acid strength. Sites for electron donation were believed to be O^{2-} and those for electron acceptance Me^{n+} ions. The ESR spectra of 7,7,8,8-tetra-cyanoquino-dimethane (TCNQ) have also been used to estimate the electron donor properties of different oxides [53,54]. TCNQ anion radicals were formed as a result of electron transfer to TCNQ from the oxide surface, and the electron donor strength was estimated from the radical concentration on the surface. The radical-forming activity (electron donor strength) decreased in the order MgO > ZnO > Al_2O_3 > TiO_2 > SiO_2 > NiO [53]. As these powders were heated only to 100°C in vacuum it was suggested that both surface hydroxyl groups and the nature of the semiconductor were responsible for the electron donating properties. At this low temperature, the number of surface hydroxyl groups is only slightly affected. Adsorption of TCNQ onto Al_2O_3 and TiO_2 from solvents with different acidic and basic properties showed large variations in the concentration of TCNQ radicals formed [54]. When the basicity of the solvent increased, the concentration of TCNQ radicals decreased considerably, as the solvent competed with TCNQ for the electron donor sites on the surface. The concentration of TCNQ radicals did not change greatly with increasing acidity of the solvent.

Nuclear magnetic resonance (NMR) spectroscopy can also be used to study surface species. Surface hydroxyl groups have been studied by NMR, and an early study by Hall et al. [55] used the absolute intensities of the proton magnetic resonance spectra to evaluate the surface densities of hydrogen. Within ±20% they obtained the same values using PMR and deuterium exchange measurements. In a recent study, the surface reactivity of hydroxyl groups on silica with trimethylchlorosilane (TMCS) was studied by ^{29}Si solid state cross-polarization (CP) magic angle spinning (MAS) NMR [56]. Using this technique it was possible to distinguish between geminal (two hydroxyl groups on the same silicon atom) and single silanols. It was found that geminal hydroxyl groups are much more reactive.

Raman spectroscopy is another technique that can be used to study species adsorbed to a surface. Hendra and Fleischman [57] have reviewed the use of Raman and especially stressed the use of laser Raman spectroscopy.

IV. INTERACTION AT THE SOLID-GAS INTERFACE

Adsorption at the solid-gas interface onto oxides has been extensively studied in relation to heterogeneous catalysis. Here we limit ourselves to studies more specifically directed toward advanced ceramics. Some of these have indeed been studied in view of their potential interest as catalyst supports. The main methods used have been IR spectroscopy and chromatography. Characterization by inverse gas chromatography seems to be of particular interest.

A. Infrared Spectroscopy

In investigations of the interaction between solid surfaces and sorbate molecules in the gas phase, the cell design is of great importance. It must be possible to control the atmosphere and the temperature. Cell requirements have been described extensively by Parkyns and Bradshaw [37].

When a sorbate molecule interacts with a surface group, the IR absorbing frequency is usually shifted. Hair [43] reviewed the possibility of using IR spectroscopy to identify the type of surface group involved in the adsorption process and to construct adsorption isotherms on silica. For example, by measuring the decrease in intensity of the freely vibrating silanol group as a function of the equilibrium gas pressure, adsorption isotherms on this surface group could be constructed [58]. Comparison between isotherms determined by volumetric and spectroscopic techniques showed very good agreement.

However, due to overlapping of absorption bands and problems in determining accurately the intensity of the IR bands, the usual method of determining the strength and nature of the sorbate-surface interaction is to measure the perturbation of the absorption band. On oxides, the frequency shift of the hydroxyl surface group is observed. An alternative method is to measure the frequency shift of the interacting group in the sorbate molecule. This more qualitative method provides information about the adsorption mechanisms and the different types of available surface groups. As the relative amounts and types of surface groups change with the pretreatment, several studies have been concerned with temperature effects.

An infrared study of surface species arising from ammonia adsorption on oxide surfaces (including SiO_2, MgO, ZrO_2, and Al_2O_3) at various stages of dehydroxylation showed different types of adsorption and dissociation reactions [38]. Using the NH bending (1100–1300 cm^{-1}) and stretching (3380–3410 cm^{-1}) absorption bands, several types of adsorption were found (Fig. 7). It was concluded that the mechanism of ammonia adsorption essentially depends on the state of hydroxyl coverage of the oxide. When the surface is highly hydroxylated, the ammonia adsorbs through hydrogen bonding between hydroxyl groups and the nitrogen atoms of ammonia (mechanism 2).

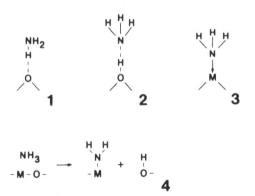

FIG. 7 Surface species arising from ammonia adsorption on oxide surfaces. (1) Hydrogen bonding via one of the hydrogens in ammonia to a surface oxygen atom. (2) Hydrogen bonding via the nitrogen atom in ammonia to the hydrogen of a surface hydroxyl group. (3) Coordination to an electron-deficient metal atom (Lewis acid site). (4) Dissociation of ammonia with the formation of surface NH$_2$ (or NH) and OH species. (From Ref. 38.)

When the powder surface is gradually dehydroxylated by increasing the temperature of evacuation, there is an increase in ammonia hydrogen-bonded to oxygen atoms in the surface (mechanism 1) and ammonia coordinated to electron-deficient metal atoms (mechanism 3). On oxides treated in vacuum at high temperatures, ammonia tends to dissociate (mechanism 4).

A thorough study on silica subjected to vacuum degassing at temperatures above 400°C showed that a new active site is generated that dissociatively adsorbs ammonia [59]. This active site has been presented as an asymmetrically strained siloxane bridge. Butylamine absorption onto SiO_2, Al_2O_3, and CaO degassed at 750°C showed two modes of adsorption: hydrogen bonding with surface hydroxyl groups or Lewis acid-base coordination to unsaturated metal ions [60]. As the frequency shift of a vibration band is considered to be proportional to the bond strength, it was concluded that the active sites on γ-Al_2O_3 are more acidic than the surface hydroxyl groups on SiO_2. Adsorption of butylamine on γ-Al_2O_3 shows a larger frequency shift ($\Delta v \approx 90$ cm^{-1}) in the NH stretching band compared to SiO_2 ($\Delta v \approx 45$ cm^{-1}).

Several attempts have been made to quantify and relate the observed frequency shifts to other physical parameters. Hertl and Hair [61] determined adsorption isotherms of several compounds on the freely vibrating hydroxyl group and calculated isosteric heats of adsorption. Comparison of these heats (ΔH) with the observed frequency shifts (Δv) of the hydroxyl band showed that the compounds studied could be divided into two groups: (i) those in which ΔH is constant and in which Δv increases with decreasing ionization potential and (ii) those in which ΔH is a function of Δv.

Pyridine adsorption has been used to provide an indication of the strength of weakly acidic surface groups. As pyridine is a significantly weaker base than ammonia, it does not react with some of the weaker sites that react with ammonia. It is also possible to differentiate Lewis and Brønsted sites, as the spectrum of adsorbed pyridine is significantly different from that of the pyridinium ion.

There has been some interest in using zirconia as a catalyst, and thus several studies have investigated the acid-base and oxidizing-reducing properties of zirconia. Nakono et al. [63] used IR spectroscopy to study the acidic properties. By varying the adsorption temperature, it was found that pyridine is coordinatively bonded to Lewis acid sites at 100°C. At higher temperatures, adsorbed pyridinium ions were observed, which was taken as an indication of the existence of Brønsted acid sites. The hydroxyl groups on zirconia are too weak to act as Brønsted acid sites at 100°C but become active at 200 to 300°C. In a thorough study, Hertl [64] used IR spectroscopy to determine and compare the different surface chemical properties of the amorphous, tetragonal, and monoclinic phases of zirconia powders. Adsorption of ammonia and pyridine showed that all three polymorphs have both Lewis and Brønsted acid activity. However, at low pretreatment temperatures most of the molecules are hydrogen-bonded to surface hydroxyl groups. In addition, tetragonal zirconia is able to dissociate ammonia and form $Zr-NH_2$ surface groups. The reaction with carbon dioxide was used as a measure of the surface basicity of the surface oxygen atoms. The results showed that the surface basicity of all three polymorphs can involve hydroxyl groups to form HCO^-_3. Reactions with 1-propanol showed that the surface hydroxyl groups of monoclinic and tetragonal zirconia are quite reactive and form alkoxyl groups on the surface.

$$Zr-OH + RCH_2OH \rightarrow ZrOCH_2R + H_2O \qquad (7)$$

The surface hydroxyl groups of the amorphous phase zirconia did not react with 1-propanol.

The surface properties of Si_3N_4, SiC, and Si_2ON_2 have also been investigated by IR spectroscopy [49,65]. The spectra of Si_3N_4 and Si_2ON_2 powders evacuated at 400°C are very similar and undergo similar modifications in response to various treatments [49]. Busca et al. [49] found that molecules of varying basicity, such as acetone and ammonia, interacted with the silanol groups on the surface, since the OH band is completely shifted to lower frequencies; but in the case of acetone they also found that part of the amino groups on the surface were perturbed. It thus seems that in this reaction, the amino groups behave as a weak acid when they interact with a relatively strong base (acetone). By adding $CHCl_3$, an acid, these authors found an interaction both with the silanol group and with a basic surface site.

Lorenzelli et al. [65] subjected the materials to a heat treatment in methanol vapor at 400°C. They found that methanol forms covalently bonded species when reacting with SiOH and Si_2NH. It was suggested that this reactivity could be utilized when chemical modification of the surface would be beneficial.

B. Inverse Gas Chromatography

Inverse gas chromatograph (IGC) is an analytical method that is an extension of conventional gas chromatography. In the IGC technique, the surface characteristics of the solid immobilized material are determined by injection of probes of known properties into the column containing the solid. The retention volume of the probe or the shape of the chromatographic peak yields information on the characteristics of the interaction between the probe and the solid material.

A recent book [66] gives an overview of IGC with attention to the characterization of polymers and other materials. Methodology and instrumentation are discussed with reference to different column materials. Special attention is paid to the interfacial characterization of materials such as carbon fibers, glass fibers, and silica. It should be noted that although IGC is a rather novel technique, there is of course extensive information on the role of the adsorbent in textbooks on conventional GC. Besides covering the theory and applications of adsorption chromatography, Snyder [67], for example, discussed the effect of adsorbent characteristics on retention volumes. Attempts were made to systematize differences between adsorbents with special reference to silica and alumina. IGC data are essentially analyzed by two methods: extrapolation to infinite dilution and direct interpretation of results at finite concentrations.

Utilizing IGC at finite concentrations, adsorption isotherms can be determined by measuring the retention volume (V_N) as a function of the sorbate concentration in the gas phase. The theory behind the conversion of V_N to adsorbed quantities is described by Dorris and Gray [68], and an example of n-decane adsorbed on short glass fibers is shown in Fig. 8 [70]. The shape of the isotherm is typical and has been observed for similar adsorbates on cellulose [68], silica [69], and short glass fibers [70]. The results can be fitted to the BET equation:

$$\Gamma = \frac{c\Gamma_m x}{(1 - x)\,[1 + (c - 1)x]} \tag{8}$$

where Γ is the surface concentration, Γ_m is the adsorbed amount required to give monolayer coverage, x is the relative vapor pressure of the adsorbate (P/P_0), and c is a constant. By conventional BET analysis, this information can be used to estimate cross-sectional areas of the adsorbed molecules and their spreading pressure. The results have been related to the dispersion part of the surface free energy [69,70]. By measuring the adsorption at different temperatures, the isosteric heat of adsorption can be determined using the Clausius-Clapeyron equation. Results on short glass fibers [70] show that the isosteric heat of adsorption for n-decane decreases with increasing surface coverage, and this method can thus be used to obtain a measure of surface heterogeneity.

The other possibility is to use IGC at infinite dilution (zero coverage). When very small amounts of solute are injected, only a small percentage of the surface area of the powder is covered by adsorbed species. Hence multilayer adsorption or effects of the degree of coverage can be neglected.

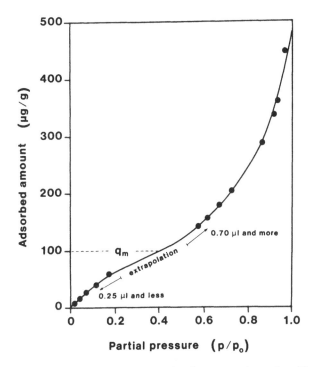

FIG. 8 Adsorption isotherm of *n*-decane on short glass fibers determined by inverse gas chromatography. (From Ref. 70. Reprinted by permission from *Ind. Chem. Prod. Res. Dev. 21*: 337, © 1982, ACS.)

The retention volume of different probes in this adsorption region can be related to the free energy of adsorption ($\Delta G°$).

$$\Delta G° = RT \ln V_N + K \tag{9}$$

where K is a constant whose value depends on the chosen reference states [71].

For homologous *n*-alkanes it has been suggested that the dispersive part of the interaction between solute and surface can be referred to the incremental increase of $\Delta G°$ per methylene group [72]. The specific component of the interaction can be evaluated by comparing the retention volume of a given probe with that of a real or hypothetical *n*-alkane with the same saturated vapor pressure [73]. The adsorption free energy calculated from the retention volume is plotted versus the saturated vapor pressure for different probes in Fig. 9 [74]. As the free energy of adsorption for the polar probes

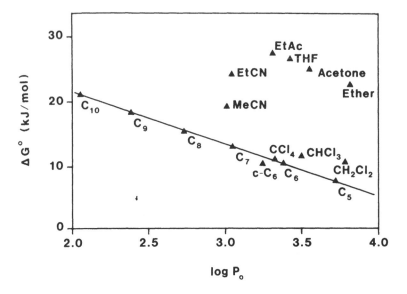

FIG. 9 Adsorption free energy versus saturated vapor pressure of different probe molecules adsorbed on a silica powder at 110°C. The increase in $\Delta G°$ for the polar molecules such as ethylacetate (EtAc) and acetone above the n-alkane line is a measure of the magnitude of the specific interaction. (From Ref. 74.)

is higher than that of n-alkanes at the same vapor pressure, this difference is a measure of the specific component of the interaction. In this case, for fumed silica, it was shown that basic probes (such as ethyl acetate) had a larger specific component than acidic probes (such as carbon tetrachloride). Thus it was concluded that the surface was mainly acidic. Attempts have been made to quantify the Lewis acid-base behavior of the powder surface using empirical scales defined by Gutman or Schmid [74,75]. IGC was, for example, applied to the characterization of the surface properties of a series of alumina powders differing in contents of impurities (Si, Ca, Na, Mg) [151]. It was shown that both the dispersive component of the surface energy and the acid-base interaction parameter are very sensitive to the presence of impurities. Furthermore, the surface characteristics could be correlated to the amount of acidic or basic polymer adsorbed from solution.

Few studies, however, have so far been made on ceramic powders, although the method seems to be convenient and accurate. One example is a study of the surface properties of α-BN [76]. It was found that α-BN is a nonspecific adsorbent, although the specific component of the interaction can increase through impurities or inhomogeneities present at the powder surface.

V. INTERACTION AT THE SOLID-LIQUID INTERFACE

Inorganic ceramic powders interacting with a liquid exhibit different types of phenomena that can be used to characterize the surface. Dispersing the powder in an aqueous solution normally results in a charge being developed on the powder surface. The magnitude and sign of this charge plays an important role in determining the properties of the sol, e.g., nonspecific ion adsorption, surfactant and polymer adsorption. The surface charge gives rise to surface forces that in many cases impart colloidal stability and strongly affect the rheological properties of the suspension, so important in ceramic processing.

When dispersing ceramic powders in nonaqueous media, the heat of immersion and quantities characterizing adsorption, such as amount adsorbed, enthalpy of adsorption, and IR shifts of adsorbed molecules, can yield information about the surface chemistry of the powders. Here we concentrate on the process of adsorption, as this is of great importance when selecting dispersants and lubricants.

A. Aqueous Media

When a powder is immersed in water, it can acquire a charge on the surface through dissociation of surface groups or binding of ions to the surface. As the suspension of the particles must be electroneutral, the surface charge σ_0 is balanced by an equal countercharge in the solution. This countercharge is created by the attraction of counterions and by the repulsion of ions of the same charge. The structure of the ion cloud outside the charged surface is usually described in terms of a two-layer model, known as the Stern-Grahame-Gouy-Chapman (SGGC) model (see, e.g., Ref. 77).

A schematic drawing of the model is given in Fig. 10. Close to the surface there is a layer of ions (or neutral species) bound at a more or less constant distance from the surface (the Stern layer). Electrically, this layer can be thought of as a condensor, in which the potential changes linearly from the surface (potential ψ_0) to the plane defining the outer boundary of the layer (potential ψ_d). Very commonly, as indicated in the figure, the charge in the Stern layer partially compensates for the surface charge, so that ψ_d has the same sign as but is smaller than ψ_0. However, specific interactions between the surface and the species bound in the Stern layer may be so

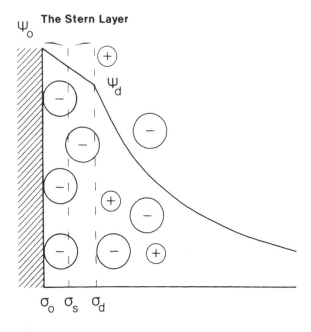

FIG. 10 Model of the charge and potential behavior at the solid-liquid interface following the Stern-Grahame-Gouy-Chapman theory.

strong that additional species with the same sign as the surface are adsorbed (in which case $\psi_d > \psi_0$) or an excess of oppositely charged species is adsorbed (in which case the sign of ψ_d is opposite to the sign of ψ_0). The total charge density in the Stern layer is denoted σ_s.

The ions adsorbed outside the Stern layer (number charge density ρ, total charge σ_d) are assumed to move freely as pointlike charges in a continuous medium with dielectric constant ε_r in the potential field created by the surface. The adsorbed layer formed in this way is called the diffuse layer. The connection between potential and distance from the surface is then given by the Poisson-Boltzmann equation

$$\nabla^2 \psi = - \frac{\rho(x, y, z)}{\varepsilon_0 \varepsilon_r} \qquad (10)$$

where ε_0 is the permittivity of vacuum. Electroneutrality demands that

$$\sigma_0 + \sigma_d + \sigma_s = 0 \qquad (11)$$

For a flat surface, this equation yields the approximate solutions

$$\sigma_d = \frac{-2\kappa k T \epsilon}{z e_0} \sinh\left(\frac{z \psi d}{2kT}\right) \tag{12a}$$

$$\kappa = \left(\frac{e_0^2 \Sigma z_i z_i^2}{\epsilon_r kT}\right)^{0.5} \tag{12b}$$

where n_i is the concentration of ions with valency z_i, e_0 is the electron charge, and κ is the so-called Debye parameter. At a distance $1/\kappa$ from the surface the potential has fallen to approximately ψ/e, and $1/\kappa$ therefore is a convenient measure of the diffuse layer thickness. The importance of these equations lies in that from the total charge and distance dependence of the potential in the diffuse layer it is possible to predict the long-range electrostatic interaction between two surfaces.

The ionic species formed by dissociation of the surface groups or associated with these are called potential-determing ions. For an oxide surface immersed in water, the potential-determining ions are H^+ and OH^-, which react with the surface hydroxyls (denoted S-OH) [79–81].

$$SOH \rightleftarrows SO^- + H^+ \tag{13}$$

$$H^+ + SOH \rightleftarrows SOH_2^+ \tag{14}$$

For each of these reactions, intrinsic equilibrium constants can be defined, and if these and the number of hydroxyl groups are known, the surface charge can be calculated.

The surface charge density σ_0 can be defined in terms of the surface concentrations (adsorption densities Γ) of OH^- and H^+:

$$\sigma_0 = F(\Gamma_H - \Gamma_{OH}) \tag{15}$$

where F is the Faraday number. When the surface concentrations are equal, $\sigma_0 = 0$. The pH at which this condition is obtained is called the point of zero charge (pzc). It may be determined by acid-base titration of the surface.

1. Solubility

Inorganic powders may also acquire a surface charge by preferential dissolution of some of the ionic species forming the material. These ionic species should also be considered as potential-determining ions. Because the species may form hydroxide complexes in solution, the preferential dissolution may be strongly pH dependent, so that the dependence of surface charge on solution conditions becomes quite complex. The thickness and charge of the diffuse layer are strongly dependent on the presence of multi-valent ions in solution (see Eq. (12)). If dissolution leads to the formation of such ions, the stability of colloids formed by the mineral may be strongly affected.

In Fig. 11, the solubility diagrams of four different ceramic oxide powders are plotted, all with a total concentration of 1 mM. The diagrams are constructed using equilibrium constants from Ref. 82.

$Al(OH)_{3(s)}$ and $Zr(OH)_{4(s)}$ both show amphoretic behavior with increasing solubility at both high and low pH-values. $Y(OH)_{3(s)}$ and $Mg(OH)_{2(s)}$ show a basic behavior with increasing solubility at low pH values.

A study of the chemical stability of different yttria-stabilized tetragonal ZrO_2 powders (3 mol % Y_2O_3) showed that yttria dissolves when the powders are suspended in acid media [83]. As the dominating species at acid pH values is Y^{3+} ions, which are very potent as a flocculating agent, this dissolution is expected to be deleterious for the colloidal stability. Another study on a mixed oxide ($BaTiO_3$) [84] showed that Ba ions were preferentially dissolved from the surface. This leaves a Ti-rich surface layer that is expected to change the electrokinetic behavior. The leached Ba can be redeposited back on the powder surface by drying.

The diagrams in Fig. 11 illustrate equilibrium solubility. In reality, the rate of dissolution is also of major importance. Segall et al. [85] have reviewed the different factors that are important in determining the reactivity of oxide surfaces in aqueous solution. The oxides were categorized into ionic, semiconducting, and covalent type, and the dissolution process was shown to follow different routes.

2. Potentiometric Acid-Base Titration

The potentiometric acid-base titration of colloidal particles was originally developed in order to investigate the silver halide system [86,87] and was later developed to cover inorganic oxide systems [88,89].

The principle is rather simple: a suspension of the colloid is titrated with the potential-determining ions (pdi) in an inert electrolyte environment. In the case of oxides and probably most ceramic powders, H^+ and OH^- are the pdi. Two titrations have to be performed, one with powder and one blank titration without powder. The difference between these two curves is then a measure of the adsorption of H^+ or OH^- on the powder surface, and hence the surface charge σ_0 can be calculated using Eq. (15). However, to measure σ_0 with reasonable accuracy a few conditions have to be fulfilled.

1. The colloid particles have to be insoluble, otherwise the soluble species compete with the surface species for the OH^- and H^+ ions.

2. The surface charge has to reach equilibrium within rather short times, otherwise the titration takes too long and the accuracy declines.
3. To be able to estimate the absolute surface charge and establish a pzc, the titration has to be performed in an inert electrolyte, i.e., an electrolyte whose ions do not adsorb at the particle surface.

The titration procedure and the different preconditions are thoroughly discussed in Ref. 78. An example of a modern setup of the titration sequence is described in Refs. 90 and 91.

These requirements limit the number of oxides that it is possible to study. Earlier studies have mainly dealt with quartz (SiO_2) [92], casserite (SnO_2) [91], rutile (TiO_2) [90], hematite (Fe_2O_3) [93], and corundum (α-Al_2O_3) [94]. A titration curve of α-Al_2O_3 is shown in Fig. 12 [94]. The result shows that NaCl is probably an inert electrolyte, as the two titration curves at different ionic strengths have a common intersection point at pH = 8.7. This pH value is identified as the pzc, and the value is consistent with earlier studies.

As was shown earlier, the solubilities of MgO, ZrO_2, and Y_2O_3 are probably too high to allow accurate determinations of the surface charge by potentiometric titrations. For silicon nitride and silicon carbide, however, a series of papers [95–97] have been published presenting surface charge data. Silicon carbide was found to behave in much the same way as silica with a pzc lower than pH = 3 [96]. Si_3N_4 showed a more complicated behavior. It was found that ammonia and silica were leached off from the surface, influencing the titration result. In order to compensate for this, supernatants from silicon nitride suspensions were used as blank electrolyte. This yielded reproducible results [95]. It was found that the different silicon nitride powders showed a significant variation in pzc. It was concluded that this variation is probably caused by the powder synthesis route. The results stress the importance of dissolution kinetics in the study of ceramic powders.

3. Electrokinetic Characterization

When electrical charges are generated at an interface, specifically a solid-liquid interface, a diffuse layer of electric charges is formed in the liquid phase. If an external electric field is applied to this system, there is a relative motion between the solid phase (including a closely bound solvent layer) and the mobile part of the diffuse layer. The relative motion is caused by the opposite signs of the charges of the solid phase and of the solvent. Conversely, an electric field will be created if the charged surface and the mobile part of the diffuse layer are made to move relative to each other.

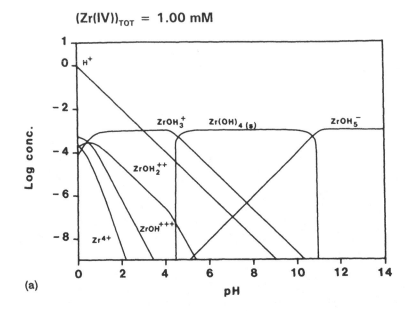

(a)

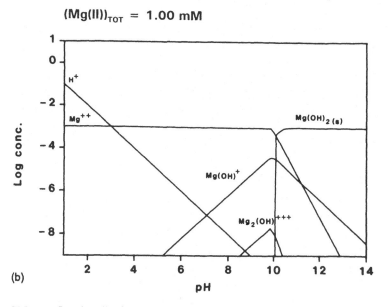

(b)

FIG. 11 Species distribution diagrams for four different cations at a total concentration of 10^{-3} M. (Equilibrium constants from Ref. 82.)

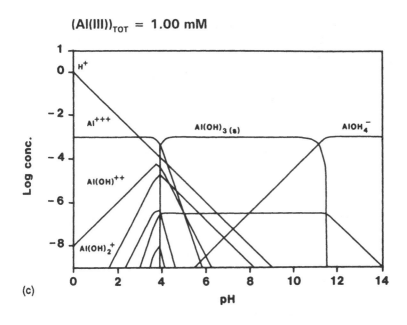

$(Al(III))_{TOT} = 1.00$ mM

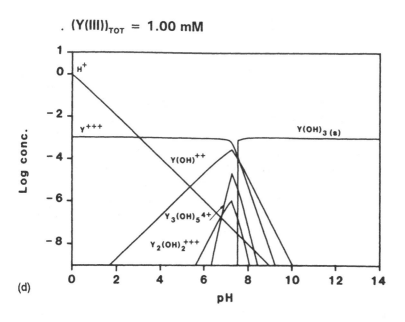

. $(Y(III))_{TOT} = 1.00$ mM

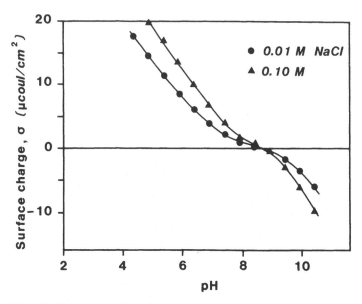

FIG. 12 Relative surface charge density (σ) versus pH for an α-Al$_2$O$_3$ powder in NaCl electrolyte determined by potentiometric titration. The apparent point of zero charge (pzc) is 8.7. (From Ref. 94. Reprinted by permission of the American Ceramic Society.)

Electrokinetic phenomena of the first kind include (a) electrophoresis, where movement of suspended particles is induced by applying an electric field across the system, and (b) electroosmosis, where the solid remains stationary and the liquid moves in response to an applied electric field. Electrokinetic phenomena of the second kind include (c) streaming potential, where a potential difference is created across a porous plug by forcing the liquid through the plug, and (d) sedimentation potential, where a potential difference is created by the sedimentation of charged particles.

Electrophoresis is by far the most extensively used method, since particles in the colloidal size range can be studied in their normal solution environment. A number of techniques are available for studying the migration of colloidal particles in an electric field, e.g., moving boundary electrophoresis, zone electrophoresis, and mass transport methods [155].

The most important and widely used method is the microelectrophoretic technique, in which the motion of individual particles is followed in an ultramicroscope (with dark-field illumination). This technique is used in very dilute suspensions. A different approach to the measurement of the velocity of the particles is to use Doppler effects in the light scattered from moving particles. Modern developments of this technique include the development of automated laser-equipped instruments allowing rapid and accurate measurements.

In more concentrated suspensions, the mass transport technique can be used. This method is based on the Hittorf method for determining ionic transport number. When the electric field is applied, the charged particles move into a collection chamber containing one of the electrodes, and the electrophoretic mobility is determined by the quantity transported.

The most recent development for the measurement of electrokinetic properties is the use of electroacoustic methods. When a suspension is subjected to an ultrasonic wave, the charge centers of the particle and of the double layer are displaced from each other, and a potential difference is created. This can be termed the colloid vibration potential (CVP) [98] or ultrasonic vibration potential (UVP) [99]. Conversely, when an alternating electric field is applied to the suspension, the coherent electrophoretic motion of the particles produces a sound, the so-called electrokinetic sonic amplitude (ESA) mode [99].

The electrophoretic mobility can be calculated from the CVP knowing the conductivity and the pressure amplitude. In a recent paper [62], the validity of the existing electroacoustic theories was tested experimentally. It was shown that a latex system provided a reliable CVP signal which could be converted to a ζ potential up to a volume fraction of solids of 3–4%. This is significantly higher than the volume fractions used in microelectrophoresis (0.01–0.1%). Experiments on polydisperse TiO_2 powders showed a significant divergence between the electrophoretic and electroacoustic results. This was said to be due to the strong influence of the larger particles on the electroacoustic results.

In the microelectrophoresis technique, the dilute particle suspension is placed in a cell of circular or rectangular cross-section. An electrode is positioned at each end of the cell, and an electric potential gradient is applied between these electrodes. The particles more toward the appropriate electrode and the velocity is measured. The mobility of a charged colloidal particle (μ_E) in the presence of a weak electric field is linearly related to the field strength. This can be expressed as

$$\mu_E = \frac{V_E}{E} \qquad (16)$$

Where V_E is the velocity of the particle and E is the field strength applied over the system.

When a charged particle (including a closely bound solvent layer) moves relative to the surrounding liquid, the zeta potential is defined as the potential difference between the slip plane and the bulk. Two simple relations between electrophoretic mobility and zeta potential have been derived by considering hydrodynamic drag and electrophoretic retardation. The two relations differ in the ratio of the double layer thickness $1/\kappa$ to the particle radius a. Provided $\kappa a > 100$ (large particles, high ionic strength), the Smoluchowski equation can be used:

$$\zeta = \frac{\eta}{\varepsilon} \cdot \mu_E \qquad (17)$$

With $\kappa a < 1$, the Hückel equation can be used:

$$\zeta = \frac{3\eta}{2\varepsilon} \mu_E \qquad (18)$$

The conditions required for the validity of the Hückel equation are, however, rarely realized in practical colloids.

For intermediate κa values, the conversion factor depends in a complex manner on particle size, ionic strength, and absolute mobility [100,101]. The morphology of the particles can also influence the conversion factor, and the existing theory [100,101] is valid only for spherical, monodisperse particles. Since most colloidal systems lie in this intermediate region and seldom consist of monodisperse, spherical particles, it is usually advisable to present the results as electrophoretic mobilities. Thus electrophoretic measurements can rarely be used to obtain reliable values of surface potentials. The isoelectric point (iep) of the surface is by definition the point at which the electrophoretic mobility of the particle is zero. In the absence of specific adsorption of ions in the Stern layer, the iep and the pzc coincide. This is usually the case for simple electrolytes such as NaCl at moderate ionic strengths. For multivalent ions it is often found that the iep depends on the electrolyte concentration and that pzc $\neq$ iep. This indicates that the electrolyte is adsorbed in the Stern layer, this layer being transported with the particle in the electrophoretic measurement so that the ζ potential (which is considered to be approximately equal to the potential in the plane immediately outside the Stern layer) is changed. The pzc or iep is not affected by hydrodynamic drag or polarization, and electrophoretic measurements can hence be used with great advantage to monitor changes in the iep due to chemical changes of the ionizable surface groups, specific adsorption, or reaction with potential-determining ions.

The effect of different ionic strengths of an inert electrolyte on the electrophoretic mobility of silicon nitride is shown in Fig. 13a [25]. Although the absolute value of the electrophoretic mobility at a given pH value decreases with increasing ionic strength due to the compression of the double layer, the iep does not shift. However, if specific adsorption takes place the iep is shifted. In Fig. 13b the iep of Si_3N_4 powders is shifted to higher pH values due to specific adsorption of $(C_2H_5)_4N^+$ ions.

Experimentally determined iep's of different ceramic powders are gathered in Table 2. Both oxide (Y_2O_3, MgO, ZrO_2, and Al_2O_3) and nonoxide (SiC, Si_3N_4) powders have been investigated. The data seem to give rather well-defined iep's for α-Al_2O_3 (pH_{iep} = 8–9), MgO (pH_{iep} = 12.5), Y_2O_3 (pH_{iep} = 9–10.6), and SiC (pH_{iep} = 2–3) but show a large variation for ZrO_2 (pH_{iep} = 5–8) and especially Si_3N_4 (pH_{iep} = 3–9). Since equilibration times and media have not in some cases been specified, it is difficult to evaluate the results. However, several of the measurements on monoclinic or tetragonal ZrO_2 compared to partially stabilized ZrO_2 containing 3 mol % Y_2O_3 [110,112] indicate that the yttria content in the latter powder increases the pH_{iep} about one pH unit above that of the former powder.

TABLE 2 Isoelectric Points for Ceramic Powders

Material	Media	pH_{iep}	Ref.
α-SiC	KNO_3	2.5±0.2	102
α-SiC	NaCl	~3	103
β-SiC	?	~3	104
Si_3N_4	?	7.5	104
Si_3N_4	KNO_3	6.5±0.5	102
Si_3N_4	NaCl	4.2–7.6	105
Si_3N_4	?	3 and 9	106
α-Al_2O_3	–	9.2±0.2	107
α-Al_2O_3	NaCl	8.8	108
α-Al_2O_3	NaCl	8.1	103
MgO	–	12.5±0.5	107
Y_2O_3	?	9	104
Y_2O_3	KNO_3	10.6	109
Y_2O_3	NaCl	8.8	110
ZrO_2	?	6.5	111
m-ZrO_2	?	5	112
t-ZrO_2	NaCl	6.5	110
t-ZrO_2^+ 3 mol % Y_2O_3	?	6	112
3 mol % Y_2O_3	NaCl	7.7	110

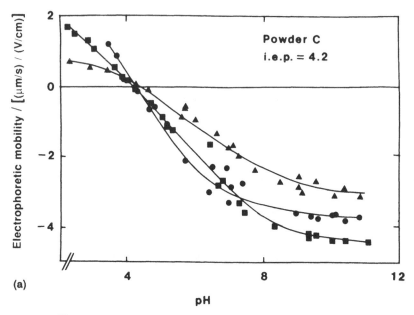

FIG. 13 Electrophoretic mobility versus pH for an Si_3N_4 powder immersed in different ionic strengths of (a) an inert electrolyte and (b) a solution containing specifically adsorbed ions ($C_2H_5N^+$). Note the shift in isoelectric point. (From Ref. 25. Reprinted by permission of the American Ceramic Society.)

As yttria has a high solubility at acidic or neutral pH-values, Fig. 11, Ref. 83, the iep can be expected to be affected by adsorption of yttrium ions. This should increase the iep and reduce the negative electrophoretic mobility. The adsorption of hydrolyzable metal ions has been extensively investigated [113–116] and has also been applied in the field of ceramic processing [103,117]. In these studies, the sorption of Mg(II) on α-SiC and α-Al_2O_3 [103] and the precipitation of $Mg(OH)_2$ and $Y(OH)_3$ on Si_3N_4 powders [117] were investigated.

Several researchers [105,107,110] have also investigated the effects of calcination and aging of the powders. Calcination of alumina [107] at temperatures above 1000°C meant that the iep was lowered to about $pH_{iep} = 6.7$. On aging the calcinated powder in water for a period of weeks, the surface was rehydroxylated and the pH_{iep} was increased. Similar results were obtained on calcinating and aging ZrO_2 powders [110]. A wide range of isoelectric points, from $pH_{iep} = 3$ to 9, has been reported for Si_3N_4. A recent study [105] shows that the pH_{iep} can be varied in a systematic way by leaching in water or annealing in vacuum. It was suggested that the surface

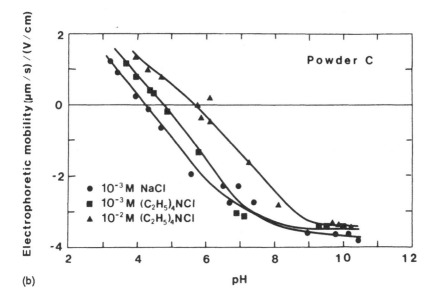

(b)

of Si_3N_4 consists of acidic silanol (Si-OH) and basic amino groups (Si_2-NH). When the relative amount of silanol groups was decreased by leaching, the pH_{iep} increased, and when the number of amino groups was decreased by annealing, the pH_{iep} decreased. The wide scatter in the pH_{iep} of Si_3N_4 in Table 2 can be explained in terms of the extent of oxidation of the surface film. Strong oxidation leads to a large number of silanol groups and hence a low pH_{iep}.

B. Organic Media

In the previous section, a description was given of how the charge on a powder surface is created in an aqueous medium. This build-up of surface charge and the associated diffuse layer of ions surrounding the particle is the most common way of creating repulsive forces in aqueous media. In nonaqueous media, surface charge formation is not considered to be an important process due to the general nondissociative nature of the solvent and the low ionic solvency. However, if the solvent is, e.g., a lower alcohol with a relatively high dielectric constant, surface charge formation probably takes place.

The most common way of creating repulsive forces on particles in non-aqueous media is by adsorbing surfactants or polymers onto the surface. As described in Chapter 4, these additives can create repulsive "steric" forces if appropriately selected. Since steric repulsion requires strongly attached and dense layers of the added surfactant or polymer, it is necessary to characterize the adsorption properties of the powder.

1. The Adsorption Process: Theoretical Background

Adsorption from solution can be described as the exchange process

$$B_{surf} + A_{sol} \rightleftarrows B_{sol} + A_{surf} \tag{19}$$

where A_{sol} and B_{sol} are the adsorbate (A) and solvent (B) molecules in solution, respectively. A_{surf} and B_{surf} represent the respective molecules on the surface.

Provided (a) that the surface is homogeneous, (b) that the volumes of the solvent and adsorbate molecules are equal, (c) that adsorption takes place only in a monolayer, and (d) that strong adsorption takes place from an ideal solution, the adsorption according to Eq. (19) is described by the so-called Langmuir isotherm

$$\Gamma_A = \frac{K_{ads} x_{A,s}}{1 + K_{ads} x_{A,s}} \tag{20}$$

where Γ_A is the surface concentration, x_A is the mole fraction of the adsorbate in solution, and K_{ads} is the equilibrium constant of (19).

However, it is obvious that the conditions listed above are rarely fulfilled [118–120]. The adsorption process is affected by many parameters. Parfitt and Rochester [120] identified the most important parameters as being (i) the surface chemistry of the solid; (ii) the nature of the solute, its chemical structure, and its interaction with the solvent; (iii) the influence of the solvent; (iv) the nature of the interactions between surface and adsorbed species; (v) the structure of the adsorbed layer; and (vi) temperature.

Although knowledge about the adsorption of different types of molecules is important as such, adsorption isotherms can be utilized as a way of characterizing powder surfaces if the system is designed so that only parameters (i) and (iv) are of predominating importance.

Fowkes [121] suggested that a useful way to interpret the interaction between a solid surface and an uncharged adsorbate is to assume that it can be divided into two parts, dispersive interactions and polar interactions. The dispersive (or London) interactions (in this context also often called nonpolar interactions) are due to the fluctuating dipole moments created by the movement of the electrons in any atom or molecule and thus occur universally between all atoms and molecules. The term polar interactions refers to specific molecular interactions between surface groups and functional groups in the adsorbate molecules. These may range from dipolar interactions to covalent binding. Although it may seem somewhat artificial to divide molecular interactions in this way, this approach has been found extremely useful from a practical point of view as a method of classifying large amounts of data on adsorbate/adsorbent interactions.

The most useful approach, also originally suggested by Fowkes, toward a systematic description of the polar interactions is to use a framework given by the Lewis acid-base or the generalized donor-acceptor concept [122,123]. In the original form of this concept, a Lewis acid is defined as a species capable of accepting a share in a pair of electrons with another species, and a Lewis base as a species capable of donating this pair of electrons. In terms of modern molecular orbital theory, this concept is generalized so that an acid is a compound with vacant molecular orbitals capable of interacting with another compound (the base) that contains filled molecular orbitals capable of acting as donors. An excellent review of modern acid-base theory is given by Jensen [124].

In the characterization of solid surfaces, the main interest is to assess the acid-base strength of the specific surface groups, such as hydroxyl groups. By adsorbing molecules with well-characterized acid-base properties and measuring, for example, the amount adsorbed, the heat of reaction, or the shifts in the IR spectra of the functional groups, systematic information for quantifying the acid-base properties of the surface groups can be acquired. Electrokinetic measurements in organic solvents have also been utilized to determine the donor-acceptor properties of solid surfaces [154]. Several different methods to quantify empirically the acid-base properties of different species have been presented. Gutman [125] defined a scale of donor numbers (DN) and acceptor numbers (AN) giving the relative acid or base strength of different compounds. The scales are based on measurements of the interaction (spectroscopically or calorimetrically) of each compound with specified reference compounds. The use of this scale has however been criticized, since it assumes that each compound can be uniquely placed on a single universal acid-base strength scale. Drago and coworkers [126,127] have presented a four-parameter equation describing the heat of reaction between an acid and a base:

$$\Delta H_{AB} = C_A C_B + E_A E_B \qquad (21)$$

where E_A and C_A are constants for the acid, and E_B and C_B are constants for the base. This theory can be viewed as a quantification of the so-called hard and soft acid-base principle introduced by Pearson [128]. Although Drago's E and C equation is theoretically more valid than the simple DN-AN concept, it is difficult to use it when characterizing amphoteric surface species, as this involves eight parameters per interaction. Thus in spite of its theoretical shortcomings, the DN-AN approach is still useful for most types of surface groups.

Recently, Riddle and Fowkes [149] presented a method of correcting AN values determined by Gutman for the van der Waals contribution to the spectroscopic shift. Using their values, several molecules such as benzonitrile, pyridine, and benzene are assigned AN $\approx$ 0 (no acidity) as opposed to the AN values assigned by Gutman.

The Brønsted acid-base properties of a surface can be characterized empirically using the so-called Hammet acidity function H_0 [129]; for a detailed discussion see [130]. A small amount of a weak neutral base (usually an indicator) is added to the solvent (which should be inert, typically benzene or toluene). The indicator may be adsorbed, and if it is a stronger base than the proton-binding surface groups it may be converted to the acid form BH^+ on adsorption. The ratio of the concentration of B to that of the protonated form BH^+ is determined spectrophotometrically. From this ratio, the quantity

$$H_0 = pK_{BH^+} + \log \frac{[BH^+]}{[B]} \tag{22}$$

is calculated. Here, pK_{BH^+} is the negative logarithm of the acidity constant of the indicator, which is taken to have the value valid for water. This procedure, which has been described in detail by, e.g., Leftin and Hobson [131], yields H_0 as a measure of the acidity of the surface. Adsorption of the base form of the indicator is expected when $pK_{BH^+} < H_0$ or when the surface coverage exceeds the number of acid sites available.

2. Adsorption Isotherms

There are several experimental methods of determining the amount adsorbed [132] and thus constructing an adsorption isotherm. Both static and dynamic measurements can be used. From the adsorption isotherms, thermodynamic quantities for adsorption may be derived, provided a proper formulation of the adsorption equilibrium can be given. Strong adsorption of simple adsorbates from organic solution can often be interpreted in terms of Eqs. (19) and (20), but care must be taken to ensure that the assumptions underlying such an interpretation are really valid. Unfortunately, this is not the case in many reports on adsorption from solution onto mineral particles.

Hammet indicators have been used extensively to characterize mainly catalysts. Benesi and Johnson [133,134] developed an amine titration method for determining the distribution of H_0 values of surface acidic sites. They used a system of indicators consisting of hydrocarbon-soluble organic bases that form brightly colored complexes with acids. However, this method has been criticized [135], and it was shown that instead of an acid site distribution, the total acidity of the sample is usually obtained due to nonequilibrium adsorption. Recently, a new method to measure the Brønsted acid and Lewis acid strength distributions using chemisorption isotherms of Hammet indicators [136] was presented. By blocking the Brønsted sites selectively with 2,6-dimethylpyridine or both Lewis and Brønsted sites with ammonia, it is possible to calculate the amount of indicator chemisorbed on each type of acid site. The structure of the sites and the adsorption states are shown in Fig. 14. It was found that the Brønsted acid sites on the silica-alumina catalysts had a relatively wide distribution of strengths, whereas most of the Lewis acid sites were distributed in the narrow range of acid strength from $H_0 = -10$ to -13. A method of characterizing surface basicity by benzoic acid titration similar to the amine titration method has also been described [137], but it is not clear whether this method may be subject to the same shortcomings as the amine titration method previously mentioned.

The Lewis acid-base properties of powder surface have been estimated by measuring the extent of adsorption of acidic or basic polymers from solvents having different acid-base properties [138]. It was found that the adsorption per unit area of polymethylmethacrylate (PMMA), a basic polymer, onto the acidic surface of silica is 50 times greater than onto the basic surface of calcium carbonate. The adsorbtion of PMMA from solvents shows a decrease in amount adsorbed with increasing basicity of the solvent, due to competition between solvent and PMMA for Si-OH sites. An increase in the acidity of the solvents also leads to a decrease in the amount adsorbed, due to competition between solvent and Si-OH for PMMA. Similar, but reversed, results were obtained when chlorinated polyvinylchloride (PVC) was adsorbed onto calcium carbonate. The extent of adsorption of PMMA and PVC from butanone onto different alumina powders was correlated to surface characterization by inverse gas chromatography [151].

Adsorption of different organic probe molecules has been used to estimate the ability of ceramic powders to interact specifically with functional groups through acidic and basic sites [152,156,157]. The various sorbate molecules chosen covered a wide spectrum of Lewis acidity-basicity (Table 3). Bergström [156] investigated the adsorption properties of silicon nitride; the adsorption data for two different silicon nitride powders (high and low pH_{iep}) are shown in Fig. 15. The results indicate distinct differences in adsorption behavior between probes with different functional groups as well as between the different powders.

The differences in adsorption behavior between the different probes were rationalized according to their acid-base strengths. The highest levels

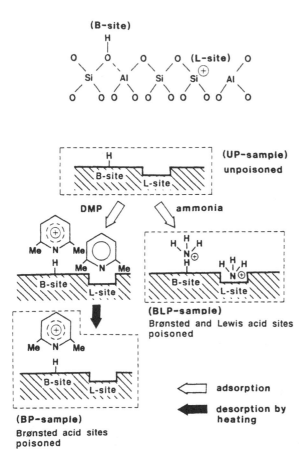

FIG. 14 Adsorption states of ammonia and 2.6-dimethylpyridine on Brønsted (B) and Lewis (L) acid sites. 2.6-dimethylpyridine can only interact strongly with Brønstedt acid sites and is hence easily desorbed from the Lewis acid sites. (From Ref. 136. Reprinted by permission from *Ind. Eng. Chem. Res. 27*: 1792, © 1988, ACS.)

of adsorption were observed with the strong acids and bases such as benzoic acid, benzylamine, and pyridine. There are also some general trends in the adsorption behavior between the different powders. The silicon nitride powder in Fig. 15 has a lower pHiep than the powder in Fig. 15. Previously, it was shown that a low pHiep of a silicon nitride powder corresponds to a high relative amount of silanol groups on the surface. A high pH_{iep}

TABLE 3 Characteristics of Organic Probe Molecules

Sorbate	Molecular mass (g/mol)	DN[a]	AN[a]
Benzoic acid	122.13	(~20)	(~50)
Phenol	94.11	(~20)	(~32)
Benzyl alcohol	108.15	(~20)	(~32)
Toluene	92.15	0.1	0.5
Ethyl benzoate	150.18	(~16)	(~5)
Pyridine	79.10	33.1	0.5
Benzyl amine	107.15	(~55)	(~15)
Cyclohexane	84.16	(0)	(0)

[a]The values in parentheses are approximate values determined from molecules with similar functional groups.

corresponds to a high relative amount of amine surface groups [105]. This difference in surface composition is for example reflected in a greater adsorption of the acidic probes (benzoic acid, phenol, and benzyl alcohol) and a lower adsorption of the basic probes on the powder with a high amount of amine groups (high pH_{iep}).

Bergström et al. [156] also used this method to characterize the effect of wet and dry milling on the surface properties of silicon nitride powders. They found a drastic difference in adsorption capacity between dry and wet milled powders. The dry milled powders show a behavior similar to that shown in Fig. 15. On the powder wet milled in isopropanol, benzoic acid was the only probe molecule that showed a high adsorption capacity. All the other molecules (benzyl amine, benzyl alcohol, and ethylbenzoate) showed very low levels of adsorption. It was suggested that the drastic difference in adsorption behavior could be explained by competitive adsorption between isopropanol coating the wet milled powder, and the added probe molecule. The strong acid benzoic acid is the only probe molecule capable of displacing isopropanol at the silicon nitride-cyclohexane interface. All the other molecules are too weak Lewis acids to displace isopropanol to any great extent.

Pugh [152] characterized the surface acid-base properties of several different ceramic powders using the adsorption method described. He found that the adsorption capacity of the acidic probes (benzoic acid and benzyl alcohol) showed a large variation between the different powders. The oxide powders α-Al_2O_3, ZrO_2, and Y_2O_3 had a high adsorption capacity for these probe molecules.

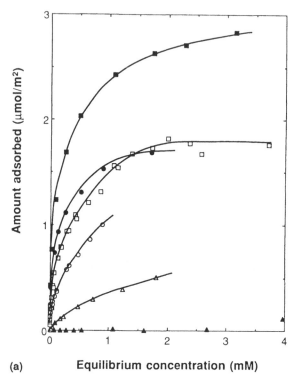

Equilibrium concentration (mM)

FIG. 15 Adsorption isotherms on two silicon nitride powders having (a) low pH_{iep}; (b) high pH_{iep}. Benzoic acid (●); phenol (a) and benzyl alcohol (b) (○); benzyl amine (■); pyridine (□); ethyl benozoate (Δ); and toluene (▲). (From Ref. 156.)

3. Calorimetry

The heat evolved on wetting or when a solute is adsorbed from a liquid solution at a solid surface provides information about the surface chemistry of the powder. The evolved heat can be determined by calorimetry. Two types of calorimetry are used for powders, batch microcalorimetry for high surface area powders (>50 m^2/g) and flow microcalorimetry for low surface area powders (>1 m^2/g). In batch microcalorimetry, the powder is immersed in the liquid by breaking a thin-walled sample bulb in the calorimeter and the temperature change is measured. The heat evolved is called the heat of immersion (ΔH_I) and is usually given in mJ/m^2. Zettlemoyer [139] has described the thermodynamic background of the wetting process.

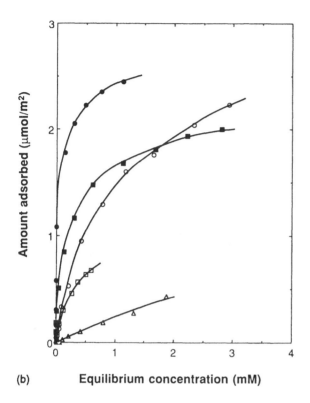

(b)   Equilibrium concentration (mM)

The heat of wetting can be measured by immersing the powder in pure liquids. Pure and oxidized laser-synthesized SiC powders have been characterized by measuring the heat of wetting in five different solvents, and they have been compared with SiO_2 [140] (Table 4). The heat of wetting was calculated per mol of solvent molecules, taking into consideration the molar surface area of each molecule. Using the Lewis acid-base concept, triethylamine (TEA) and acetone are bases; chloroform and antimony pentachloride are acids; and cyclohexane is neutral. The heat of wetting for cyclohexane was used as a measure of the dispersive interaction energy, and the heat of wetting of the other solvents, corrected for the dispersion interaction, was used as a measure of the Lewis acid-base interaction.

The heat of adsorption can be determined by immersing powders in solutions of a solute in a liquid or by introducing solute to a powder dispersion. Mills and Hockey [141] used this technique to evaluate the adsorption of methyl esters of n-fatty acids at the silica-benzene and silica-carbon tetrachloride interface. They found that the heat of displacement of one solute molecule for one solvent molecule is independent of surface coverage of the

TABLE 4 Measured Heats ΔH of Wetting of SiO_2 and SiC in Different Solvents (kJ/mol)

Solvent	SiO_2	Pure SiC	Oxidized SiC
Triethylamine	92.0	95.4	56.9
Acetone	50.6	45.6	44.6
Cyclohexane	33.4	38.9	35.3
Chloroform	46.0	61.5	38.7
$SbCl_5$	463.8	130.6	126.4

Source: Ref. 140. Reprinted by permission of the American Ceramic Society.

adsorbate. It was also found that the chain length of the methyl esters did not affect the heat of adsorption. This suggested that the heat evolved represents the difference between the localized interaction of single surface silanol groups with the carbonyl group of the solute and with the solvent.

Measurements of heat of adsorption coupled with adsorption isotherms provide more information about the surface chemistry and the solute-surface interaction. Sanders and Keweshan [142] have used this fact to study the adsorption of organic sorbents at the silica-isooctane interface. They found that although the free energy of displacement (ΔG_{disp}) obtained from the adsorption isotherm decreased with surface concentration, the heat of displacement (ΔH_{disp}) obtained by calorimetry showed a constant value up to a critical surface concentration for the N-containing rings (Fig. 16). It was suggested that this was due to steric effects involved in the formation of hydrogen bonds between the N atom in the sorbate and the silanol group. At a critical surface concentration, this bond has to rearrange to a more unfavorable orientation.

The number and character of basic and acidic surface sites on several metal oxides have been studied by microcalorimetry together with adsorption isotherms of NH_3 and CO_2 from the gas phase [153]. Although the adsorption took place at the gas-solid interface and hence the dispersive forces contributed to the heat of adsorption, the different oxide powders could be divided into three groups: basic oxides, acidic oxides, and amphoteric oxides. A relationship between the average heat of adsorption as a function of the percentage ionic character and of the charge-radius ratio was found.

In flow microcalorimetry, a known quantity of solute in a liquid is pumped through the fixed powder bed and the heat evolved is measured. Joslin and Fowkes [143] have improved this sytem using a UV concentration

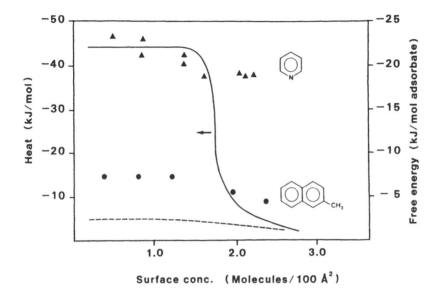

FIG. 16 Adsorption of pyridine and 2-methylnaphthalene onto silica from isooctane. The results show the free energy (ΔG) of displacement for pyridine (▲) and 2 methylnaphthalene (●) and the heat of displacement (ΔH) as a function of surface concentration. Pyridine (——) and 2-methylnaphthalene (----). (From Ref. 142.)

detector in series with a flow microcalorimeter. They used this system to study the surface acidity of ferric oxides and found that FeOH surface sites of ferric oxides are acidic using the Lewis acid-base concept and adsorb nitrogen and oxygen bases with appreciable heats of adsorption, which were correlated to the Drago E and C equation. Another important conclusion was that the moisture level strongly influences the heats of adsorption. Thorough moisture control was necessary to achieve reproducible results.

4. Infrared Spectroscopy

Infrared spectroscopy is a useful technique for obtaining information about the nature of sorbate-surface interactions in solid-liquid systems. However, compared to solid-gas studies, relatively few investigations have been performed on the solid-liquid interface. One of the main problems has been to construct a cell in which the intensity of the absorption bands of the liquid phase itself is minimized. In most of the early studies this was solved by separating the solid and liquid phases before measurement [40]. Due to the risk that the removal of the liquid phase would change the equilibrium and structure of the adsorbed molecules, cells were developed in which the IR measurement could be performed in situ. Rochester [144] has reviewed different cell designs.

In the same monograph [144], IR studies of the adsorption behavior at the solid-liquid interface are reviewed. The perturbation of surface hydroxyl groups by immersion of oxide powders in liquids and adsorption of molecules containing specific functional groups is discussed with special reference to silica.

Several attempts have been made to quantify the adsorbate-surface interaction by IR spectroscopy. The spectral shift Δv of a surface group or of the adsorbate group has been correlated to donor and acceptor numbers of the adsorbed molecule [125].

Fowkes et al. [145] used the frequency shift of the carbonyl group in esters and ketones such as ethylacetate and acetone to characterize the surface acidity. By correlating the frequency shift to measured heat of acid-base complexation, the shift can be used as a measure of $-\Delta H^{AB}$ (Fig. 17). If the C_B and E_B values of the different probe molecules are known, this information can be used to construct a C and E plot according to

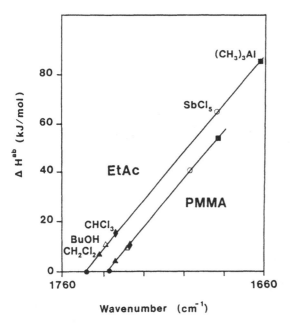

FIG. 17 Relation of measured heats of acid-base complexation (ΔH^{AB}) to the carbonyl stretching frequency (v) in ethyl acetate and in polymethylmethacrylate. (From Ref. 145. Reprinted by permission of John Wiley & Sons, Inc.; copyright © owner *J. Polym. Sci., Polym. Chem. Ed.*, 1984.)

$$E_B = \frac{-\Delta H^{AB}}{E_A} - \frac{C_B C_A}{E_A} \qquad (23)$$

Using probe molecules with different C-to-E ratios, an accurate intersection can be determined and used to characterize the surface acidity [145].

The great advantage of IR spectroscopy in solid-liquid systems is that it can be used to study more applied surfactant systems relevant to ceramic processing in situ. Sokoll and Hobert [146] studied the adsorption of octadecylamine at the MgO-to-CCl$_4$ interface. The powder was dehydroxylated by pretreatment in vacuum at 700°C, and it was found that octadecylamine adsorbs through a coordinative interaction between amine and Lewis acidic surface sites.

The adsorption of stearic acid on silica and alumina, and of decanoic acid on magnesia from CCl$_4$ solutions [147], has also been measured. Silica degassed at 800°C adsorbed stearic acid predominantly as monomers at low concentration. At higher coverages, more weakly bound acid dimers were found. The adsorption of stearic acid on alumina powder with a high amount of surface hydroxyl groups resulted in the formation of an aluminium stearate layer. The main reaction of magnesia with decanoic acid solution was the formation of a surface decanoate layer.

In a similar study by Low and Lee [148] on silica, it was found that octadecanol adsorbed as dimers mainly through interaction with surface silanols. However, an interaction probably involving hydrocarbon chain interaction was also found. Octadecylamine interacted very strongly with surface silanols and resulted in a large spectral shift ($\Delta v_{OH} \approx 750$ cm^{-1}).

VI. SUMMARY

This chapter has been concerned with the methods used for the surface chemical characterization of ceramic powders. A wide variety of methods is available allowing the ceramist to characterize the elemental composition of powder surfaces, the chemistry of functional surface groups, and the interactions at the solid-gas and solid-liquid interface. Some of these methods, such as infrared spectroscopy and electrokinetic characterization, have existed for many years. Others, such as SIMS, have evolved during the last 10–15 years. However, almost all the methods described have experienced a considerable improvement in, for example, detection limit, accuracy, and spatial resolution during the last 10 years, mainly through the implementation of microelectronics and computers.

Generally speaking, the number of studies on the surface characteristics of ceramic powders is rather limited. In most of these studies, the surface characterization constitutes only a minor part, usually related to the behavior during processing. There is a great need for more systematic knowledge on the surface chemistry of ceramic powders, particularly of the nonoxide powders, which in many cases are more difficult to process. A fruitful approach would be to employ a variety of methods, for example, ESCA, infrared spectroscopy, electrophoretic mobility, and adsorption isotherms, on different powders. The combined knowledge from the different methods would make it possible to predict and control the behavior during processing.

REFERENCES

1. P. G. Harris and A. D. Trigg, *Materials and Design 9*: 127 (1988).
2. D. Briggs and M. P. Seah, eds., *Practical Surface Analysis.* John Wiley, New York, 1983.
3. M. N. Rahaman, Y. Boiteux, and L. G. de Jonghe, *Am. Ceram. Soc. Bull. 65*(8): 1171 (1986).
4. W. Braue, H. J. Dudek, and G. Ziegler, in *Ceramic Powders* (P. Vinzenzini, ed.). Elsevier, Amsterdam, 1983, pp. 661–670.
5. H. J. Möller and G. Welsch, *J. Am. Ceram. Soc. 68*: 320 (1985).
6. S. Ichimura, H. E. Bauer, H. Seiler, and S. Hofmann, *Surf. Interface Anal. 14*: 250 (1989).
7. S. Hofmann, *Surf. Interface Anal. 9*: 3 (1986).
8. A. Benninghoven, F. G. Rüdenauer, and H. W. Werner, *Secondary Ion Mass Spectroscopy.* John Wiley, New York, 1987.
9. A. Brown and J. C. Vickerman, *Surf. Interface Anal. 6*: 1 (1984).
10. C. G. Pantano, C. A. Houser, and R. K. Brow, in *Better Ceramics Through Chemistry* (C. J. Bringer, D. E. Clark, and D. R. Ulrich, eds.). Elsevier, Amsterdam, 1984, pp. 255–266.
11. J. Chabala, R. Levi-Setti, S. A. Bradley, and K. R. Karasek, *Appl. Surf. Sci. 29*: 300 (1987).
12. A. Benninghoven, W. Sichtermann, and S. Storp, *Thin Solid Films 28*: 59 (1975).
13. J. D. Andrade, ed., *Surface and Interface Aspects of Biomedical Polymers*, Vol. 1. Plenum Press, New York, 1985, p. 105.
14. T. N. Taylor, *J. Mater. Res. 4*: 189 (1989).
15. K. R. Karasek, S. A. Bradley, J. T. Donner, H. C. Yeh, J. L. Schienle, and H. T. Fang, *J. Am. Ceram. Soc. 72*: 1907 (1989).
16. M. P. Seah, *Surf. Interface Anal. 2*: 222 (1980).
17. J. H. Scofield, *J. Electron Spectroscopy Rel. Phenom. 8*: 129 (1976).
18. M. P. Seah, *Surf. Interface Anal. 9*: 85 (1986).
19. C. D. Wagner, L. E. Davis, M. V. Zeller, J. A. Taylor, R. H. Raymond, and L. H. Gale, *Surf. Interface Anal. 3*: 211 (1981).

20. M. P. Seah and M. T. Anthony, *Surf. Interface Anal. 6*: 230 (1984).
21. M. P. Seah, M. E. Jones, and M. T. Anthony, *Surf. Interface Anal. 6*: 242 (1984).
22. R. J. Pugh, *J. Coll. Interface Sci. 138*: 16 (1990).
23. L. E. Firment, H. E. Bergna, D. G. Swartzfuger, P. E. Bierstedt, and M. L. van Kavelaar, *Surf. Interface Anal. 14*: 46 (1989).
24. M. Peuckert and P. Greil, *J. Mater. Sci. 22*: 3717 (1987).
25. L. Bergström and R. J. Pugh, *J. Am. Ceram. Soc. 72*(1): 103 (1989).
26. J. E. Fulghum and R. W. Linton, *Surf. Interf. Anal. 13*: 186 (1988).
27. V. G. Aleshin, A. N. Sokolov, M. G. Chudin, and A. A. Shulzhenko, *Poroshkovaya Metallurgiya 12*: 76 (1986).
28. E. Paparazzo, *Surf. Interface Anal. 12*: 115 (1988).
29. P. R. Griffiths and J. A. de Haseth, *Fourier Transform Infrared Spectroscopy*. John Wiley, New York, 1986.
30. T. Theofhanides, ed., *Fourier Transform Infrared Spectroscopy*. Reidel, Dordrecht, 1984.
31. R. J. Bell, *Introductory Fourier Transform Spectroscopy*. Academic Press, New York, 1972.
32. J. R. Ferraro and L. J. Basile, eds., *Fourier Transform Infrared Spectroscopy — Applications to Chemical Systems*, Vols. 1-4. Academic Press, New York, 1979.
33. J. L. Koenig, *Appl. Spectrosco. 29*: 293 (1975).
34. H. Ishida, *Rubber Chem. and Tech. 60*: 497 (1987).
35. M. P. Fuller and P. R. Griffiths, *Anal. Chem. 50*: 1906 (1978).
36. A. Taboudoucht and H. Ishida, *Appl. Spect. 43*: 1016 (1989).
37. N. D. Parkyns and D. I. Bradshaw, in *Laboratory Methods in Vibrational Spectroscopy* (H. A. Willis, J. H. van der Maas, and R. G. J. Miller, eds.). John Wiley, New York, 1987, pp. 363-410.
38. A. A. Tsyganenko, D. V. Pozdnyakov, and V. N. Filimonov, *J. Molec. Struc. 29*: 299 (1975).
39. K. J. Nilsen, R. E. Riman, and S. C. Danforth, in *Ceramic Powder Science II* (G. L. Messing, E. R. Fuller, Jr., and H. Hausner, eds.). American Ceramic Society, Westerville, Ohio, 1988, pp. 469-477.
40. L. H. Little, *Infrared Spectra of Adsorbed Species*. Academic Press, New York, 1966.
41. M. L. Hair, *Infrared Spectroscopy in Surface Chemistry*. Marcel Dekker, New York, 1967.
42. A. V. Kiselev and V. I. Lygin, *Infrared Spectra of Surface Compounds*. John Wiley, New York, 1975.
43. M. L. Hair, in *Vibrational Spectroscopies for Adsorbed Species* (A. T. Bell and M. L. Hair, eds.). American Chemical Society, Washington, D.C., 1980, pp. 1-11.
44. R. K. Iler, *The Chemistry of Silica*. John Wiley, New York, 1979.
45. W. A. P. Claesson, H. A. J. Th. v. d. Pol, A. H. Goemans, and A. E. T. Kuiper, *J. Electrochem. Soc. 133*: 1458 (1986).

46. T. M. Kramer, W. E. Rhine, and H. K. Bowen, *Adv. Ceram. Mater. 3*: 244 (1988).
47. A. Celikkaya and M. Akinc, in *Ceramic Powder Science II* (G. L. Messing, E. R. Fuller, Jr., and H. Hausner, eds.). American Ceramic Society, Westerville, Ohio, 1988, pp. 110–118.
48. P. A. Agron, E. L. Fuller, Jr., and H. F. Holmes, *J. Colloid Interface Sci. 52*: 553 (1975).
49. G. Busca, V. Lorenzelli, M. I. Baraton, P. Quintard, and R. Marchand, *J. Mol. Struct. 143*: 525 (1986).
50. R. A. McFarlane and B. A. Morrow, *J. Phys. Chem. 92*: 5800 (1988).
51. J. E. Wertz and J. R. Bolton, *Electron Spin Resonance – Elementary Theory and Practical Applications*. McGraw-Hill, New York, 1972.
52. D. Cordischi, V. Indovina, and M. Occhiuzzi, in *Magnetic Resonance in Colloid and Interface Science* (J. P. Fraissard and H. A. Resing, eds.). D. Reidel, Dordrecht, 1980, pp. 461–466.
53. H. Hosaka, T. Fujiwara, and K. Meguro, *Bull. Chem. Soc. Jap. 44*: 2616 (1971).
54. K. Esumi, K. Miyata, F. Wuki, and K. Meguro, *Colloids Surf. 20*: 81 (1986).
55. W. K. Hall, H. P. Leftin, F. J. Cheselske, and D. E. O'Reilly, *J. Catal. 2*: 506 (1963).
56. R. W. Linton, M. L. Miller, G. E. Maciel, and B. L. Hawkins, *Surf. Interface Anal. 7*: 196 (1985).
57. P. J. Hendra and M. Fleischman, in *Topics in Surface Chemistry* (E. Kay and P. S. Bagus, eds.). Plenum Press, New York, 1978, pp. 373–402.
58. M. L. Hair and W. Hertl, *J. Phys. Chem. 73*: 4269 (1969).
59. B. A. Morrow and I. A. Cody, *J. Phys. Chem. 80*: 1998 (1976).
60. R. Sokoll, H. Hobert, and I. Schmuck, *J. Chem. Soc., Faraday Trans. 1, 82*: 3391 (1986).
61. W. Hertl and M. L. Hair, *J. Phys. Chem. 72*: 4676 (1968).
62. R. W. O'Brien, B. R. Midmore, A. Lamb, and R. J. Hunter, *Faraday Discuss. Chem. Soc. 90*: 301 (1990).
63. Y. Nakono, T. Hattori, and K. Tanabe, *J. Catal. 57*: 1 (1979).
64. W. Hertl, *Langmuir 5*, 96 (1989).
65. V. Lorenzelli, G. Ramis, G. Busc, and P. Quintard, *Proceeding 31st Int. Symp. Pure and Appl. Chem.*, Section 5, Sofia, Bulgaria, 1987, pp. 40–52.
66. D. R. Lloyd, T. C. Ward, and H. P. Schreiber, eds., *Inverse Gas Chromatography*, 391 ACS Symposium Series. ACS, Washington, D.C., 1989.
67. L. R. Snyder, *Principles of Adsorption Chromatography*. Marcel Dekker, New York, 1968.
68. G. M. Dorris and D. G., Gray, *J. Colloid Interface Sci. 71*: 93 (1979).
69. E. Papirer, A. Vidal, W. M. Jiao, and J. B. Donnet, *Chromatographia 23*: 279 (1987).
70. C. L. Flour and E. Papirer, *Ind. Eng. Chem. Prod. Res. Dev. 21*: 337 (1982).
71. J. Schultz, L. Lavielle and C. Martin, *J. Adhesion 23*: 45 (1987).
72. G. M. Dorris and D. G. Gray, *J. Colloid Interface Sci. 77*: 353 (1980).
73. C. S. Flour and E. Papirer, *J. Colloid Interface Sci. 91*: 69 (1983).

74. A. Vidal, E. Papirer, W. M. Jiao, and J. B. Donnet, *Chromatographia 23*: 121 (1987).
75. C. S. Flour, E. Papirer, *Ind. Chem. Prod. Res. Dev. 21*: 666 (1982).
76. T. B. Gavrilova, A. V. Kiselev, I. V. Parshina, and T. M. Roshchina, *Kolloidnyi Zhurnal 48*: 421 (1986).
77. R. O. James and G. A. Parks, in *Surface and Colloid Science* (E. Matijevic, ed.). Plenum Press, New York, 1982, pp. 119–216.
78. R. J. Hunter, *Zeta Potential in Colloid Science*. Academic Press, London, 1981.
79. T. W. Healy and L. R. White, *Adv. Coll. Inter. Sci. 9*: 303 (1978).
80. J. A. Davis, R. O. James, and J. O. Leckie, *J. Coll. Inter. Sci. 63*(3): 480 (1978).
81. D. E. Yates, S. Levine, and T. W. Healy, *J. Chem. Soc. Faraday Trans. I, 70*: 1807 (1974).
82. A. E. Martel and R. M. Smith, *Critical Stability Constants*, Vol. 5, *Inorganic Complexes*, 1st Suppl. Plenum Press, New York, 1982.
83. J. C. Farinas, R. Moreno, J. Requeno, and J. S. Moya, *Mat. Sci. Eng. A109*: 97 (1989).
84. D. A. Anderson, J. H. Adair, D. Miller, J. V. Biggers, and T. R. Shrout, in *Ceramic Powder Science II* (G. L. Messing, E. R. Fuller, Jr., and H. Hausner, eds.). American Ceramic Society, Westerville, Ohio, 1988, pp. 485–492.
85. R. L. Segall, R. St. C. Smart, and P. S. Turner, in *Surface and Near-Surface Chemistry of Oxide Materials* (J. Nowotny and L. C. Dufour, eds.). Elsevier, Amsterdam, 1989, pp. 527–579.
86. E. L. Mackor, Thesis, Utrecht, 1951.
87. J. A. W. van Laar, Thesis, Utrecht, 1952.
88. G. H. Bolt, *J. Phys. Chem. 61*: 1166 (1957).
89. G. A. Parks and P. L. de Bruyn, *J. Phys. Chem. 66*: 967 (1962).
90. E. A. Barringer and H. K. Bowen, *Langmuir 1*: 420 (1985).
91. M. R. Houchin and L. J. Warren, *J. Coll. Inter. Sci. 100*: 278 (1984).
92. Th. F. Tadros and J. Lyklema, *J. Electroanal. Chem. 17*: 267 (1968).
93. A. Breeuwsma and J. Lyklema, *Discuss. Faraday Soc. 52*: 324 (1971).
94. J. Cesarano III, I. A. Aksay, and A. Bleier, *J. Am. Ceram. Soc. 71*: 250 (1988).
95. P. K. Whitman and D. L. Feke, *Adv. Ceram. Mat. 1*: 366 (1986).
96. P. K. Whitman and D. L. Feke, *J. Am. Ceram. Soc. 71*: 1086 (1988).
97. D. L. Feke, NASA Contractor Report 179634 (1987).
98. B. J. Marlow, D. Fairhurst, and H. P. Pendse, *Langmuir 4*: 611 (1988).
99. A. J. Balchin, R. S. Chow, and R. P. Sawatzky, *Adv. Coll. Inter. Sci. 30*: 111 (1989).
100. P. H. Wiersema, A. L. Loeb, and J. Th. G. Overbeek, *J. Coll. Inter. Sci. 22*: 78 (1966).
101. R. W. O'Brien and L. R. White, *J. Chem. Soc. Faraday Trans. II, 74*: 1607 (1978).
102. M. J. Crimp, R. E. Johnson, Jr., J. W. Halloran, and D. L. Feke, in *Science of Ceramic Chemical Processing* (L. L. Hench and D. R. Ulrich, eds.). John Wiley, New York, 1986, pp. 539–549.
103. R. J. Pugh and L. Bergström, *J. Coll. Inter. Sci. 124*: 570 (1988).

104. M. Persson, L. Hermansson, and R. Carlsson, in *Ceramic Powders* (P. Vincenzini, ed.). Elsevier, Amsterdam, 1983, pp. 735–742.
105. L. Bergström and E. Bostedt, *Coll. Surf. 49*: 183 (1990).
106. Y. Imamura, K. Ishibashi, and H. Shimodaira, *Ceram. Eng. Sci. Proc. 7*: 828 (1986).
107. M. Robinson, J. A. Park, and D. W. Fuerstenau, *J. Am. Ceram. Soc. 47*: 516 (1964).
108. D. Ballion and N. Jaggrezic-Renault, *J. Radio-Analytical Nuc. Chem. 92*: 133 (1985).
109. Y. Chia, in *Ceramic Powder Science II* (G. L. Messing, E. R. Fuller, Jr., and H. Hausner, eds.). American Ceramic Society, Westerville, Ohio, 1988, pp. 511–519.
110. M. Kagawa, M. Omori, Y. Syono, Y. Imamura, and S. Urui, *J. Am. Ceram. Soc. 70*: C-212 (1987).
111. M. Kagawa, Y. Syono, Y. Imamura, and S. Usui, *J. Am. Ceram. Soc. 69*: C-50 (1986).
112. R. Moreno, J. Requena, and J. S. Moya, *J. Am. Ceram. Soc. 71*: 1036 (1988).
113. R. O. James and T. W. Healy, *J. Coll. Inter. Sci. 40*: 42 (1972).
114. R. O. James and T. W. Healy, *J. Coll. Inter. Sci. 40*: 53 (1972).
115. R. O. James and T. W. Healy, *J. Coll. Inter. Sci. 40*: 65 (1972).
116. K. J. Farley, D. A. Dzombak, and F. M. M. Morel, *J. Coll. Inter. Sci. 106*: 226 (1985).
117. T. M. Shaw and B. A. Pethica, *J. Am. Ceram. Soc. 69*: 88 (1986).
118. A. Dabrowski, M. Jaroniec, and J. O'scik, in *Surface and Colloid Science*, Vol. 14 (E. Matijevic, ed.). Plenum Press, New York, 1987, pp. 83–214.
119. A. Derylo-Marczewska and M. Jaroniec, in *Surface and Colloid Science*, Vol. 14 (E. Matijevic, ed.). Plenum Press, New York, 1987, pp. 301–380.
120. G. D. Parfitt and C. H. Rochester, eds., *Adsorption from Solution at the Solid/Liquid Interface*. Academic Press, New York, 1983.
121. F. M. Fowkes, in *Chemistry and Physics of Interfaces*. American Chemical Society, Washington, D.C., 1965, pp. 1–12.
122. F. M. Fowkes, in *Surface and Colloid Science in Computer Technology* (K. L. Mittal, ed.). Plenum Press, New York, 1987, pp. 3–26.
123. W. B. Jensen, in *Surface and Colloid Science in Computer Technology* (K. L. Mittal, ed.). Plenum Press, New York, 1987, pp. 27–59.
124. W. B. Jensen, *The Lewis Acid-Base Concepts*. Wiley-Interscience, New York, 1980.
125. V. Gutman, *The Donor-Acceptor Approach to Molecular Interactions*. Plenum Press, New York, 1978.
126. R. S. Drago and B. B. Wayland, *J. Am. Chem. Soc. 87*: 3571 (1965).
127. R. S. Drago, G. C. Vogel, and T. E. Needham, *J. Am. Chem. Soc. 93*: 6014 (1971).
128. R. G. Pearson, *Science 151*: 172 (1966).
129. L. P. Hammett and A. J. Deyrup, *J. Am. Chem. Soc. 54*: 2721 (1932).
130. C. H. Rochester, *Acidity Functions*. Academic Press, London, 1970.

131. H. P. Leftin and M. C. Hobson, Jr., *Adv. Catal. 14*: 115 (1963).
132. C. E. Brown and D. H. Everett, in *Colloid Science*, Vol. 2 (D. H. Everett, ed.). Chemical Society, London, 1975, pp. 52–100.
133. H. A. Benesi, *J. Phys. Chem. 61*: 970 (1957).
134. O. Johnson, *J. Phys. Chem. 59*: 827 (1955).
135. M. Deeba and W. K. Hall, *J. Catal. 60*: 417 (1979).
136. K. Hashimoto, T. Masuda, H. Sasaki, *Ind. Eng. Chem. Res. 27*: 1792 (1988).
137. K. Tanabe, M. Misono, Y. Ono, and H. Hattori, eds., *New Solid Acids and Bases*. Elsevier, Amsterdam, 1989.
138. F. M. Fowkes and M. A. Mostafa, *Ind. Eng. Chem. Prod. Res. Dev. 17*: 3 (1978).
139. A. C. Zettlemoyer, *Chemistry and Physics at Interfaces*. American Chemical Society, Washington, D.C., 1965, pp. 139–148.
140. M. Okuyama, G. J. Garvey, T. A. Ring, and J. S. Haggerty, *J. Am. Ceram. Soc. 72*: 1918 (1989).
141. A. K. Mills and J. A. Hockey, *J. Chem. Soc. Faraday Trans. I, 71*: 2392 (1975).
142. N. D. Sanders and C. F. Keweshan, *J. Coll. Inter. Sci. 124*: 606 (1988).
143. S. T. Joslin and F. M. Fowkes, *Ind. Eng. Chem. Res. Dev. 24*: 369 (1985).
144. C. H. Rochester, *Adv. Coll. Inter. Sci. 12*: 43 (1980).
145. F. M. Fowkes, D. O. Tischler, J. A. Wolfe, L. A. Lannigan, C. M. Ademu-John, and M. J. Halliwell, *J. Polymer Sci., Pol. Chem. Ed. 22*: 547 (1984).
146. R. Sokoll and H. Hobert, *J. Chem. Soc. Faraday Trans 1, 82*: 1527 (1986).
147. M. Hasegawa and M. J. D. Low, *J. Coll. Inter. Sci. 30*: 378 (1969).
148. M. J. D. Low and P. L. Lee, *J. Coll. Interf. Sci. 45*: 148 (1973).
149. F. L. Riddle, Jr., and F. M. Fowkes, *J. Am. Chem. Soc. 112* 3259 (1990).
150. L. -S. Johansson, *Surf. Interface Anal. 17*: 663 (1991).
151. E. Papirer, J. -M. Ressin, B. Siffert, and G. Philippeneau, *J. Coll. Interf. Sci. 144*: 263 (1991).
152. R. J. Pugh, in *Ceramic Powder Science III* (G. L. Messing, S.-I Hirano, and H. Hausner, eds.). American Ceramic Society, Westerville, Ohio, 1990, pp. 375–382.
153. A. Auroux and A. Gervasini, *J. Phys. Chem. 94*: 6371 (1990).
154. M. E. Labib and R. Williams, *J. Coll. Interf. Sci. 97*: 356 (1984).
155. D. J. Shaw, *Electrophoresis*. Academic Press, London, 1969.
156. L. Bergström, *Colloids Surf. 69*: 53 (1992).
157. L. Bergström, M. Ernstsson, B. Gruvin, R. Brage, B. Nyberg, and E. Carlström, in *Ceramics Today – Tomorrow's Ceramics* (P. Vincenzini, ed.). Elsevier, Amsterdam, 1991, p. 1005.
158. T. A. Dang and R. Gnanasekaran, *Surf. Interface Anal. 15*: 113 (1990).
159. T. A. Dang and R. Gnanasekaran, *J. Vac. Sci. Technol. A9*: 1406 (1991).
160. T. A. Dang, R. Gnanasekaran, and D. D. Deppe, *Surf. Interface Anal. 18*: 141 (1992).

4

Dispersion and Stability of Ceramic Powders in Liquids

ROBERT J. PUGH   Institute for Surface Chemistry, Stockholm, Sweden

I. INTRODUCTION

The homogenization, dispersion, and stability of particles in liquids are primary important steps in the processing of high-performance ceramics produced by conventional slurry consolidation methods such as tape casting and slip casting. In industry today, it is generally accepted that the driving force and efficiency of sintering is influenced by the basic powder properties such as purity, grain size, and chemical heterogeneity. However, more recent research studies suggest that the packing and distribution of particles throughout the green body control the porosity and microstructure and play an important role in determining the reliability of the final product [1-4].

Essentially, the aim of the dispersion process is to achieve a high-solids (in some cases today >60 vol %) homogeneous suspension with a well-defined rheological behavior and possibly an intermediate (but at present not well-defined) degree of stability. This enables the ceramic to be formed into complex geometrics with a defect-free microstructure after sintering. In single-component systems, the uniformity of packing generally depends not only on the particle size distribution, but also the number of secondary agglomerates which should be reduced to a minimum, since they cause defects such as voids and cracks in the composite. In other cases, where two, three, or even four different colloidal components are mixed together in the fluid, the situation becomes more complex, and it is difficult to obtain a homogeneous dispersion of particles. For example, a typical slurry may contain matrix material plus one or two sintering aids, plus fibers. Under such circumstances, segregation, coagulation, and separation of components —or worse, some degree of heterocoagulation—can occur during processing, and this is obviously detrimental, generating cracks and warping during sintering.

From a fundamental point of view, the process of homogenization and dispersion of a powder in a liquid may be dealt with in several stages [5]. Initially, wetting must occur, in which the liquid phase wets and spreads on the powder's external surface, and also air must be displaced from the internal pores during immersion. At a later stage, secondary units (clusters or agglomerates) must be broken down or fractured into essentially primary units. Finally, the primary particles must *remain* dispersed through the liquid medium and reagglomeration prevented by some type of stabilization mechanism.

In the case of the wetting of a high-energy ceramic powder surface (such as a metal or oxide) with a lower-energy liquid (such as water or a hydrocarbon) then the wetting and immersion processes occur almost simultaneously. Generally, these processes are both irreversible and rapid. However, the dispersion and stability steps (*in the absence of any form of stabilization barrier*) are more complex and may occur on different time scales, depending on the particle concentration in suspension.

For example, in a concentrated colloidal suspension, the dispersion of the components in a ball mill may occur relatively *fast*, but after the slip is poured into the mold this "breakup step" may be rapidly *reversed* by coagulation. This can be illustrated from the classical theory of second-order perikinetic coagulation kinetics as described by Smoluchowski [6], dealing with the *rate of rapid coagulation* of monodispersed colloidal particles. The rapid coagulation rate that occurs in the absence of any stabilization mechanism can be expressed in terms of the half life $t_{1/2}$, or the time to reduce the number of particles in the system to half the original value; it may be represented by

$$t_{1/2} = \frac{3\eta}{4k_B T N_0} \approx \frac{2 \times 10^{11}}{N_0(\text{cm}^{-3})} \quad \text{seconds in water at } 20°C \tag{1}$$

where N_0 is the original particle concentration of the fully dispersed system expressed in particles/cm^3, η is the viscosity of the liquid medium, k_B is the Boltzmann constant, and T is the temperature.

For a *dilute* colloidal suspension in water, where for example $N_0 \approx 10^7$ particles/cm^3, then at room temperature $t_{1/2}$ would be several hours, such that the system could be regarded as being relatively stable, whereas for a *concentrated* suspension, then $N_0 \approx 10^{14}$ particles/cm^3, and $t_{1/2}$ would be reduced to milliseconds (the process would be reversed extremely rapidly).

In ceramic powder dispersions where the size and density of the particles can be fairly well defined, values of $t_{1/2}$ can be easily determined and compared to the sedimentation rate. Generally, the concentration, particle size, and viscosity (as controlled by the addition of binder) can all have a pronounced influence on stability and sedimentation. This is clearly illustrated in Table 1. In concentrated suspensions, it is essential therefore to introduce some type of stabilizing agent, especially in highly concentrated suspensions, to prevent reagglomeration from occurring. In the case of *slow* coagulation, the process is essentially controlled by the interparticle interactions that have the net effect in controlling the overall state of the dispersion. In systems where *attractive* forces dominate the interaction, the system become unstable, and the particles coagulate. This causes an increase in viscosity and sedimentation, and in concentrated dispersions it may produce a structured green body. In cases where *repulsive* forces are strong, then a relatively stable, well-dispersed (lower viscosity) suspension of individual particles will be formed.

In recent years, the importance of colloidal processing has been strongly emphasized, and research studies suggest that defect-free composites with structural reliability can be produced from a highly concentrated suspension, provided the stability is critically controlled [7–9]. For example, the suspension must be *sufficiently* colloidally stable and free from agglomerates and unwanted foreign bodies; yet it must be sufficiently fluid to be easily formed into the desired shape. Also, the packing should be sufficiently dense to give a minimum pore volume. This can only be achieved by control of the particle size distribution and a balance of the interparticle surface forces. *Too strong an interparticle attraction* must be avoided, since

TABLE 1 A Comparison Between the Half-Life of Rapid Coagulation $t_{1/2}$ and the Time of Sedimentation t_{sed} for α-Al$_2$O$_3$ Particles (Size 0.1 to 10 μm) in Dilute Suspensions (1 and 10 vol %). Calculations Are for a Binder System (η = 1 poise and ρ = 1.0 g/cm^3)

(a) 1 Vol % Suspension in Binder Solution

Particle size (μm)	$t_{1/2}$	t_{sed}/cm (assuming no hindrance)
10	$\approx$12 days	1.7 hr
1	17 min	7.1 day
0.1	1 s	1.9 year

(b) 10 Vol % Suspension in Binder Solution

Particle size (μm)	$t_{1/2}$	t_{sed}/cm (assuming no hindrance)
10	29 hr	1.7 hr
1	1.7 min	7.1 day
0.1	0.1 s	1.9 year

particles tend to stick on contact, causing flocs or open structures; but also *strong repulsion* causes a packed structure with a low density, which causes difficulties in removing the liquid during heat treatment. It would therefore seem reasonable to aim for an intermediate degree of stability.

By use of surface active processing chemicals for controlling the inter-particle forces, it would appear feasible to meet this requirement. In practice, however, the situation is often more complicated, since if a suitable dispersant is found, it may not stabilize all the particles to the same extent in a multicomponent system. In the case of not finding a single dispersant, then several dispersants may be added; but this may cause competitive adsorption to occur on the different components. Before these types of problems can be discussed further, it is necessary to review the theoretical background to the wetting, dispersion, and stability steps in the following sections. However, to begin it is important to devote some effort to considering the types of particles and organizations that may occur in ceramic powders.

II. THE STRUCTURE OF CERAMIC POWDERS

Before one can discuss the dispersion of powders in liquid, some consideration must be given to the initial powder structure and the size of the individual particles. In ceramic powders, the primary discrete units are usually considered to be grains, which for synthetic powders usually lie in the submicron range. These can easily adhere to each other, forming secondary clusters, which are often formed during precipitation or processing. Clusters can build up to larger units or agglomerations, which can extend several microns in size and usually are composed of a network of interconnected pores. A simplified illustration of the structure of ceramic powders is shown in Fig. 1. The dimension, density, pore size, distribution, and surface area

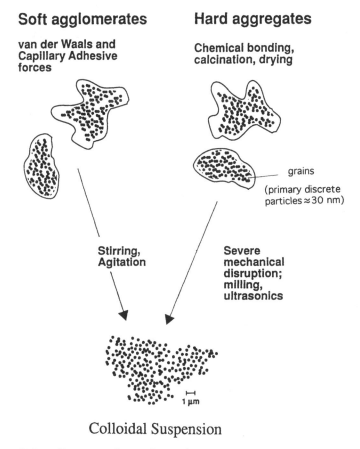

Soft agglomerates

van der Waals and
Capillary Adhesive
forces

Hard aggregates

Chemical bonding,
calcination, drying

grains

(primary discrete
particles ≈ 30 nm)

Stirring,
Agitation

Severe
mechanical
disruption;
milling,
ultrasonics

1 μm

Colloidal Suspension

FIG. 1 Structure of ceramic powders.

of the agglomerate depend on the size, shape, microstructure, spatial arrangement, and coordination number of the primary particles within the agglomerate [10].

The agglomerates may be soft due to attractive van der Waals forces or capillary forces and fairly easy to break down into primary particles. Other types of high-strength agglomerates (known as aggregates) are hard due to bonding caused by chemical reactions, fusion, calcination, or drying during manufacture. For example, synthetic powders produced from metal alkoxides tend to produce strongly agglomerated powders (agglomerate size ≈ 0.5 μm) yet comprising very small primary particles (≈ 30 nm) [11]. These hard types of aggregates are usually difficult to break down except by severe mechanical methods (i.e., milling). Powder agglomeration is probably the most difficult property to control throughout the processing cycle.

Most particle-size-determining techniques, such as light scattering and centrifugation, give an accurate assessment of the degree of dispersity of the powder. In addition, with modern instrumentation, it is often possible to differentiate between primary and secondary units. This may be achieved using laser diffraction instruments, which use a combination of different light-scattering or light-diffraction techniques. An instrument such as the Master Sizer (Malvern Instruments) can deal with size ranges from 0.1 to 600 μm covered by two overlapping ranges from 0.1 to 90 μm and from 1.2 to 600 μm.

Barringer and Bowen [12] defined an ideal powder for high-performance ceramics as having a particle size between 1 and 0.1 μm, where the particles have a narrow size distribution and are equiaxed and nonagglomerated [12]. The upper limit (1 μm) depends on the requirements of reactivity (surface area) to sintering, and the lower limit (0.1 μm) is a critical particle size below which problems with forming techniques occur. In reality, however, most commercial powders contain between 0.1 and 10% large hard aggregates above this size range, and it is essential to remove these by colloidal processing techniques such as milling or ultrasonic treatment. Also, sedimentation or filtration helps before sintering. For example, a typical titania powder (grain size 0.5 μm) was reported to contain 80% of aggregates > 1 μm; but after milling and dispersing, few aggregates > 1 μm were found to remain in suspension [13].

Most powders used in high-technology applications are produced by chemical routes and contain a wide range of agglomerates, whereas particles produced by grinding usually consist of larger particles with a much wider size distribution. For example, α-SiC (UF-15 grade) produced by the Acheson process, followed by wet pulverization, gives particles having a size distribution of 90% < 1 μm and 50% < 0.7 μm [14]. Roosen and Hausner

[10] recently reviewed the various techniques to suppress agglomerate formation that occurs during the synthesis and processing of wet chemically prepared powders.

In dealing with submicron-size grains, consideration must also be given to the relationship between particle size and surface energy. For small colloidal particles, as the surface-area-to-volume ratio increases, there is a significant increase in the proportion of molecules lying in the interfacial region. These usually have a higher surface energy and are more reactive than bulk molecules, and they can cause interfacial reactions to occur during processing. For example, leaching can occur for fine powders in water [15]. Also reprecipitation, hydrolysis, and chemical complexing can affect the stability of the system. In addition, large surface area powders can adsorb large quantities of chemicals (dispersants, processing aids, and impurities).

For crystalline grains the surface energy is not isotropic and the edges and corners have a higher surface energy. This has been explained by the atoms at the edges having fewer neighbors than those in a plane surface. This is an important factor to consider with small α-SiC particles (≈ 0.1 μm), which usually preserve their crystallinity and have angular rather than spherical shapes, whereas other particles, such as ZrO_2, form single crystalline grains [16]. Additional problems can result from fine powders due to their interfacial reactivity when exposed to the environment during storage (in air or humid atmospheres), prior to dispersion. This may change the composition of the surface layer due to redox reactions and/or hydrolysis.

In recent surface chemistry studies, different Si_3N_4 powders produced by different manufacturing methods were shown (from electrokinetic measurements) to have surface layers with different compositions [17-19]. For Si_3N_4 powders, it has been shown that mechanical grinding in air produces reactive surface sites [20].

In many cases during the production of powders, aggregates are produced deliberately by the use of binders or other additives during pelletization or spray drying. This enables easy handling of the powders and prevents demixing of homogeneously mixed powders, but it can also cause difficulties, since the powders must be dispersed during processing. Attempts have been made, occasionally, to produce disagglomerated powders, but these are more difficult to deal with in press molding due to increased fluffiness [21].

III. WETTABILITY OF POWDERS

Before the powder can be dispersed, it must be initially wetted by the liquid. In systems containing a solid, a liquid, and a vapor phase, three types of wetting phenomena can occur, which may be classified as adhesional wetting, spreading wetting, and immersional wetting. Adhesional wetting is the process of formation of a three-phase contact; spreading wetting is the process of displacement of the vapor by the liquid on the surface of the solid; finally, immersional wetting is the process of transfer of the particles from vapor into the liquid phases. The free energy changes ΔG describing the processes can be formulated as

Adhesion $\quad \Delta G_A = \gamma_{SL} - \gamma_{SV} - \gamma_{LV}$ (2)

Spreading $\quad \Delta G_S = \gamma_{SL} - \gamma_{SV} + \gamma_{LV}$ (3)

Immersion $\quad \Delta G_I = \gamma_{SL} - \gamma_{SV}$ (4)

where γ_{SL}, γ_{SV}, and γ_{LV} are the surface free energies of the solid-liquid, solid-vapor, and liquid-vapor interfaces, respectively. The more negative the value of ΔG, the greater the ease of the process occuring.

In addition, γ_{SL}, γ_{SV}, and γ_{LV} are related by the Young's equation

$$\gamma_{SV} = \gamma_{SL} + \gamma_{LV} \cos \theta \tag{5}$$

where θ is the contact angle or equilibrium angle measure through the liquid phase at which the liquid surface contacts the solid.

If complete wetting occurs (i.e., spreading), $\gamma_{SV} - \gamma_{SL} > \gamma_{LV} \cos \theta$, and a spreading coefficient defined by $S = \gamma_{SV} - (\gamma_{SL} + \gamma_{LV})$ will be positive, indicating a spontaneous process.

Under practical conditions, γ_{SL} and γ_{SV} are difficult to measure, although θ can be fairly easily determined. For the three wetting processes, the ratio of ΔG to γ_{LA} can be plotted as a function of the contact angle (Fig. 2), and it indicates that, for a given θ value, the ease of probability of the processes occurring follow the order

spreading < immersion < adhesion

An important aspect of the plots is the critical value of θ (90°). As the curve passes through this value, the cosine changes sign from positive to negative. According to these curves, it can be seen that the adhesion process is energetically more favorable. However, adhesion cannot occur before spreading, such that the adhesion process must be considered as an activated phenomenon involving an activation energy barrier. Also, changes in the various interfacial tensions due to adsorption of surfactant can significantly affect the wetting behavior. These effects have recently been discussed by Chander and Hogg [22].

Under practical conditions, poor wetting in ceramic slips causes cavitation, foaming on shearing, and increased viscosity. There are several meth-

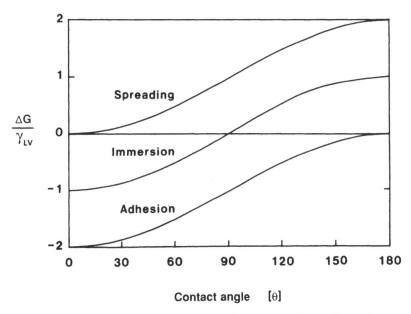

FIG. 2 The effect of contact angle on the free energy change for various wetting processes. (From Ref. 22.)

ods developed to determine the wetting characteristics of powders, which include immersion-sink time measurements [23], capillary flow techniques that determine the rate of penetration of liquids into cylindrical pore channels [24], induction time methods [25], etc. In some cases, difficulties can be encountered for wetting ceramic powders under processing conditions. For example, poor wetting often occurs in slips prepared from SiC with polyphenylene added (to produce free carbon), since the powder can become coated with an organic film. In other cases, the particle surface may become hydrophobic due to contamination by rubber or plastic from mill linings or residual organic solvent. When the particles are only partially wetted, they adsorb at the L-V interface and may stabilize foam lamellae. In order to break the foam, wetting agents such as high alcohols or silicones can be added to reduce γ_{SL} and γ_{LV}. Obviously, the foaming problem can be avoided by initially adding a wetting agent such as a naphthalene sulfonic acid condensate. These problems have recently been studied by Carlström et al. [26].

For nonaqueous tape casting systems, wetting agents such as alkylaryl polyether alcohols [27], ether-modified polyethylene glycols [28], polyoxyethylene acetates [27], and esters [27] have been used.

IV. INTERPARTICLE INTERACTIONS IN LIQUIDS

The dispersion of ceramic powders is usually carried out in vapor or in bulk liquid. In vapor, consideration must be given to capillary forces and electrostatics, which are influenced by the humidity. Usually, under high humidity conditions the particles become bonded together by condensed water bridges. This causes the electrostatic forces to be reduced. Low humidity conditions are therefore the best for dispersing powders in air, since the particles are easily charged and there are fewer possibilities of condensed water bridges being formed.

In dispersing powders in liquids, after the wetting processes have been completed, consideration must be given to the particle size and the amount of energy dissipated in the breakdown and mixing processes. Large aggregates require large amounts of energy for breakdown (i.e., dry or wet milling or grinding), while soft agglomerates can be broken down more easily. During stirring, pumping, piping, filtration, consolidation, etc., the slurry is subject to shear and capillary forces, and the degree of turbulence can vary according to the type of equipment. In order to control the state of the suspension, consideration must also be given to the particle size range involved with respect to characteristics of the liquid flow. For powders containing some noncolloidal size fractions, it is important to define the influence of particle size on the interparticle interactions with respect to Brownian (perikinetic), hydrodynamic, (orthokinetic), gravitational, or inertial effects resulting from shear flow.

Warren [29] emphasized that surface forces such as van der Waals forces, electrostatic forces, and Brownian motion can dominate *only* over gravitational and inertial forces for particles in the 0.1 μm size range, as indicated in Fig. 3. In Table 2, the energies associated with some of these forces are given for particles in the 0.1–10 μm size range. For large particles (1–10 μm), the inertial and gravitational forces become extremely important, and the fluid flow motion can confer enormous energies to the particles. Under certain conditions, the mechanical forces can overcome the repulsive potential energy between particles, causing coagulation. In other cases, the particles can be forced apart if they are weakly agglomerated.

Overall for the colloidal particle fraction, the suspension stability can be increased by the use of a dispersing agent. In an aqueous environment, this may simply involve the change in solution conditions such as pH; in other

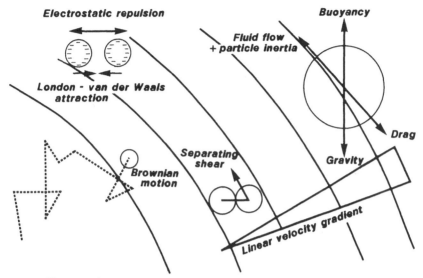

FIG. 3 Forces acting on and between particles in laminar flow. (From Ref. 29.)

cases, adsorption of a specific ion on the particle may occur, causing an increase in the repulsive charge. However, processing at low or high pH in the slip can cause corrosion attacks on both the plaster mold and other raw materials. Hence, frequently, more complex chemicals are used to stabilize ceramic slips. In both aqueous and nonaqueous media, chemical disper-

TABLE 2 Energies of Particles in Suspension due to Various Interactions

Type of interaction	Energy (in units of kT) for particles of given size		
	0.1 μm	1 μm	10 μm
van der Waals attraction	10	100	1000
Electrostatic repulsion	0–100	0–1000	0–10,000
Brownian motion	1	1	1
Kinetic energy of sedimentation	10^{-13}	10^{-6}	10
Kinetic energy of stirring	1	1000	10^6

Source: Ref. 29.

sants can produce a strong electrostatic repulsive force or reduce the van der Waals attractive potential between interaction particles. In addition, particles may also be stabilized by an adsorbed polymer (steric stabilization) or by adsorption of a strongly hydrophilic film, causing structural hydration forces. Also, the masking of the van der Waals forces can be achieved by selecting a suitable dispersing medium to match the ceramic and "depletion stabilization" using high concentrations of nonadsorbing polymer are well established methods of stabilizing ceramic suspensions. Finally, in many cases stabilization can be achieved by a combination of steric and electrostatic stabilization. Some of these mechanisms are illustrated in Fig. 4.

To proceed, we need to discuss these mechanisms of stability in some detail, and we begin by considering electrostatic stability, which is controlled by electrostatic charge, as described by Derjaguin, Landau, Verwey, and Overbeek (DLVO) theory [30a]. First, it is important to discuss the mechanisms that cause the buildup of charge on a ceramic particle in an aqueous environment.

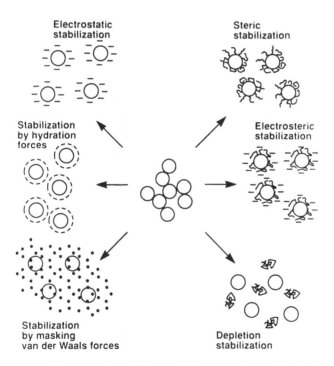

FIG. 4 Methods of stabilizing colloidal ceramic particles in liquids.

V. THE DEVELOPMENT OF SURFACE CHARGE ON CERAMIC OXIDE POWDERS DISPERSED IN WATER

On immersing a ceramic powder in water, the suface usually acquire a charge by interaction with ionic solution species. The charge is affected by pH and the electrolyte concentration of the water. Hydrolysis, surface dissociation, or ion adsorption or desorption can occur. On dealing with simpler oxide ceramic, the surface-charging mechanism has been explained by the Lewis acid-base concept involving direct proton transfer with the surface (by the surface hydroxyl groups). For example, if the electron density of the oxygen atom in the $-MOH$ group is *low*, then the strength of the hydrogen bond formed with the polarized hydrogen atom will be reduced, and ionization with a water molecule via the following dissociation process may occur:

$$-MOH_{(surface)} + H_2O \overset{K_{a1}^s}{\rightleftarrows} -MO^-_{(surface)} + H_3O^+ \qquad (6)$$

If the electron density of the oxygen is *high*, then protons may become bound to the $-MOH$ groups, causing dissociation via

$$-MOH_{(surface)} + H_2O \overset{K_{a2}^s}{\rightleftarrows} -MOH^+_{2(surface)} + OH^- \qquad (7)$$

where M represents a ceramic cation such as Al^{3+}, Ti^{4+}, Zr^{4+}, or Si^{4+}. To compensate for the surface charge, ions form bulk solution (counterions) concentrate at the interface. The counterions are electrostatically attracted to the surface, and due to thermal motion they become evenly distributed through the solution. The charge in the particle surface, together with the ionic atmosphere of counterions (which consist of an excess of opposite-charge ions), constitutes an electrical double layer (d.l.). From the balance between the coulombic forces (the attraction and repulsion between like and unlike charges) and the entropic requirements of the ions in solution (which causes mixing uniformly throughout the available space), the Gouy-Chapman theory [31] was developed, which was mathematically described by the Poisson-Boltzmann equation. This model has several weaknesses; for example, the ions are considered as point charges, and the solvent is considered as a structureless continuum. Also, the charge is considered to be spread uniformly on the surface with no discreteness.

Another weakness of this simple acid-basic dissociation model for oxides involves the definition of the surface acidity constants K_{a1}^S and K_{a2}^S in Eqs. (6) and (7). These are microscopic equilbrium constants and differ from true intrinsic acidity constants because each loss of a proton reduces the charge on the ceramic polyacid surface, and the acidity of the neighbor groups will be affected to some extent. The free energy of deprotonation as expressed by K_{a1}^{S} is a summation of the free energy of dissociation of H^{+} as defined by K_{a}^{S} (intr) and a free energy of removal of the proton from the site of dissociation into the bulk of solution, as expressed by the Boltzmann (or electrostatic) factor. Thus

$$K_{a1}^{S} = K_{a1}^{S} \text{ (intr) } \exp\left(\frac{F\psi_1}{RT}\right) \tag{8}$$

where ψ_1 is the effective potential difference between the binding sites and bulk solution, K_{a1}^{S} (intr) is the acidity constant of an acid group in a hypothetically completely chargeless surrounding, and F is the Faraday constant. To obtain K_{a1}^{S} (intr), K_{a1}^{S} values can be determined experimentally and the data linearly extrapolated to the point of zero surface charge [32, 33].

The proton conditions of pH where the surface charge is zero is commonly known as the zero point of charge (pH $_{zpc}$) and may be expressed by

$$pH_{zpc} = \frac{1}{2} [pK_{a1}^{S} \text{ (intr) } + pK_{a2}^{S} \text{ (intr)}] \tag{9}$$

Also the difference in surface ionization constants is given by

$$\Delta pK_a = pK_{a1}^{S} \text{ (intr) } - pK_{a2}^{S} \text{ (intr)} \tag{10}$$

ΔpK_a is an important parameter that quantifies the amount of charge at the interface at pH$_{zpc}$. Negative values of ΔpK_a correspond to a surface with a large (but equal) number of positive and negative sites at pH$_{zpc}$. As ΔpK rises to positive values, the number of charged sites (present at pH$_{zpc}$) becomes smaller. For oxides, ΔpK_a has usually values of about three.

Determination of the equilibrium constants is useful for estimating the surface charge on the particles and interpreting adsorption data in association with appropriate ion exchange reactions. This information can be useful in controlling the dispersibility of powders in water. Unfortunately, the constants are not amenable to direct experimental measurement because the ion activity in the surface plane cannot be directly measured. However, from potentiometric titrations, values of K_{a}^{S}(int) can be estimated, as will be discussed later in this section.

According to conventional Debye-Hückel theory, the thickness of the electrostatic d.l. (κ^{-1}) can be expressed according to

$$\kappa^{-1} = \left(\frac{\varepsilon_r\varepsilon_0 kT}{e^2\Sigma_i n_i Z_i^2}\right)^{1/2} \tag{11}$$

where e is the electronic charge, n_i is the number of ions per cm^3, and Z_i is the valency of the ionic species; ε_0 is the dielectric permittivity of a vacuum, and ε_r is the relative permittivity (or dielectric constant).

Various models have been used to describe the spatial distribution of charge near the surface and have been recently reviewed by Hunter [33]. In one of the simplest models, the double layer of charge can be divided into two sections. The first region, the adjacent to the surface, is known as the Stern layer and contains a layer of counterions with the distance extending into solution determined by the size of the hydrated counterions adsorbed on the surface. Beyond the Stern layer, the diffuse layer of counterions extends further into solution, and the boundary between the two layers is known as the Stern plane. With regard to colloid stability, it is important to attempt to quantify the electrical potential value near this boundary.

From d.l. theory, the Stern layer potential may be equated to the ζ potential, which is the potential drop across the mobile part of the double layer that surrounds a colloidal particle and is responsible for the electrokinetic behavior of the particle under an electric field. From ζ potential measurements, a basic understanding of the types of ions and species that determine the structure of the d.l. can often be derived. This type of information is important in relation to the control of the stability, rheology, and sedimentation characteristics of the dispersion.

With regard to the counterions, a distinction is made between indifferent and specifically adsorbing ions. Indifferent ions are usually simple ionic species such as K^+ or Cl^- that constitute the diffuse part of the d.l. The term usually implies that they do not adsorb on specific surface sites. Specifically adsorbing ions have a "chemical" affinity for the surface. In addition to coulombic forces, other specific forces, such as hydrophobic interactions and complex formation, can alter the charge on the surface and hence influence the ζ potential and the colloidal stability of the suspension.

VI. ADSORPTION OF SIMPLE IONIC SPECIES AT THE PARTICLE INTERFACE

The adsorption of simple ions as dispersing agents at the particle liquid interface can change the interfacial electrical properties (charge and potential). In fact, most dispersing agents act by the specific adsorption of surface active counterions. In the opposite case of nonspecific adsorbing ions such as inert electrolyte, then the overall effect is simply to reduce the thickness of the double layer causing coagulation.

For oxide systems, protons and hydroxy ions are specific adsorbed as illustrated by Eqs. (6) and (7) and hence are potential determining. The fixed surface charge (σ_0) can be expressed by the difference in adsorption density between H^+ and OH^- adsorbed on the surface and can be formulated as

$$\sigma_0 = F(\Gamma_{H^+} - \Gamma_{OH^-}) \tag{12}$$

where Γ_{H^+} and Γ_{OH^-} are the surface excess of H^+ and OH^-. Hence by increasing or decreasing the pH, the electrostatic charge on the particles can be controlled.

The surface charge became *less* negative in the presence of specific adsorbed metal cations M^{2+}, such that

$$\sigma_0 = F(\Gamma_{H^+} - \Gamma_{OH^-} + z\Gamma_{M^{2+}}) \tag{13}$$

In the presence of specifically adsorbed anions $H_nA^{(z-n)-}$ the surface charge becomes *more* negative.

$$\sigma_0 = F(\Gamma_{H^+} - \Gamma_{OH^-} - z\Gamma_{A^{2-}}) \tag{14}$$

where $\Gamma_{M^{2+}}$ and $\Gamma_{A^{2-}}$ are the adsorption densities of metal cations and deprotonated anions, respectively. The relationship between net charge and pH established by the proton balance and their complexes and specifically adsorbed cations or anions is illustrated in Fig. 5.

Simple models that relate interfacial potential (ψ_0) to pH for oxides were originally developed from conventional electrochemical theory. The basic assumption is that the oxide, in the absence of any specifically adsorbing ions, can be theoretically treated as a reversible electrode (to H^+ and OH^- ions) in a galvanic cell. The total double layer potential can hence be described by the Nernst equation

$$d\psi_0 = \frac{RT}{F} \, d \ln a_{H^+} \tag{15}$$

$$= \frac{RT}{F} \, d \ln a_{OH^+} \tag{16}$$

or

$$\frac{d\psi_0}{d \ln a_{H^+}} = \frac{RT}{F} = 26 \text{ mV at } 25°C \tag{17}$$

However, for most oxide systems, values for $d\psi_0/d \ln a_{H^+}$, when experimentally determined, generally give a lower result than 26 mV/pH unit and appear to be dependent on the concentration of indifferent electrolyte. For α-SiC powder (containing an oxide film), a value of 9 mV/pH unit was recently reported [35].

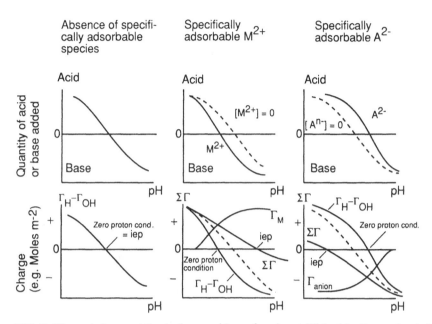

FIG. 5 The net charge at the hydrous oxide surface is established by the proton balance (adsorption of H^+ or OH^- and their complexes at the interface and specifically bound cations or anions). Specifically adsorbed cations (anions) increase (decrease) the pH of the isoelectric point (pHiep) but lower (raise) the pH of the zero proton condition. (From Ref. 34.)

In recent years, the validity of the treatment of oxides as classical reversible electrodes has been reassessed. A more realistic picture was developed, by taking into consideration the number and type of surface sites on the oxide and relating the charge to the site density by the equation

$$\sigma_0 = eN_s(n_+ - n_-) \qquad (18)$$

where N_s is the total number of sites, n_+ is the number of MOH^+ sites, and n_- is the number of $MO-$ sites.

This leads to the development of a modified Nernst equation

$$|d\psi_0| = \frac{RT}{F}\,d\ln a_{H^+} - \frac{RT}{F}\,d\ln\left(\frac{n_+}{n_-}\right) \qquad (19)$$

At pH_{zpc}, $n_+ = n_-$ and there is no contribution from the sites to the chemical potential. However, at other pH values, n_+ and n_- play an important role and need to be taken into consideration. For the oxide-water interface in the region of pH_{zpc}, this effect can explain the lower $d\psi_0/d \ln a_{H^+}$ values. However, it was later shown that the difference between the surface potential and the Nernst value was also dependent on the background electrolyte concentration. Further advancement of the amphoteric dissociation model as described by Eqs. (6-10) resulted in the introduction of equilibria to take account of the weak binding of the supporting electrolyte such as NaCl, which can be expressed in the form

$$MO^-_{(surface)} + Na^+ \overset{K^{Cl^-}_{(intr)}}{\rightleftarrows} MO^- Na^+ \tag{20}$$

$$MOH^+_{2(surface)} + Cl^- \overset{K^{Cl^-}_{(intr)}}{\rightleftarrows} MO^+_2 Cl^- \tag{21}$$

The more sophisticated triple layer model of the charged interface distinguishes two separate adsorption planes: the inner Helmholtz plane, in which specifically adsorbed ions lie (with ψ_i and σ_i values), and the outer Helmholtz plane of closest approach of the non-specifically adsorbed ions (with ψ_d and σ_d values). Detailed descriptions of the structure and properties of these d.l. models are available [33]. Site-dissociation-site-binding models were developed by Yates et al. [36] and Chan et al. [37]. These have been reviewed by James and Parks [38]. Mathematically, ionization and complexation models can be described by a set of simultaneous equations that can be solved mathematically by using appropriate analytical solutions. Six unknowns, ψ_0, ψ_i, ψ_d, σ_0, σ_i, and σ_d, can be solved in terms of the independent variables, the electrolyte concentration, and pH, providing suitable values are chosen for the parameters pK^{int}_{Na}, pK^{int}_{Cl}, pK^{int}_{a1}, pK^{int}_{a2}. N_s, pK_{a1}, and pK_{a2}. Complete amphoteric site-dissociation-site-binding models for the oxide-solution interface can be solved using computer programs.

From potentiometric experiments, values of complexation constants and intrinsic ionization constants for several oxide powders, such as SiO_2, Al_2O_3, TiO_2, ZrO_2, and ThO_2, have been determined and recently summarized by Schindler and Stumm [39] and James [38].

Since the pH controls the fundamental electrical properties of the powder, it is essential to have some knowledge of the pK values and pH_{zpc} and also the ψ versus pH relationship in order to predict the stability of the dispersed powder in water. In fact, the magnitude of the charge and the degree of stability depend on the departure of the pH from pH_{zpc}.

The surface charge and electrokinetic ζ potential can be determined experimentally. In practice, values of pH_{zpc} can be determined by titration with potential-determining ions. The condition under which the ζ potential is zero is called the isoelectric point (pH_{iep}). For oxide minerals, when the ζ

potential measurements are carried out in the presence of nonspecific adsorbing ions, $pH_{iep} \approx pH_{zpc}$. However, in the presence of specifically adsorbing species such as polyvalent or surface active dispersant species, the pH_{zpc} may not be identical to the pH_{iep} (as illustrated in Fig. 5). Hence the difference between these two values can sometimes be useful for studying the specific adsorption of surfactants, which will influence the stability.

Most ζ potential measurements today are carried out on small particles in *dilute* suspensions using the electrophoresis technique. For *concentrated* ceramic suspensions, instruments have been developed to use electric and ultrasonic impulses to determine ζ potential values. Earlier measurements on concentrated dispersions were made using moving boundary methods, electrophoresis mass transfer, and instruments built to determine the mass of particles deposited on an electrode. However, these types of instruments have today been replaced by acoustic techniques. The original principle of the acoustic method is based on sound waves passing through a suspension of charged particles, which can generate a force causing an electric field, due to a difference in relative motion of the two phases.

Several new instruments have recently been introduced on which it is possible to measure both surface charge (by titration) and the ζ potential (by acoustophoretic measurements) on concentrated dispersions. This enables both the pH_{iep} and pH_{zpc} to be determined on the same sample. In cases where the two values differ, it becomes possible to understand the role of the potential-determining ion, indifferent ion, and chemically adsorbed ion on the stability of the system. The principles of the techniques have been recently reviewed by Babchin et al. [40], who compare acoustophoretic mobility results to electrophoretic mobility data for several dispersed systems.

Essentially, the suspension will show colloidal instability in the region of the pH_{iep}, where the rate of coagulation has a maximum value indicating that there is no repulsive energy barrier to prevent rapid coagulation. As the charge and potential increase at pH values either side of the pH_{iep}, the system becomes more stable. In general, stability requires a repulsive potential of at least 30–40 mV. By convention, both charge and potential are negative above and positive below the pH_{iep}. In Fig. 6 the stability of an alumina powder (as determined by the turbidity of the suspension) can be correlated to the ζ potential, so that at negative or positive potentials, an electrostatic repulsion barrier to coagulation exists, and the suspension remains turbid.

From DLVO theory, the stability of a charged suspension can be quantified by the critical coagulation concentration (ccc) or the concentration of various monovalent ions added on each side of the pH_{iep}, which will coagu-

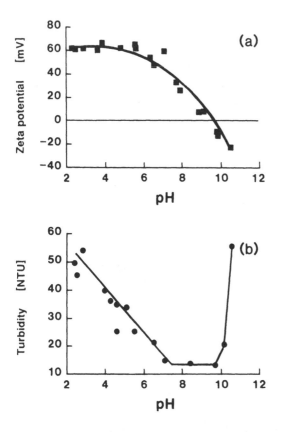

FIG. 6 The dispersion of α-Al_2O_3 particles in water at constant ionic strength (10^{-2} M NH_4Cl) as a function of pH. (a) Zeta potential measurements. (b) Stability as determined from turbidity. (From Ref. 121.)

late the suspension by collapsing the d.l. For example, on the negatively charged powder surface, a cationic adsorption sequence can be derived for Li^+, Na^+, K^+, and Cs^+. On the positively charged surface, the adsorption of anions IO_3^-, F^-, BrO_3^-, and NO_3^-, etc. can be determined to give another sequence.

According to the Stern d.l. theory, the cations maintain their hydration shell on adsorption. The higher the ion valency, the stronger the adsorption and the lower the concentration of electrolyte required to coagulate the system. For ions having the same valency, the smaller the size of the hydrated ion, the greater the adsorption capacity. Generally, in the absence of specific chemical effects, both the size of the ion and its charge can have an influence on the adsorption process and d.l. properties.

However, several coagulation experiments have shown that in addition to valency and size of hydrated ions, other parameters may also influence the adsorption process. For example, recent models have been proposed by Giest et al. [42], Bérubé and de Bruyn [43], and Stumm [44]. These workers have discussed the surface-ion interaction in terms of the modification of the local water structure and introduced an additional short range entropic effect that could be superimposed on the electrostatic contribution. Essentially, it was suggested that structure-maker ions will be preferentially adsorbed on a structure-maker surface in preference to structure-breaker ions and vice versa. This idea was based on the generalization of the ion-surface interactions from the Gurney concept [45] of the ion-ion interactions and was recently found effective in explaining the adsorption sequences on many rutile hydrosols [46].

Other explanations for the role of ion size in coagulation involve differences in diffusion and friction coefficients around the charged particles, which are affected by the d.l. around them. It was also reported by Jan and Fuerstenau [47] that the interaction strength of alkaline-earth cations on TiO_2 followed the sequence $Ba^{2+} > Sr^{2+} > Ca^{2+} > Mg^{2+}$. In this case, it was suggested that the cations adsorbed as a bidentate complex, and a surface-induced hydrolysis model was proposed.

For oxide ceramic powders, values of the pH_{zpc} (as traditionally determined in water) are fairly reproducible and therefore serve as a general guide to the ability of the surface to lose or attract protons. Generally, more acidic oxides have low pH_{zpc}, and basic oxides will have high pH_{iep} values: for example, α-SiC, $pH_{iep} \approx 3$; ZrO_2, $pH_{iep} \approx 7$; α-Al_2O_3, $pH_{iep} \approx 8$, and Y_2O_3, $pH_{iep} \approx 9$. For other powders, such as nitrides, a wide range of values have been reported, which can be explained by considering the complicated effects associated with the composition of the surface layer.

There are two theoretical methods of predicting pH_{iep} of simple oxides and hydroxides. One method is to use the minimum solubility theory [48], provided no specific adsorption occurs. The pH_{zpc}, which frequently corresponds to the pH of minimum solubility, can be determined by constructing solubility equilibria diagrams.

An alternative method of predicting pH_{zpc} is to use the relationship developed by Parks [49], in which pH_{zpc} can be directly related to Z/R where Z is the formal charge on the cation and R is the sum of the cationic radius r_+ and the oxygen diameter 2.8 Å (Fig. 7). Values of Z and r_+ are

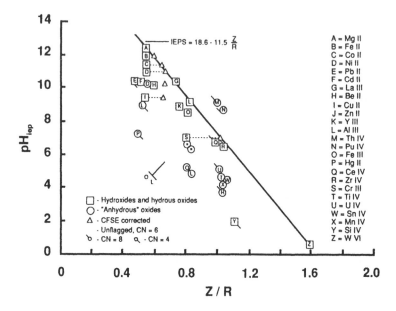

FIG. 7 The relationships between pH_{iep} and Z/R for oxides, where Z is the formal charge on the cation and R is the sum of the cationic radius (r_+) and the oxygen diameter (2.8 Å). (From Ref. 49.)

readily available from crystallographic data for most oxides/hydroxides. This relationship also explains why tetravalent cations bonded to surface hydroxyls are very acidic, with low pH_{iep} values. Yoon et al. [50] have also developed an improved version of Parks' relationship.

VII. ADSORPTION OF DISPERSANT SPECIES AND POLYMERS

Although there is an abundance of literature covering the area of adsorption, there still remains much to be uncovered on the subject, particularly with complex systems. A series of recent publications edited by Parfitt and Rochester [51], Anderson and Rubins [52], Tewari [53], and Ottewill et al. [54] give a good overview of the field.

One must take into account an enormous range of simple and complicated dispersant ions and molecules and then apply some type of subdivision. A simple classification of the dispersant types could be as follows: neutral species, simple ions, complex ions, simple surface active agents, polyelectrolytes, charged and uncharged polymers. All dispersants are surface active and may be generally classified according to the surfactant functional group (as illustrated in Table 3).

However, in order to comprehend the process in some detail, certain molecular characteristics of the dispersant must also be taken into consideration as well as the type of surfactant functional head group, for example,

TABLE 3 Classification of Some Types of Anionic Dispersants According to Functional Groups

(a) size and shape, (b) ratio of polar to nonpolar segments, (c) chemical properties of its ligand for complex species, and (d) its orientation at the interface, etc.

A simplified theoretical approach to the overall problem of adsorption of dispersant on powders can be achieved by isolating the major variables that influence the adsorption step. For example, the total free energy of adsorption ΔG_{ads}^0 can be separated into several individual contributions. These may be due (for example) to coulombic attraction, hydrogen bond intractions, dipole-dipole interaction, etc. This general approach may be expressed as

$$\Delta G_{ads}^0 = \Delta G_{elect}^0 + \Delta G_{vdw}^0 + \Delta G_{chem}^0 + \Delta G_{hb}^0 + \Delta G_{solv}^0 \qquad (22)$$

where ΔG_{elect}^0 is the electrostatic contribution, ΔG_{vdw}^0 is the London or van der Waals interaction, and ΔG_{chem}^0 includes specific "chemical" interactions ranging from hydrogen bonding to covalent bonding. Also, ΔG_{solv}^0 is the change in free energy of the hydrated sphere of an ion as it adsorbs and becomes partly desolvated in the adsorption process. Finally, ΔG_{hb}^0 refers to the contributions to adsorption due to hydrophobic interactions.

Depending upon the nature of the adsorbing species, some of these terms may or may not be important. For example, for the adsorption of simple inorganic ions on an oxide surface, ΔG_{ads}^0 will be mainly due to ΔG_{elect}^0. However, as earlier discussed, certain ions exhibit a specific affinity for the surface and adsorb in the inner region of the d.l., and in these cases chemical effects and solvation have to be taken into consideration. For example, phosphate and silicate dispersant ion species specifically adsorb on many surfaces. Under such circumstances, the adsorption process is usually associated with a shift in pH_{iep}.

In dealing with the special case of the adsorption of more complex multivalent cations, i.e., Mg^{2+}, Co^{2+}, Cu^{2+}, Ca^{2+}, etc., it has been found relevant to relate the adsorption process in aqueous media to their hydrolysis/precipitation behavior. Several models have been developed in order to interpret the pH dependence of the adsorption process. This is usually characterized by an "abrupt pH adsorption edge," where a rapid uptake of cations occurs over a narrow pH range depending on the properties of the cation and the surface.

Early models, such as ion exchange and surface complex formation, have been proposed to explain this behavior [55]. In addition, James and Healy [56,57] have developed a solution chemistry and hydrolysis model by considering ion-solvent interactions in terms of coulombic, solvation, and specific chemical energies of interactions of the hydrolyzed solution species

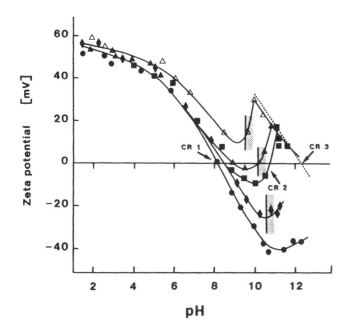

FIG. 8 Zeta potential versus pH plot of α-alumina ceramic powder dispersed (0.01 w/w % solids) in ($\triangle$) 10^{-2} M $MgCl_2$; ($\blacktriangle$) 10^{-3} M $MgCl_2$; ($\blacksquare$) 5×10^{-4} M $MgCl_2$; ($\blacklozenge$) 10^{-4} M $MgCl_2$; ($\bullet$) without $MgCl_2$; --- $Mg(OH)_2$ dispersion, at constant ionic strength (10^{-2} M NaCl). ($-$) represents the precipitation edge of bulk magnesium hydroxide. CR indicates the points of change reversal. (From Ref. 58.)

with the surface. This theory has recently been used to explain the charge reversal (CR) and change in stability associated with the uptake of Mg(II) on α-SiC and α-Al$_2$O$_3$ ceramics as determined by electrophoresis [58]. In Fig. 8 the experimental zeta potential measurements for α-Al$_2$O$_3$ in Mg(II) is shown as a function of the solution pH. It is also interesting to note that Shaw and Pethica [59] achieved a uniform blend of MgO and Si_3N_4 by controlled hydrolysis/precipitation of Mg^{2+} species from aqueous solution onto the surface of an Si_3N_4 powder.

Organic ions and molecules are rather complex due to their amphipatic nature, and in addition to the electrical and chemical contributions to ΔG^0_{ads} from the polar head group, hydrophobic chain interactions must also be considered. When the organic species adsorbs in the Stern layer, association through interactions of their hydrocarbon chains must be taken into account. The basis of the understanding of this mechanism was due to

Gaudin and Fuerstenau [60] and Somasundaran and Fuerstenau [61]. It was proposed that a so-called two-dimensional surface aggregate or hemimicelle was formed on the surface. The free energy associated with hemimicelle formation can be expressed by

$$\Delta G^0_{ads} = N\phi \tag{23}$$

where N = the number of CH_2 groups in the molecule and ϕ is the standard free energy for removing one CH_2 group from the water phase. In experimental studies, $\Delta G^0_{CH_2}$ was related to chain length for a series of alkyl sulphonate surfactants adsorbed on alumina [62].

Dick et al. [63] have studied the relationship between the hydrophobic and electrostatic interaction, for the adsorption of homologs of linear alkylbenzene sulphonates on alumina particles ($pH_{zpc} \approx 8$). From Fig. 9a, it can be seen that the branching of the dodecyl groups reduce the adsorption

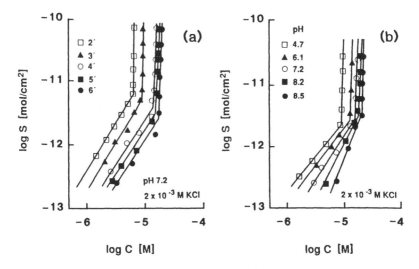

FIG. 9 Effect of structure of hydrophobic group and pH on adsorption of anionic surfactants on alumina. (a) The logarithm of the surface concentrations as a function of the logarithm of solution concentration is shown for a series of 4-(n'-dodecyl) benzenesulfonate. The numbers in the legend indicate the position on the dodecyl chain at which the benzene ring is attached. Increased branching of the dodecyl group leads to weaker adsorption. (b) The logarithm of the surface concentrations as a function of the logarithm of solution concentration is shown for 4-(2'-dodecyl) benzenesulfonate as a function of pH. An increase in pH leads to weaker adsorption, consistent with the onset of electrostatic repulsion at the surface, of which the pH_{zpc} (in the absence of surfactant) is 8.9. (From Ref. 63.)

density of this negatively charged molecule in a systematic way. At first sight, it would appear that the adsorption of this group is primarily hydrophobic. However, further inspection of the data for one homolog at a series of different pH values (Fig. 9b) suggests a systematic decrease in adsorption with increase in pH. Also, above the pH_{zpc} only low concentrations of surfactant are adsorbed. This complementary data indicates that although the adsorption of the surfactant on the oxide is influenced by ΔG_{hb}, ΔG_{elect} is extremely important and on varying pH conditions has a greater control of the process.

The adsorption of polymers on surfaces differs drastically from the adsorption of simpler small molecules. This is due to the large number of configurations that the macromolecule can achieve in solution and on the interface. The subject has been extensively studied in recent years and has recently been reviewed [64]. Polyelectrolytes are usually dealt with as a special case of polymer adsorption, and the adsorption process is usually controlled to some extent by ΔG_{elect}; positive surfaces tend to adsorb negative electrolytes and vice versa. However, there are many cases where negatively charged polymers adsorb on negative surfaces and neutral polymers adsorb on charged surfaces.

For polyelectrolytes noncoulombic interactions are therefore also extremely important and must be taken into account. One simplified approach is to define a free energy of polymer adsorption dealing with both types:

$$\Delta G_{ads}^0 = \bar{z}e\bar{\psi}_\delta + \bar{n}_{CH_2}\phi_{CH_2} \tag{24}$$

where $\bar{z}$ is the effective charge on the polymer, $\bar{n}_{CH_2}$ is the equivalent number of methylene groups per polymer, ϕ_{CH_2} is the hydrophobic interaction energy/CH_2, and $\bar{\psi}_\delta$ is the mean interfacial potential, which is negative for the process of transfer of a CH_2 group from water to the interface.

However, since coiling and uncoiling of the polymer will change $\bar{z}$ and $\bar{n}_{CH_2}$, ΔG_{ads}^0 will also be drastically affected. An alternative approach is to define ΔG_{ads}^0 in terms of the enthalpy ΔH_{ads}^0 and entropy ΔS_{ads}^0 changes:

$$\Delta G_{ads}^0 = \Delta H_{ads}^0 + T\Delta S_{ads}^0 \tag{25}$$

For adsorbing polymers, ΔG_{ads}^0 is usually large and positive in relationship to ΔH_{ads}^0. However, since it is more likely that both coulombic and hydrophobic energies contribute to the polymer adsorption process, a more appropriate equation for the process may be

$$\Delta G_{ads}^0 = \bar{z}e\bar{\psi}_\delta + \bar{n}_{CH_2}\phi_{CH_2} - T\Delta S_{ads}^0 \tag{26}$$

where the coulombic and hydrophobic contributions are combined in ΔS_{ads}^0.

Adsorption from solution is usually quantitatively expressed in terms of the surface excess, and adsorption isotherms usually have different shapes and magnitudes for different systems. From the isotherms, the adsorption capacity and adsorption affinity of the dispersant for the powder can be determined.

Generally, today in ceramic processing less expensive homopolymers are almost universally used. Unfortunately, there may be subject to desorption, which is a complex problem depending on several different parameters related to both solvent and particle. Homopolymers such as polyacrylates are fairly simple molecules that can provide both electrostatic and steric stabilization, provided the polymer adsorbs on the surface to give full coverage in a good solvent. Strong attraction to the surface can occur when the polymer has multisegments available with an affinity for adsorption. However, in cases where very strong adsorption of segments occurs, the polymer may stick to the first area of contact and prevent further polymer from adsorbing in the vicinity, producing only weak coverage. Weaker adsorption generally allows mobility to occur in the surface, rearrangement, and better packing. Also, there must be sufficient polymer present in solution to give a high coating density. In cases of low polymer concentration in a poor solvent, it is possible for adsorption to occur at more than one particle surface at a time, forming an extended link. This can cause bridge flocculation to occur.

Anionic polyelectrolyte such as polyacrylic acids (PAA) are frequently used to disperse oxide particles. This has been well documented for alumina [65,66], rutile, and hematite [67,68]. The mechanism appears to be mostly electrostatic and hydrogen bonding. Gebhardt and Fuerstenau [67] showed that PAA strongly adsorbed on positively charged hematite and rutile from pH 4 up to their pH_{iep} values. Above pH_{iep}, the particles became positively charged and start to repel the polymer. It was also shown that PAA did not absorb on a negatively charged silica ($pH_{iep} \approx 2.5$). Cesarano et al. [65] obtained similar results for the adsorption of Na^+ salt of polymethacrylic acid (PMAA) adsorbing in alumina powder. These workers also showed that at a given pH the transition from the flocculated to the dispersed state corresponded to the adsorption saturation limit. As the pH was decreased the adsorption saturation limit increased until charge neutralization (of the PMAA) was approached. It has been shown by several workers that the carboxyl groups adsorb in increasing amounts on the surface as the pH of the suspension is lowered from pH_{iep}.

As the pH is reduced, the degree of ionization of the polymer is decreased and also the number of positive surface sites on the oxide surfaces is increased. From adsorption studies of PMAA on Al_2O_3, Cesarano et al. [65] constructed stability diagrams that enabled the optimal addition of PMAA to create a stable slip to be determined. Similar studies were made by Persson [9] for PAA on Al_2O_3 from viscosity measurements. Also Shashidhar et al. [69] have studied adsorption of polyacrylate on Al_2O_3 powders.

PAA does not adsorb on acidic particles (i.e., with low pH_{iep}) such as Si, SiO_2, SiC. To stabilize slips produced from these powders, cationic polymers such as cationic polyethyleneimines (PEI) have been shown to be effective [70]. Also lignosulphates have been shown to stabilize silicon nitride powders at neutral pH [71]. Unfortunately, a complete understanding of electrostatic and steric stabilization is a difficult goal, but the practical use of these polymers in many industrial systems has been developed to a fine art.

VIII. ELECTROSTATIC STABILITY OF DISPERSIONS (DLVO THEORY)

The DLVO theory was developed initially from the early theory of doublet formation between submicron particles under rapid coagulation conditions (no repulsive energy barrier) as described by Smoluchowski theory [6]. The rate of disappearance of primary particles in the initial stages of the coagulation process can be written as

$$- \frac{dN}{dt} = kN_0^2 \qquad (27)$$

where k is the coagulation rate constant. For rapid coagulation (in the absence of an energy barrier), $k = k_0 = 8\pi DR$, where D = the diffusion coefficient of the particle and R = the collision radius. Fuchs introduced an arbitrary interaction potential into the theory that slowed down the coagulation process and expressed the relative rates of slow and fast coagulation in terms of a stability ratio W, where k can be put equal to k_0/W and equation (1) rewritten as

$$\frac{-dN}{dt} = \frac{k_0}{W} N_0^2 \qquad (28)$$

Since the rate remains constant in the rapid coagulation region, and making the assumption that $W = 1$ leads to a value of $k_0 = k$.

The DLVO theory provided the quantitative explanation of coagulation by equating the total interaction potential equation V_T term as the summation of the dispersion attraction V_A and the electrostatic repulsion V_R:

$$V_T = V_R + V_A \tag{29}$$

To calculate V_R we need to consider the interaction between approaching d.l. as the particles come together. This causes the repulsive force that has been treated by several different methods and at various levels of refinement. Although a complete mathematical solution to the electrostatics has not yet been achieved, the equations that describe V_R have been derived [30] by using mathematical approximations and physical refinements.

In the case of V_A, the intermolecular attractive forces must be taken into consideration. Originally, three types of molecular attractive forces were postulated by van der Waals. These were classified as permanent dipole interactions, dipole-induced dipole interactions, and attraction forces between nonpolar molecules. London [30] explained the latter type by the differences in polarizability of molecules in the solid phase and the intervening thin film. General London van der Waals forces are always present and are known as dispersion forces. They account for nearly all the van der Waals forces.

Although London attractive forces between individual molecules are extremely short range for an assembly of molecules they can be approximated to be additive causing a longer decay length.

For identical particles, the forces are always attractive, and for $H_0 \gg a$, where H_0 is the interparticle separation distance and a is the radius of the particles, then it scales as A/H_0^6, where A is the Hamaker function, which is related to the polarizability mismatch and is weakly dependent on the interparticle distance.

The magnitude of A can be determined by the microscopic approach, in which the values are evaluated for individual atomic polarizabilities and atomic densities of the materials involved. However, values calculated by this method are not precise.

An alternative macroscopic approach known as the Lifshiftz method can be used to give more accurate values of A. In this method, the interaction particles and the intervention medium are treated as a continuous phase. Unfortunately, the calculations are complex, and data of the bulk optical/dielectric properties of the materials over a wide frequency range is required.

Generally, A values for a single material in vacuum (air) have been reported in the 10^{-19} to 10^{-20} J range. For oxide powders across a water film, values about $0.5-1 \times 10^{-20}$ J have been estimated. (30b).

The complete mathematical expressions for V_A and V_R are given in several well-known texts [30a,30b]; however, a useful practical form of expressing V_A and V_R to allow for dissimilar particles (with surface potentials ψ_1 and ψ_2) with unequal sizes (radii a_1 and a_2) has been formulated by Hogg, Healy, and Fuerstenau [72] and may be expressed as

$$V_R = F \times \frac{\varepsilon}{4} \left\{ 2\psi_1\psi_2 \ln\left[\frac{1 + \exp(-\kappa H_0)}{1 - \exp(-\kappa H_0)}\right] + (\psi_1^2 + \psi_2^2) \ln(1 - \exp 2\kappa H_0) \right\} \quad (30)$$

and

$$V_A = -F \frac{A_{123}}{6H_0} \quad (31)$$

where A_{123} is the net Hamaker-London function for materials 1 and 2 in medium 3 and is given by $A_{123} = A_{12} + A_{33} - A_{13} - A_{23}$; F is the form factor or size $a_1a_2/(a_1 + a_2)$; κ is the Debye-Hückel reciprocal length parameter, which is related to the ionic strength of the medium; and ε is the dielectric constant of the liquid. These equations are particularly useful when dealing with multicomponent systems with particles of different charges and sizes. In the case of monodispersed particles, $a_1 = a_2$, and for a system with only one type of particle, $\psi_1 = \psi_2$.

In using Eqs. (30) and (31), it is usually assumed that the surface potentials ψ_1 and ψ_2 are relatively low and can be replaced by the zeta potential of the ceramic particles. Also, two limiting cases are implied: (a) completely relaxed double layers, and (b) total unrelaxed double layers as the particles approach. The first corresponds to constant potential of the interacting surfaces and the second to constant charge density. Recent experiments suggest that the rate of d.l. relaxation is not sufficiently rapid to guarantee constant potential conditions, and that for many cases the real situation is somewhere intermediate between these two extremes. Also interaction between approaching particles can cause distortion of the d.l. and desorption of ions, and this is not considered in the theory.

The total potential energy of interaction V_T, interparticle separation, H_0 and the curve when plotted, show several characteristic features that describe the stability of the system. At close distances of separation, a "primary minimum" exists and a "primary maximum" height V_{MAX} that acts as a barrier to prevent coagulation. This maximum must be surmounted before the particles fall into the "primary minimum" and make contact (coagulate). In principle, when V_{MAX} is sufficiently high in terms of k_BT (i.e., at least 5 k_BT), the particles will not be able to surmount the barrier and the dispersion will be stable.

At larger interparticle separation distances, a secondary net attraction or "secondary minimum" V_{MIN} may exist, which becomes deep enough with particles larger than a few microns to cause interparticle association. In Fig. 10, the total potential energy of interaction versus interparticle distance

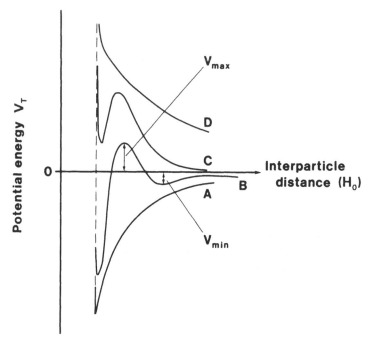

FIG. 10 Total potential energy versus interparticle distance curves between two particles, showing four different types of interactions. (A) Fast coagulation into primary minimum. (B) Weak secondary minimum coagulation. (C) No primary minimum coagulation due to high V_{MAX} energy barrier. (D) Spontaneous dispersion of particles. (---) indicates particle contact.

curves show four classes of shapes with large variations in stability. In the first case, curve A, the particles experience no repulsive forces and fall directly into the deep "primary minimum." Under these circumstances "fast coagulation" occurs and the system is completely unstable. In case B, coagulation in the primary minimum may be prevented by a fairly high V_{MAX} energy barrier, but "weak coagulation" may occur in the secondary minimum. This "secondary minimum" effect is more pronounced with particles of large radii and for flat particles than with spherical ones. In Fig. 10, curve C, the V_{MAX} value is sufficiently high to prevent coagulation; and finally curve D represents only strong repulsive forces producing a dispersed system.

Generally, the primary minimum stability will be reduced by lowering V_{MAX} (by reducing the electrostatic repulsion term). In practice, this may be achieved either by increasing the ionic strength of the solution, which decreases κ^{-1} (the double layer thickness) and "screens" out the surface charge, or by decreasing the surface potential ψ of the particles.

The general picture of the DLVO theory has been confirmed by Horn [73,74] by direct measurements of surface forces between sapphire crystal platelets in aqueous solution of NaCl at pH 6.7–11. The surface force apparatus, using mica as a substrate, has over the past 20 years been used to measure fundamental equilibrium or static interactions such as (I) attractive van der Waals forces, (II) repulsive electrostatic "double layer" forces, (III) solvation and hydration forces, (IV) hydrophobic interactions, and (V) steric forces. The principle of the method is based on measuring the separation (down to an accuracy of ±0.02 nm) between two mica sheets glued to cross cylinders of silica using optical interference fringes while the corresponding force is measured (down to a sensitivity of 1 µN) by the deflection of a cantilever spring (Fig. 11). The results on sapphire crystal were the first

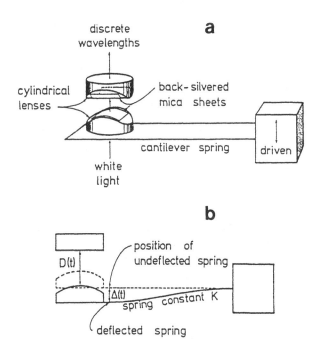

FIG. 11 The surface force apparatus. (a) Optical interference fringes are used to measure the separation between two mica sheets glued to cross cylinders of silica. (b) The force between the mica sheets is measured by the deflection of a cantilever spring as indicated by Δ, where D is the separation.

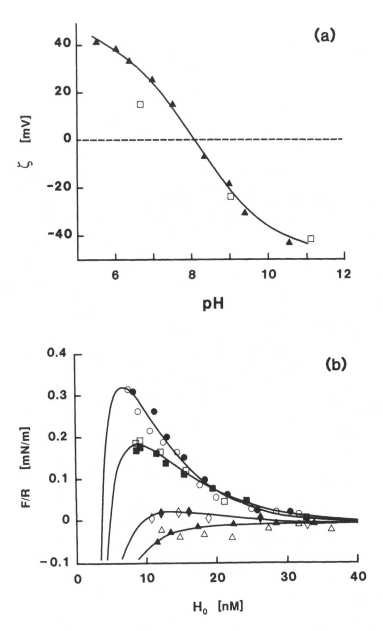

FIG. 12 (a) Surface potentials determined from double layer force measurements between sapphire crystals immersed in 0.001 M NaCl at various pH values (squares) compared with independent measurements of zeta potentials (triangles) of γ-alu-

measurements using the Israelachvili [75] force apparatus (replacing mica as a substrate) and demonstrate the possibilities of using the technique to explore the surface chemistry of other ceramic materials. From the double layer force measurements, the surface potential was determined from DLVO theory and showed good agreement with experiments determined from ζ potential measurements (Fig. 12).

Following an extension of the theory, it was also shown that the repulsive barrier could reduce the coagulation rate by a perikinetic stability factor $W_{(1,2)}$ between particles 1 and 2, which could be related to the interparticle potential V_{12} by the equation

$$W_{(1,2)} = 2\bar{a} \int_{2\bar{a}}^{\infty} \exp \frac{V_{12}}{k_B T} \frac{dr}{r^2} \tag{32}$$

where r is the distance between the centers of the two particles, i.e., $r = a_1 + a_2 + H_0$, and $\bar{a} = (a_1 + a_2)/2$.

Equation (32) can be presented in an approximate form as

$$W_{(1,2)} \sim \frac{1}{2\kappa\bar{a}} \exp \frac{V_{MAX}}{k_B T} \tag{33}$$

where V_{MAX} is defined as the maximum height of the potential barrier. This equation enabled a coagulation rate to be predicted that could be compared to experimental measurements. In many cases, good agreement has been reached, as reported, for example, by Barringer [76] for TiO_2 powders (Fig. 13). In addition, from DLVO theory, the coagulation rate in Eq. (1) as expressed by $t_{1/2}$ can be redefined as

$$t_{1/2} \approx \frac{2 \times 10^{11} W}{N_0 (\text{cm}^{-3})} \quad \text{in water at } 20°C \tag{34}$$

Although the DLVO theory has been fairly successful over the past 35 years in describing stability of colloidal particles under perikinetic conditions, there are certain restrictions in the use of the theory which need to be considered.

FIG. 12 (continued) mina. (From Ref. 73.) (b) Force F (divided by the radius of curvature) between the basal planes of two sapphire crystal platelets immersed in 10^{-3} M NaCl, plotted as a function of surface separation. The continuous lines are theoretical DLVO forces computed at constant ψ_0 with A = 6×10^{-20} J. Experimental data: $\triangle\blacktriangle$ measured at pH 6.7, ψ_0 = 13 mV; $\diamond\blacklozenge$ measured at pH 9, ψ_0 = 22 mV; $\square\blacksquare$ measured at pH 10, ψ_0 = 35 mV; $\bullet\circ$ measured at pH 11, ψ_0 = 41 mV. (From Ref. 74.)

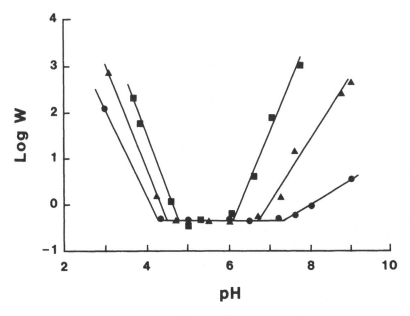

FIG. 13 Coagulation rates for TiO_2 particles expressed in terms of stability ratio (W), as a function of pH in KCl electrolyte concentrations of 0.025 M (■), 0.015 M (▲), and 0.0075 M (●). (From Ref. 76.)

Initially, it must be stressed that it only applies to monodispersed electrostatically charged sols that collide to form doublets (or the collision between single sols and a cluster); a quantitative theory of polydispersed ceramic powder systems is far too complex to be derived at present. Even with model colloids, the Smoluchowski approach applies only to early stages, i.e., the interaction between two particles; but in high-solid suspensions several interactions may occur simultaneously. Also, as discussed earlier, orthokinetic collisions with particles become more important as the early aggregates grow over 1 μm in size; the perikinetic stage is only a minor process in the coagulation of moderately concentrated slurries. The major process is the orthokinetic stage, where sweeping up of the small particles by hydrodynamic collisions with smaller ones occurs. Also, the theory only applies to colloids with smooth, well-characterized surfaces. It is sometimes difficult to apply this knowledge to ceramic powders with rough and less-well-defined surfaces.

In the DLVO theory, van der Waals and double layer interactions only are considered. "Structural" hydration and hydrophobic interactions are not included. At present, there is no general theory to deal with these additional forces, because such forces often depend on specific features of the surface. Derjaguin et al. [77] have recently presented a review of the present-day state of surface forces in which special attention is given to the structural forces of hydrophilic repulsion and hydrophobic attraction. More recently, theoretical studies have also demonstrated that nonuniform charge distribution in interacting double layers can also correlate with each other, giving additional attractive contributions to the force; or they may reduce the double layer repulsion. Such phenomena are known as charge correlation effects and are important when the surfaces are highly charged and the counterions are polyvalent [78]. Also, although DLVO theory can often be used to predict primary minimum coagulation for model colloidal systems, the theory has been found to be less successful in explaining secondary minimum effects.

However, a rough evaluation of stability of many ceramic systems under perikinetic conditions can be obtained by calculating W. In addition to predictions from the theory, several obvious predictions can be made to give a well-dispersed powder, which can be summarized as follows: the particles should have (a) a high electrostatic repulsive potential V_R, (b) a large particle size, and finally (c) a high Hamaker constant. As discussed earlier, cases where ψ_1 and ψ_2 have opposite sign must be avoided, since heterocoagulation will occur.

As previously mentioned and illustrated in Fig. 6, the stability can be easily achieved for many ceramic powders dispersed in water by simply adjusting the pH so that it is well above or below the pH_{iep}. Also from the zeta potential versus pH plot, it becomes fairly easy to predict in which range of pH values the powders will remain stable. For example, in the pH range 4 to 7, a mixture α-Al_2O_3 and ZrO_2 could be easily dispersed in water. However, at pH > 7 the particles will be oppositely charged and heterocoagulation can occur. It has been recently shown that a suspension of Al_2O_3 and Si_3N_4 can be stabilized by pH control producing a consolidated compact with narrower than average pore size [79]. Recent studies also showed clearly that good dispersions of SiC whiskers in a Si_3N_4 matrix can be achieved by manipulation of the system pH [80]. The zeta potential versus pH plots, indicating the most stable region of the multicomponent system, are illustrated in Fig. 14.

Obviously, when working with multicomponent systems for which the particles have different pH_{iep}, it can be extremely difficult to electrostatically stabilize all particles in the system. For example, to find a suitable pH range for electrostatic stabilization, except at extremes of low and high values, may be difficult. Also, to develop a compatible dispersant for the dif-

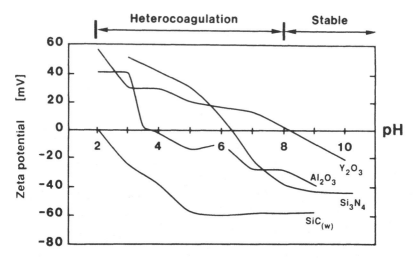

FIG. 14 Plots of zeta potential versus pH for the 10 wt % SiC whiskers/Si₃N₄ (plus 4 wt % Y₂O₃ and 2 wt % Al₂O₃) composite. (From Ref. 80.)

ferent powders may be time-consuming. To simplify the situation, several workers have attempted to modify the surfaces of the different particles with alkoxides to match the surface chemistry [81].

However, to some extent the stability of different types of ceramic powders can be roughly predicted from the theoretical treatment. From equations (30), (31), and (33) a series of stability curves have been computed relating $W_{(1,2)}$ to the radii a_1 and a_2, and the surface potentials ψ_1 and ψ_2 of two species for various values of the Hamaker constant A, and the Debye-Hückel reciprocal length parameter κ [82a]. In Fig. 15, a plot is shown for data computed under conditions of constant potential. Other plots have been computed at constant charge, and in addition a rough guide to a secondary minimum can be evaluated from similar curves. These plots give a useful general guide to the stability of electrostatically charged colloidal dispersions.

A coagulated sol can sometimes be redispersed or peptized by washing or removing counterions that cause coagulation, reestablishing the d.l. Obviously, it is almost impossible to peptize a sol that has fallen into the primary minimum, since this requires an energy of several hundred kT. However, in practice it is often easy to redisperse many ceramic powder systems, and this can be explained by tightly bound hydration layers that prevent the particles entering V_{min}. In the case of silica, the hydration effect is so strong that the

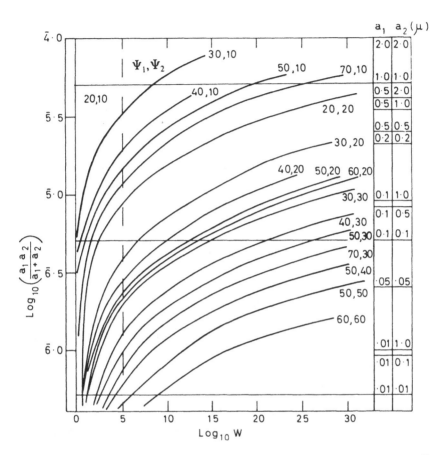

FIG. 15 Stability curves designated ψ_1, ψ_2 at constant potential with $A = 3 \times 10^{-20}$ J and $K = 10^6$ cm^{-1}. (From Ref. 82.)

colloid is stable at pH_{iep}. To destabilize the aqueous sol it is necessary to reduce the degree of hydration by adding salt to the suspension.

Although DLVO theory has been found useful to predict primary minimum coagulation, the theory has been found to be much less successful in explaining secondary minimum effects, particularly with concentrated monodispersed colloidal systems. For example, in DLVO theory a decrease in electrolyte concentration increases κ^{-1} and decreases the depth of the secondary minimum or removes it altogether. However, with several concentrated monodispersed systems a decrease in electrolyte has been shown to produce a stabilized ordered system that is believed to be caused by an increase in the range of the electrostatic repulsive forces. At high concentrations, the particles can maintain repulsive contact with all particles interacting simultaneously with several neighboring particles. Although these ordered-disordered transitions and repulsive structures have been found mostly with polymer colloids, they have also been detected with monodispersed SiO_2 and CeO_2 [82b].

IX. STERIC STABILITY OF DISPERSIONS WITH POLYMERS

High-molecular-weight natural polymers such as gums and starches have a long history of stabilizing colloidal particles. The advantages of these steric stabilizers in comparison to electrostatic stabilizing agents is that they are less sensitive to ionic strength (in aqueous media) and are more effective at high solids under extreme flow conditions in both aqueous and nonaqueous media. In fact, many natural and synthetic polymers act as good dispersants, since they provide a combination of both *steric* and *electrostatic* repulsive forces (so-called *electrosteric*).

In recent years possibly the most effective synthetic polymers used to stabilize colloids in both aqueous and nonaqueous media are block or graft copolymers that contain an insoluble functional head group that can be strongly attached to the surface. The tail must have a high affinity for the solvent, so that it can then extend into solution, producing the steric effects required for stabilization.

A polymer chain that is very soluble in the dispersion medium can in some cases be chemically attached to the surface via functional groups that react with specific sites. The reactive groups eliminate the need for anchor segments. There are several examples in the literature; for example, Laible and Haman [83] reviewed the formation of chemical bonded polymers with oxide surfaces using polymerization reactions or reaction of the surface hydroxyl group with chemically active chlorosilane or alkoxysilane groups. Green et al. [84] more recently used a similar approach with both short- and long-chain Ti-alkoxide dispersants, capable of reacting with the surface of ceramic oxide powders. In some cases, organosilane compounds were used, and in other cases commercially available polymers were modified to react with the particle surface. For example, reactive dispersants for both Al_2O_3 and $BaTiO_3$ powder in organic media were prepared by treating Ti-tetra-isopropoxide with both simple-chain and branched-chain carboxylic acids (Fig. 16). The organotitanate could then be chemically bonded to the surface hydroxyl species on the powder. In addition to the bonding of polymeric dispersant to oxide powders, the bonding to clean covalent nonoxide

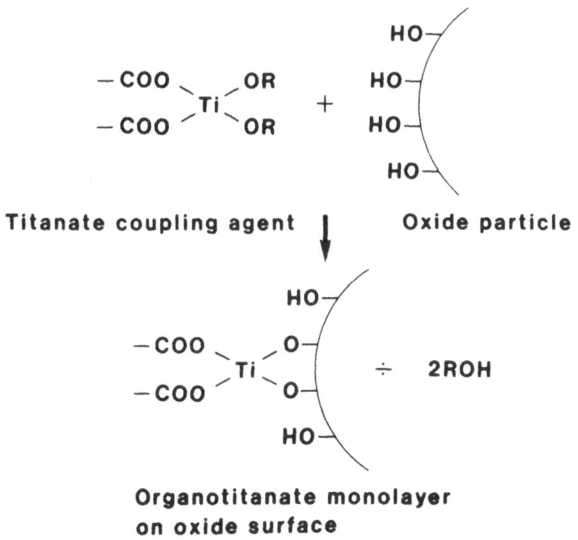

$$-COO \diagdown Ti \diagup OR \\ -COO \diagup \diagdown OR \quad + \quad \begin{array}{c} HO- \\ HO- \\ HO- \\ HO- \end{array}$$

Titanate coupling agent   Oxide particle

$$-COO \diagdown Ti \diagup O- \\ -COO \diagup \diagdown O- \quad \begin{array}{c} HO- \\ \\ HO- \end{array} \quad \div \quad 2ROH$$

Organotitanate monolayer
on oxide surface

FIG. 16 Reaction of an organotitanate coupling agent with an oxide. (From Ref. 85.)

powders such as laser-formed silicon or silicon carbide particles (with a hydrogenated surface) were also studied. The attachment of alkene molecules to these surfaces was achieved by hydrosilylation reactions.

Aizawa et al. [85] have studied the dispersability of TiO_2 powders treated with titanate coupling agents containing caprate and lower carboxylates. The powders were successfully dispersed in an organic matrix. It was reported that the lower the value of the pK_a of the carboxylic acid functional group of the titanate dispersant, the lower the viscosity of the corresponding dispersion.

These techniques generally proved effective and reliable means of dispersing and stabilizing the ceramic powders in nonaqueous media, since the bound polymer layers are resistant to desorption. Additional advantages of chemically bonded dispersants are that they reduce competitive adsorption in ceramic slips where dispersants, binders, plasticizers, etc. have been added. Also, excess of dispersant can be washed away, which is beneficial to the consolidation process. Another advantage is that the metal part of the selected metal alkoxide can be chosen to fit the requirements for sintering agents. This ensures a control on the homogeneous distribution of components.

Although extensive progress has been made in the development of theories to explain the mechanism of steric stabilization over the last few decades, there is still a need to apply this theory to practical systems. In fact, it is only recently that experimental methods such as NMR and neutron scattering have been used to test some of these theories. Several different theoretical approaches have been developed to describe steric stabilization, but most of these are based on the change in the Gibbs free energy ΔG as the particles approach. These have been extensively reviewed in several books [86-88] and fall into two groups which may be classified as (a) elastic or volume restriction stabilization theories and (b) osmotic or mixing stabilization theories.

The elastic or entropic interactions cause an increase in ΔG_{el}, resulting from the loss of configurational freedom of the adsorbed macromolecules as the particles approach and the chains become compressed (no penetration). Since the total volume available to each chain is reduced, the configurational entropy of the chain is also reduced, causing an increase in ΔG_{el}. In the case of the mixing interaction, the change in Gibbs free energy ΔG_M arises from the interpenetration of adsorbed layers causing a buildup in segment concentration in the interaction zone. This leads to a decrease in configuration entropy and free energy.

Earlier studies dealing with the elastic theory used simple bond models consisting of inflexible rods, terminally adsorbed but freely jointed [91]. Using this crude model it was assumed that the number of configurations Ω available to the molecule was proportional to the surface area of the hemisphere swept out by the free rod. The decrease in free entropy ΔS was then calculated from the loss of the number of configurations using the Boltzmann equation and expressed by

$$\Delta S = k \ln \Omega \qquad (35)$$

where k is the Boltzmann constant. The free energy of repulsion was then determined from

$$\Delta G_{el} = N_S k T \theta_\infty \ln \Omega \qquad (36)$$

where θ_∞ is the fraction of surface covered by the adsorbed layer when the surfaces are at infinite separation and N_S is the number of adsorbed rods per unit area. Later computer simulation methods were developed to calculate the reduction in configuration entropy of flexible terminally adsorbed macromolecules with several links. In these models, segment-solvent interactions were not considered. Later, random flight statistics were used to calculate ΔG_{el} for terminally adsorbed polymers on flat surfaces using restricted boundary conditions.

In the case of the mixing interaction, early theories calculated ΔG_M by assuming that the segment concentration in the adsorbed layer was uniform or that the segment concentration in the overlap region was equal to the sum of the individual concentrations from both adsorbed layers. The free energy of mixing of the adsorbed layers was then calculated from dilute polymer solution theory.

Recent theories of steric interaction have tended to avoid this artificial separation between the elastic and mixing contributions. However, several theories retain this separation in an attempt to simplify the main parameters involved in the mechanism.

Napper [91] for example gives the following equation for the steric interaction V_s between two particles with adsorbed polymer layer of thickness δ as

$$V_S = \underbrace{\frac{2\pi akTV_2^2\Gamma_2^2}{V_1}(\frac{1}{2}-\chi)\,S_{mix}}_{\text{mixing term: } V_{S,mix}} + \underbrace{2\pi akTS_{el}}_{\text{elastic term: } V_{S,el}} , \qquad (37)$$

where a ($\gg \delta$) is the particle radius, V_1 the solvent molecular volume, V_2 the polymer molecular volume, Γ_2 the adsorbed amount of polymer (number of chains/area), and χ the Flory polymer-solvent interaction parameter; S_{mix} and S_{el} are geometric functions that depend on the form of the segment concentration profile. Analytical expressions for S_{mix} and S_{el} have been derived by Napper [91].

A more approximate but practical expression for the interparticle steric interaction is given by

$$V_S(D) = \frac{2\pi akT}{\bar{V}_1}\,\bar{\phi}_2\Big(\frac{1}{2-\chi}\Big)\,(2\delta + 2a - D)^2 \qquad (38)$$

This expression is valid in the interpenetrational domain $\delta < (D - 2a) < 2\delta$, where D is the distance between particle centers.

This model assumes a constant density radially of the adsorbed steric layer, which is a reasonable approximation for thin, dense layers. The precise choice of the values for the volume fraction of chains in the adsorbed layer $\bar{\phi}_2$, and for the adsorbent-solvent interaction parameter χ, is not so important, since the steric repulsive interaction increases rapidly when the distance of closest approach is below 2δ.

Generally, although these theoretical aspects concerning the mechanism of the steric stabilization need further refinement, it is obvious that an additional short range repulsive term V_S (which rises rapidly at short interparticle distances) must be introduced to Eq. (29). Hence the total interaction must be modified to

$$V_T = V_R + V_A + V_S \tag{39}$$

In fact, V_S may be regarded as a combination of the mixing interaction and the entropic interaction. Also, both V_R and V_A need to be modified in the presence of the adsorbed polymeric layer. Potential energy diagrams showing the affect of the V_S on V_A and V_R are shown in Fig. 17.

It is also important to consider instability in a sterically stabilized dispersed system, which is governed by the solubility requirements of the polymer. In this case, a change in stability can be induced by reducing the solvency (i.e., changing the temperature or pressure or by adding a miscible nonsolvent). Usually a sharp transition point is observed from long-term stability to fast flocculation, but the process is usually reversible. The point at which flocculation occurs is known as the critical flocculation pressure (CFP) or critical flocculation volume (CFV) or critical flocculation temperature (CFT).

The *temperature dependence of stability* is one of the main features that distinguishes a *sterically* stabilized system from an *electrostatically* stabilized dispersion. Some sterically stabilized dispersions flocculate on heating,

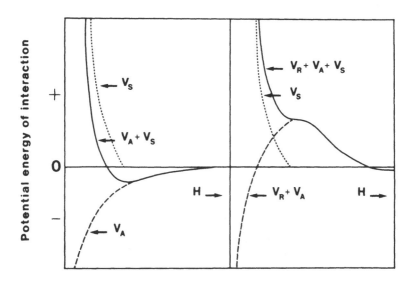

FIG. 17 Plots of the potential energies of interaction versus the interparticle distances H for sterically stabilized particles (a) without electrical double layer repulsion ($V_T = V_A + V_S$); (b) with electrical double layer repulsion ($V_T = V_R + V_A + V_S$).

others on cooling; in some cases some suspension may flocculate on either cooling or heating. However, other sterically stabilized systems may remain stable under such temperature changes, and in such cases it has been suggested that the CFT is not readily accessible.

Napper and coworkers [89] have studied the *sensitivity of temperature* (particularly the θ temperature) on the stability of different polymeric stabilized systems near the CFT, and this has enabled a basic thermodynamic understanding of the process to be developed. The temperature dependence can be related to the Gibbs free energy ΔG_T and the corresponding entropy ΔS_T and enthalpy ΔH_T by the equation

$$\Delta G_T = \Delta H_T - T\Delta S_T \tag{40}$$

and also

$$\frac{\delta(\Delta G_T)}{\delta T} = -\Delta S_T \tag{41}$$

On crossing from a stable to an instable region, ΔG_T must change sign (from positive to negative) and must be dependent on the sign and magnitude of ΔH_T and ΔS_T. This approach enabled Napper [89] to identify different types of steric stabilization phenomena to be identified from their change in state on heat treatment. Essentially, enthalpically stabilized dispersions are characterized by flocculation on heating, for example, systems stabilized by poly(ethylene oxide) moieties in aqueous $MgSO_4$ solution, where turbidity can be observed as the systems become unstable on increasing the temperature. Entropically stabilized dispersions are generally flocculated by cooling, for example, polyacrylic acid in 0.2 HCl.

Generally, however, entropic stabilization is more common in nonaqueous dispersions. However, the combined effect of entropic and enthalpic stabilization can occur in both environments. In fact, many dispersions that do not flocculate at any normally accessible temperature are usually stabilized by a combination of the two effects. In Table 4, a comparison of the critical flocculation temperature (cft) and the θ temperature for several different types of polymer systems is shown. A strong correlation for both flocculation by heating (upper critical flocculation temperature, ucft) and cooling (lower critical flocculation temperature, lcft) can be seen.

Sato and Ruch [90] and Napper [91] give good overviews of the historical development of ideas on steric stabilization and also give a good collection of free energy expressions that can be computed to determine the extent of steric stabilization for different systems. To date, the important results to derive from these theoretical studies are crude potential energy diagrams for sterically (thermodynamically) stabilized systems. Generally, these have

TABLE 4 Comparison of Theta Temperatures with Critical Flocculation Temperature

Stabilizer	Molecular weight	Dispersion medium	cft/K	u/l	θ/K
Poly(ethylene oxide)	10,000	0.39 M MgSO$_4$	318 ± 2	ucft	315 ± 3
Poly(ethylene oxide)	96,000	0.39 M MgSO$_4$	316 ± 2	ucft	315 ± 3
Poly(ethylene oxide)	1,000,000	0.39 M MgSO$_4$	317 ± 2	ucft	315 ± 3
Poly(acrylic acid)	9,800	0.2 M HCl	287 ± 2	lcft	287 ± 5
Poly(acrylic acid)	51,900	0.2 M HCl	283 ± 2	lcft	287 ± 5
Poly(acrylic acid)	89,700	0.2 M HCl	281 ± 2	lcft	287 ± 5
Poly(vinyl alcohol)	26,000	2 M NaCl	302 ± 3	ucft	300 ± 3
Poly(vinyl alcohol)	57,000	2 M NaCl	301 ± 3	ucft	300 ± 3
Poly(vinyl alcohol)	270,000	2 M NaCl	312 ± 3	ucft	300 ± 3
Polyacrylamide	18,000	2.1 M (NH$_4$)$_2$SO$_4$	292 ± 3	lcft	—
Polyacrylamide	60,000	2.1 M (NH$_4$)$_2$SO$_4$	295 ± 3	lcft	—
Polyacrylamide	180,000	2.1 M (NH$_4$)$_2$SO$_4$	280 ± 3	lcft	—
Polyisobutylene	23,000	2-methylbutane	325 ± 1	ucft	325 ± 2
Polyisobutylene	150,000	2-methylbutane	325 ± 1	ucft	325 ± 2
Polyisobutylene	760,000	2-methylbutane	327	ucft	318
Polyisobuylene	760,000	2-methylpentane	381	ucft	376
Polyisobutylene	760,000	2-methylhexane	423	ucft	426
Polyisobuylene	760,000	3-ethylpentane	463	ucft	458
Polyisobutylene	760,000	cyclopentane	455	ucft	461
Poly(α-methyl styrene)	9,400	n-butyl chloride	254:1	lcft	263
Poly(α-methyl styrene)	9,400	n-butyl chloride	403:1	lcft	412
Polystyrene	110,000	cyclopentane	410	ucft	427
Polystyrene	110,000	cyclopentane	280	ucft	293
PDMS[a]	3,200	n-heptane/ethanol	340	lcft	340 ± 2
PDMS	11,200	(51:49 v/v)	340	lcft	340 ± 2
PDMS	23,000		341	lcft	340 ± 2
PDMS	48,000		338	lcft	340 ± 2

[a]Poly(dimethylsiloxane)
Source: Ref. 87.

been shown to be radically different from electrostatic (meta-)stable systems. For example, owing to the elastic effect, aggregation into a deep primary minimum does not occur, and redispersion takes place on reverting to better than θ solvent conditions. In Fig. 18, the potential energy curve for two approaching particles coated with polymer is shown for conditions in a

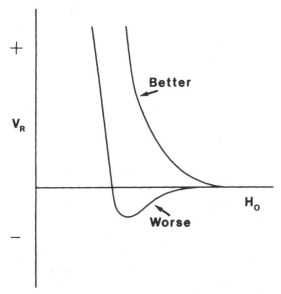

FIG. 18 Schematic representation of the potential energy curves for both better than and worse than θ solvents. (From Ref. 87)

dispersion medium that is (a) better than θ solvent and (b) worse than θ solvent.

Also, other types of stability/instability transitions can occur in special cases where the polymer is in an extremely good solvent (i.e., the segments are nonadsorbing on the particle surface). Under these conditions, two competing effects may occur as two particles approach under Brownian motion. In the first case (which occurs at high polymer concentrations), as the polymer chain between the surfaces becomes compressed, the configuration and entropy of the chains are reduced, causing an increase in free energy and repulsion. This situation is known as *depletion stabilization.* In the second case, the system loses less entropy if the polymer is evicted from the region between the particles, causing a loss in translational entropy but no configurational entropy charges. With more polymer outside the region between approaching surfaces and fewer inside, there is an osmotic pressure difference causing attraction. This behavior is known as *depletion flocculation,* when the separation distance between surfaces approaches the radius of gyration of the polymer.

In recent years, extensive amounts of research have been done on the thermodynamics and solubility requirements of polymeric coated particles.

Unfortunately, the preoccupation with solubility requirements has led many investigators to conclude that precipitation of stabilizers is the main mechanism of colloidal instability and to neglect the dispersion force interactions between particles. In a simplified form, the dispersion force interaction potential, dealing with polymer coated particles, can be described in the same way as in electrostatic stability.

In order to stabilize a ceramic slip, it is essential to begin by choosing a polymer that does not precipitate when the system is expected to remain stable (i.e., temperature, solvent composition, etc.). Once this has been accomplished, then attention should be paid to the dispersion force attraction between particles and the *critical* separation distance between approaching particles, which is needed to prevent the particles from coming within the range of van der Waals forces, which can cause coagulation.

For practical purposes, an estimate of the thickness of the polymer film necessary to mask the dispersion force can be estimated, and this value may be used to describe the stability of the system. This idea may be developed by selecting a simple model, where there is only London van der Waals attraction to consider (the electrostatic repulsion may be neglected in a low-dielectric-constant organic liquid), and the repulsion becomes infinite as soon as the adsorbed layers come into contact (i.e., assuming "nonpenetration" of adsorbed polymer layers). Under these circumstances, the approaching particles will be held in an energy minimum, the depth of which V_M depends on the thickness of the adsorbed layer.

If we represent the potential energy of attraction between two equal-size spherical particles of the same ceramic (1) in water (2) (when the polymer layers coating the two approaching particles touch) by V_{121}^d, then this value should be greater than $k_B T$, the average kinetic energy per particle, in order that sticking will occur on contact (Fig. 19). If the value is less than $k_B T$, the particles will not stick but will redisperse.

An approximate form of the potential energy of attraction at short interparticle distances may be represented by

$$V_{121}^d = \frac{a A_{121}}{12 H_0} \times f \qquad (42)$$

where a is the particle radius, f is the retardation force, and H_0 is the interparticle distance.

The shortest distance of separation H_0^{min} for the particles to remain stable when the particles are touching (assuming $V_{121} \approx k_B T$ at this point) can be given by

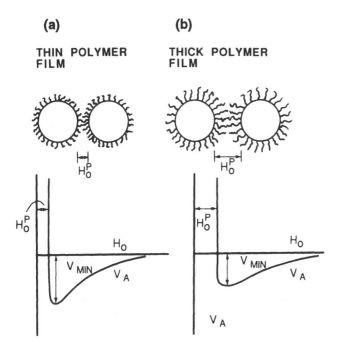

FIG. 19 Systematic representation of the potential energy of interaction versus interparticle distance curves for two approaching ceramic particles with a film thickness of adsorbed dispersant of $H_0^P/2$. In order for sticking to occur, $V_{MIN} > k_B T$. Two situations are considered, with a thin (a) and a thick (b) adsorbed film.

$$H_0^{min} = \frac{a A_{121}}{12 k_B T} \times f \qquad (43)$$

The requirement for stability is that a particle surface must have a minimum film thickness of adsorbed dispersant of $H_0^P/2$. The retardation effects on the London forces f may be calculated from the equations derived by Schenkel and Kitchener [92] so that Eq. (43) may be reexpressed as

$$H_0^{min} = \frac{a A_{121}}{12 k_B T} \left(\frac{1}{1 + 1.77 p_0} \right) \quad \text{for} \quad 0 < p_0 < 2 \qquad (44)$$

and

$$H_0^{min} = \frac{a A_{121}}{12 k_B T} \left(\frac{+2.45}{5 p_0} - \frac{2.17}{15 p_0^2} + \frac{0.59}{35 p_0^3} \right) \quad \text{for} \quad 0.5 < p_0 < \infty \qquad (45)$$

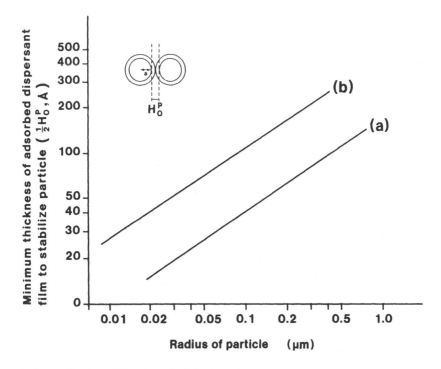

FIG. 20 Steric stabilization of mineral particles with adsorbed dispersant film. The minimum thickness of the film ($H_0^P/2$) to achieve stability versus the radius of the particle. (a) Silica particles in water with A_{121} = 0.8 × 10^{-20} J; (b) Al_2O_3 in water with A_{121} = 4 × 20^{-20} J.

where A_{121} is the Hamaker function of the ceramic in water and p_0 = $2\pi H_0/\lambda$, where λ is the wavelength corresponding to the intrinsic electronic oscillation of atoms (= 1000 Å for the ceramic in water). Using these equations, the minimum film thickness of the dispersant (to establish stability) can be related to the particle radius for different types of ceramic particles dispersed in liquids (see Fig. 20).

Generally, van der Waals forces for colloidal particles operate over the range of 5–10 nm. In addition, the root-mean-square end-to-end distance $\langle r^2 \rangle^{1/2}$ of linear polymers is according to Napper [91] an estimate of the diameter of the random coil conformed polymer. Also, $\langle r^2 \rangle^{1/2}$ is proportionate to $M^{1/2}$, where M is the molecular weight of the polymer. From this approach, it can be estimated that polymers with molecular weight over 10^3 can neutralize the attraction potential.

A more sophisticated analysis for the approach of two polymer coated spheres was more recently developed by Russel [93]. In this model, criteria for both (a) the dispersion force and (b) the phase separation of the sterically stabilized particle were taken into consideration.

Direct measurement of steric interactions have been mostly restricted to measurement carried out in two-dimensions using, for example, the surface force measurements with both homopolymers and segmented copolymers adsorbed on mica. These have been recently reviewed by Patel and Tirrell [94] and Luckham [95]. Klein [96] has studied polystyrene-coated mica in cyclohexane and confirmed the general shape of the potential energy diagram shown in Fig. 18. Other experiments, using a compression cell to determine three-dimensional repulsion forces with poly(12-hydroxystearic acid), adsorbed on a polymer latex colloid, have demonstrated the short-range nature of the steric barrier [97], and these experiments suggest that no significant repulsion occurs until the steric barriers interpenetrate.

However, most of these surface force and compression experiments deal with the slow approach of polymer layers. Also the theories of steric interaction are equilibrium theories. However, in practice, it would appear unlikely that two rapidly approaching particles under Brownian motion or shear will not respond to equilibrium conditions. Steric forces are time dependent and rate dependent due to the rheological behavior of the adsorbed polymer.

In addition, although a considerable effort has been devoted to promoting the use of electrostatic and steric stabilizing polymers to produce stable ceramic slips, recent studies appear to suggest that this may not be the whole picture. In fact, attention should also be paid to studies by Yin et al. [98] that suggested that the desired goal in the dispersion process is to achieve low viscosity at high solids, rather than a colloidal stability per se. They demonstrated that an Al_2O_3 powder could be packed to high density (58.7 vol %) during slow sedimentation in heptane by use of a poly(octadecyl) methacrylate polymer. It was suggested that the polymer acted by a lubrication mechanism, reducing friction between particles and allowing flow to occur by reducing the van der Waals forces. This results in a fairly low viscosity and high packing density. This polymer lubrication mechanism was also recently studied by Lannutti et al. [99].

Also Velamakanni et al. [100] and Lange et al. [101] exploited an alternative method, utilizing short-range hydration repulsion caused by specifically adsorbed hydrated counterions to prevent contact between aluminum particles in concentrated aqueous electrolyte. These forces were found to be of sufficient magnitude to counteract van der Waals forces and prevent flocculation into a primary minimum as indicated in Fig. 21. Again the par-

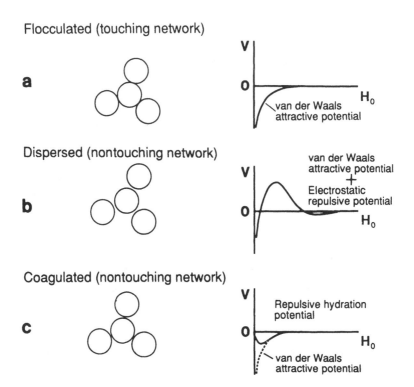

Flocculated (touching network)

a

Dispersed (nontouching network)

b

Coagulated (nontouching network)

c

FIG. 21 Schematic of the three networks and their respective interparticle potentials; V = interparticle potential and H_0 = the separation distance between particle surfaces. (From Ref. 101.)

ticles are only weakly aggregated (due to a lubricating weak film), allowing rearrangement to occur during sedimentation. Also the weak floc structure formed as a fairly high viscosity, preventing sedimentation. Unfortunately, at present it is not clear what type of surface layers are required to give the optimum lubrication or viscosity reductions for ceramic systems. Also a clearer understanding of the mechanism is required before further progress can be achieved in this direction.

X. DISPERSION AND STABILITY OF POWDERS IN ORGANIC LIQUIDS

Tape casting is generally performed in organic solvents and is difficult to carry out in water. Also, slip casting can be applied in organic solvents as in water, but water is obviously preferable for environmental and economic reasons. In the injection molding process, the powder, prior to shaping, is dispersed in an organic polymer melt.

A popular misconception in the processing of ceramic powder in organic media with a *low dielectric constant* is that good stability can only be obtained by using a dispersant with a sufficiently long chain length to give steric repulsion. However, numerous studies extending over 50 years have clearly demonstrated that stability in organic media may be controlled as in aqueous media by *London dispersion forces* and *charge*.

In the simplest situations, we may consider the stability of *uncharged* particles in hydrocarbons with low dielectric constants where the London van der Waals forces play a dominant role. Then, for two spherical particles suspended in the liquid, the potential energy of interaction rising from the van der Waals forces may be approximated from Eq. (31); in the case of equal-size particles (i.e., $a_1 = a_2 = a$),

$$V_A = -A_{S/L/S} \left[\frac{a}{12 H_0} \right] \text{ for } H_0 \ll a. \tag{46}$$

where $A_{S/L/S}$ is the Hamaker constant of the solid in the liquid. Although values of $A_{S/L/S}$ can be accurately evaluated from spectral data, it can be shown [106] that

$$A_{S/L/S} = A_S + A_L - 2\sqrt{A_L A_S} \tag{47}$$

where A_S and A_L refer to the values of solids and liquids in vacuum.

The value of A_S for ceramic powders can be estimated using static dielectric constant data for the materials using the formula

$$A_S \text{ } (kT \text{ units}) = 113.7 \frac{(\varepsilon_S - 1)^2}{(\varepsilon_S + 1)^{3/2}(\varepsilon_S + 2)^{1/2}} \tag{48}$$

where ε_S is the dielectric constant of the powder. An analogous expression can be used for liquids by replacing ε_S by ε_L.

From these theoretical equations (provided the sizes of the particles are fairly well defined) it is possible to calculate the magnitude of V_A for many ceramic powders in organic liquids. For the calculation, the value of H_0 is required; it is usually assumed to be twice the molecular diameter of the liquid.

The quantification of V_A can be used as a simple means of estimating the dispersibility of the powder in the hydrocarbon, in the *absence* of any type of dispersing agent. *Low* V_A values indicate a fairly *high* level of dispersibility, while *high* V_A values indicate the presence of *strong* interparticle attractive forces that *prevent* the particles from becoming completely dispersed.

Blier [102,103] calculated $A_{S/L/S}$ and V_A values for Si, SiC, and Si_3N_4 powders in a range of organic liquids with well-defined ε_L values. From Eq. (48), values of $A_{Si} = 78.3kT$, $A_{SiC} = 71.9kT$, and $A_{Si_3N_4} = 39.4kT$ were reported. Values of V_A from $-5kT$ to $-10kT$ were selected, as criteria for the ease of dispersion of the powders in the liquids. V_A values below this range indicated poor dispersibility, and greater values indicated good dispersibility. A series of dispersion experiments were carried out and the extent of agglomeration of the powders was visually estimated and compared to the theoretical predictions.

Typical results are presented in Table 5 for a series of organic liquids including aliphatic hydrocarbons, aromatic hydrocarbons, alcohols, ketones, ethers, aliphatic acids, and aldehydes. These studies generally indicated that the calculated V_A values, could in many cases, adequately account for the observed dispersion behavior. However, some anomalies were reported in the cases of Si_3N_4 and SiC powders. These were explained by the presence of small amounts of surface impurities that could obviously influence the powder dispersibility. From these studies, it could be suggested that an easy means of obtaining a well-dispersed system was to choose a hydrocarbon and powder with similar ε values in order to *mask* the van der Waals dispersion interaction. To stabilize suspensions in hydrocarbons in which $|V_A| \geq 10kT$, however, required some type of added dispersant, a surfactant or polymer to generate an electrostatic charge to provide steric repulsion.

The dielectric constants of nonaqueous solvents cover a wide range of values from about 2 for some hydrocarbons to about 180 for certain types of amides, and for some of these systems surface charge becomes important.

In tape casting formulations, a range of organic solvents can be considered. Mistler [104] in a recent review lists fifteen different solvents.

Also, in organic solvents as in aqueous solvents, the influence of the repulsive potential must be taken into consideration. This may be expressed by the Derjaguin equation [30]

$$V_R = 2\varepsilon_r\varepsilon_0 a\psi_0^2 \ln(1 + e^{-\kappa H_0}) \tag{49}$$

This indicates that the electrostatic repulsion depends on the product of a dielectric constant of the liquid and the square of the surface potential. Therefore in ketones with $\varepsilon = 20$, ψ_0 would need to be double the value in water to provide the same repulsive force.

The dielectric constant also plays an important role in determining the electric field gradient in the diffuse double layer that also influences the

TABLE 5 Van der Waals Potential Energy of Interaction (V_A in kT Units) for Si_3N_4 and SiC Particles and Predicted Dispersion Quality at 20°C

Liquid	Si_3N_4		SiC	
	V_A^a	Dispersion quality	V_A^a	Dispersion quality
Aliphatic hydrocarbons				
n-Hexane	-5.1	Good-poor	-23	Poor
Cyclohexane	-4.5	Good	-21	Poor
n-Octane	-4.4	Good	-20	Poor
Aromatic hydrocarbons				
Benzene	-3.2	Good	-18	Poor
Toluene	-2.6	Good	-16	Poor
Xylene	-2.4	Good	-15	Poor
Alcohols				
Methanol	-9.9	Poor-good	-2.5	Good
Ethanol	-7.8	Poor-good	-1.6	Good
n-Propanol	-6.4	Good-poor	-1.1	Good
Isopropanol	-6.0	Good-poor	-0.91	Good
n-Butanol	-5.4	Good-poor	-0.74	Good
n-Octanol	-2.5	Good	-0.0097	Good
Benzyl alcohol	-3.9	Good	-0.22	Good
Ethylene glycol	-9.3	Poor-good	-2.5	Good
Ketones				
Acetone	-6.5	Good-poor	-1.1	Good
2-Pentanone	-4.6	Good	-0.46	Good
2-Heptanone	—b	Good	b	Good
Ethers				
Isopentyl ether	-3.8	Good	-18	Poor
Aliphatic acids				
Propionic acid	-0.39	Good	-8.4	Poor-good
Butyric acid	-0.90	Good	-10.4	Poor
Aldehydes				
Benzaldehyde	-5.2	Good-poor	0.71	Good
Inorganics				
Water	-16	Poor	-5.3	Good-poor

[a]Separation is twice the molecular diameter of the liquid.
[b]Not estimated; quality was deduced by analogy with 2-pentanone.
Source: Refs. 107, 108.

electrostatic interparticle forces. If we assume that the concentration of counterions in solution n_i, expressed in ions/cm^3, remains constant, then the Debye length is inversely proportional to the square root of ε_r according to Eq. (11).

For a given n_i value, $1/\kappa$ in a ketone may be twice the value in water. This approach suggests that the main influence of low ε_r is to restrict the dissociation of ionizable substance, which causes a reduction in n_i and hence in the electric field. This has led many investigations to assume that only low concentrations of ions can occur in liquids with low ε_r values, but this is not the general case. It has been shown by several workers that ionic species can be produced by alternate interfacial processes other than by dissociation and ionization of solubilized dispersants. In such cases, these ions can form in the double layer regions surrounding the particles and may also contribute toward the ionic strength of the system. Such effects become particularly pronounced in suspensions containing high concentrations of small particles where the surface-area-to-volume ratio provides fairly high concentrations of counterions.

Fowkes and coworkers [105,106] have described these interfacial mechanisms in terms of an acid-base reaction. In aqueous solutions where surfactants ionize, the surface active ions adsorb onto the surface giving electrostatically charged surfaces. However, in organic media, a basic dispersant can adsorb as a neutral molecule onto acidic surface sites where proton transfer can occur, and the adsorbed base becomes positive. If the concentration of dispersant dissolved in the organic phase is sufficiently high, then dynamic adsorption and desorption occur, and some proton-carrying dispersants will desorb into solution and provide the counterions for the negative charges remaining on the surface. This mechanism explains why acidic dispersants give basic inorganic particles positive charges in neutral hydrocarbon liquids (as determined by zeta potential measurements). In addition, basic dispersants give acid particles negative charges. In other systems, where dispersants or additives are not present in the solution, then electrophoretic measurements of inorganic particles with basic surface sites give positive zeta potential when dispersed in acidic organic solvents (chloroform or methylene chloride), while particles with acidic surface sites show positive zeta potentials when dispersed in basic organic solvents (ketones or ethers). Solvents such as nitriles have both acidic and basic functionality, and when powders are dispersed in acetonitrile the particles may show a positive or negative charge depending on the predominant acidity or basicity of the powder surface. Solvents such as methylisobutylketone and methanol show similar behavior.

Other common oil-soluble dispersants that have proved effective in neutral hydrocarbons are basic amines or nitrogen bases such as the polyisobutene type that have a basic anchoring group. Chevron OLOA 1200 is a basic commercial product that strongly adsorbed on acidic surface sites. Other acidic type dispersants that adsorb on basic sites include partial esters of phosphoric acid and poly(12-hydroxy stearic acid), which is produced commercially by ICI. Inverse micellar soaps or sulphonates of polyvalent metals such as calcium, zinc, and barium also act as dispersants. In addition, oil-soluble high-molecular-weight alkyl lauryl sulphonate anions are weak bases, whereas the calcium or zinc ions are weak acids. Barium and calcium sulphonates react with hydroxide surfaces, providing basic species that can confer a high negative charge. In selecting a dispersant for an organic system, it is essential to choose an acidic type of additive for powder surfaces that have a predominance of basic sites and vice versa. Since many of the dispersants such as polyisobutene and the poly-hydroxy stearic acid types are short chain polymers with molecular weight in the range of 2000 to 3000, there is also the possibility that they act to some extent as steric stabilizers.

Other dispersants in hydrocarbon media include calcium octanoate, which was earlier studied by Minne and Hermanie [107]. More recent work by Ottewill [108] demonstrated the effectiveness of this dispersant in dodecane. On addition of the additive to the dispersion, the conductivity of the dodecane was found to increase, and the colloidal particles became charged. Also, in concentrated suspensions, an increased ordering of the structure was detected by small-angle neutron scattering.

The stability of many oxide powders in liquids with low dielectric constant has also been achieved in many early studies with fatty acid dispersants. Verwey and Boer [109] used oleic acid and Koelmans and Overbeek [110] increased the stability of Al_2O_3 and Fe_2O_3 in xylene suspensions with oleic acid. However, in these early studies no ionization effects were considered, since the dispersant was regarded as a nonionic molecule. Many recent studies have verified the use of short-chain acids including oleic, stearic, and palm oil as effective dispersants for particles with basic surface sites.

The most effective and economical dispersant for ceramic oxide systems is Menhaden fish oil, which consists of triglycerides or glycerol trioleate, which are used in preference to fatty acids. In fact, fish oil is almost universally used in dispersing alumina or ferrite powders in tape casting. Numerous investigations have been made to illuminate the mechanism of the dispersion process with this product. In most cases, it has been explained by electrostatic charge with an additional steric contribution from the long-chain hydrocarbon that extends from the surface into the nonaqueous solution. Another commercial additive, denoted KD-2 and produced by ICI, consists of a basic ethoxylated amine and was shown to stabilize oxidized silicon particles in ethanolic media [111]. Mikeska and Cannon [112] screened over 70 commercial dispersants for tape casting $BaTiO_3$ in an ethanol-methyl ethyl ketone solution. A phosphate ester, a fish oil, a fatty acid, and an ethoxylate were found to give the best performance. Double layer repulsion and steric hindrance was suggested as the stabilizing mechanism.

Measurement of acidic and basic sites on ceramic particles requires some current understanding of Lewis acid-base theory. The adsorption of organic acidic-basic probes from cyclohexane onto α-SiC, ZrO_2, α-Al_2O_3, and Y_2O_3 powders has recently been studied [113]. The probes (benzoic acid, benzyl alcohol, pyridine, benzyl amine, and ethyl benzoate) covered a wide range of acceptor numbers (AN) and donor numbers (DN) as defined by Gutmann [114]. The results showed a wide variation in the adsorption capacity of the probes for the different powders that appeared to depend on the surface heterogeneity of the powders. The results (for example as shown in Fig. 22 for α-SiC, ZrO_2, α-Al_2O_3, and Y_2O_3) indicate that the amount of benzoic acid (strong acid) adsorbed on the powders is much higher than for the other probes. Also the amount adsorbed on the basic Y_2O_3 powder is about three times greater than for the acidic α-SiC. In fact, the adsorption density of the probes generally follows the increase in pH_{iep} as determined in the aqueous environment. A similar behavior trend is shown also for the benzyl alcohol, but the adsorption levels are considerably less for all powders. The pyridine and ethyl benzoate probes adsorb on the powders but at a reduced level. However, it is interesting to observe that the benzyl amine (strong base with acid functionality) is more strongly adsorbed on Y_2O_3. For the benzoic acid (with the carboxylic acid function group) these results are not altogether surprising.

It has also been demonstrated by Calvert et al. [115] that oxidized fish oil that degrades to carboxylic acids was a much superior dispersant for Al_2O_3 than toluene. Also from IR studies, it was found that the trioleate ester groups were strongly adsorbed to the powder surfaces in preference to the ester groups.

In recent years the presence of surface charge on oxide powders dispersed (in the absence of additives) in polar type hydrocarbons such as chloroform, methylene chloride, dichloroethane, nitrobenzene, ethyl acetate, ethylene diamine, etc. has been reported. In these experiments the buildup of surface charge as determined from zeta potential measurements was

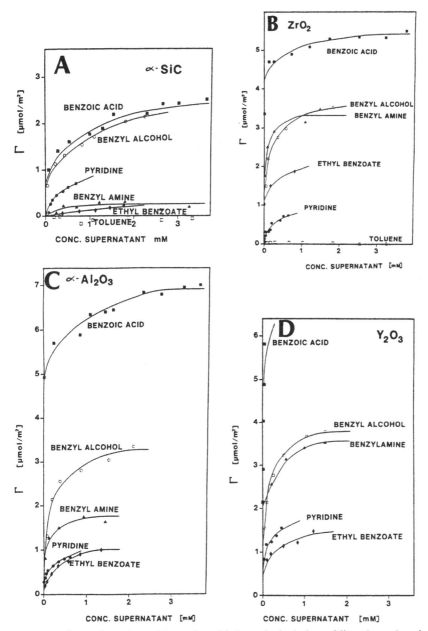

FIG. 22 The adsorption of benzoic acid, benzyl alcohol, pyridine, benzyl amine, ethyl benzoate, and toluene probes from cyclohexane on α-SiC, ZrO_2, α-Al_2O_3, and Y_2O_3 ceramic powders. (From Ref. 123.)

again explained by an acid-base mechanism. In cases where the surface charge is of sufficient magnitude, it can be anticipated that it may contribute to the electrostatic stability of the systems.

For example, Labib and Williams [116] reported surface charge values as high as 80 mV on several oxide powders (SiO_2, Al_2O_3, TiO_2, and MgO) dispersed in a series of organic liquids. These values were determined by zeta potential measurements. The liquids had a wide range of electron donor properties as defined by the donor number. It was reported that the zeta potential varied systematically with the DN of the liquid. Additional experiments were carried out with several other nonoxide powders such as gold, mica, diamond, zinc sulfide, etc., which represented several different types of bonding including metallic, covalent, and ionic [117]. The value of the DN of the liquid corresponding to the pH_{zpc} of the system was determined and an attempt made to relate the DN scale to a pH_{zpc} scale.

Stability and adsorption data for many metal powders in organic media have been reported. However, parts of this early data must be treated with some caution since many metals are easily oxidized, even in cases where the surfaces are freshly prepared. In fact, it is extremely difficult to produce a clear surface for a reactive metal powder. However, recently Satoh et al. [118] carried out an extensive investigation by preparing 18 different metals in 20 different organic solvents. The dispersibility of the powders in the organic liquids was correlated to the dielectric constant of the liquid and the electronic affinities for the metal surface for the solvent. Generally, in the same solvent, the tendency of the dispersibility of the system was found to follow the order Au > Sn > Al > Zn, Mg following the electronic affinity; the later metals are known to have a relatively low tendency to form ions in aqueous solution. This study suggested that surface atoms are important for the dispersibility of metallic sols.

XI. IN SITU SETTLING MEASUREMENTS ON DISPERSED SYSTEMS

A γ-ray densitometry apparatus has recently been developed by Pacific Northwest Laboratories, USA, to study the degree of dispersion and transient settling characteristics of concentrated ceramic powder systems [119]. The technique enables the differences in behavior between dispersed and flocculated systems to be recorded during the sedimentation-consolidation aging process. Essentially, the γ-ray apparatus (Fig. 23) measures directly the density of the suspension in terms of the height of the settled sediment versus the volume fraction of solids.

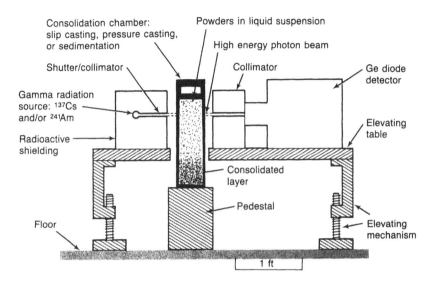

FIG. 23 Schematic of γ-ray densitometer. (From Ref. 119.)

The apparatus directs a beam of γ-radiation from a ^{137}Cs isotope source through a tube containing the suspension. The beam is collimated by passing through a small hole in lead plates positioned between the isotope chamber and a germanium diode detector. Attenuation measurements were performed with a multichannel analyzer by integrating the energy as a function of the intensity of the photoelectric peak.

At a given elevation, the average density within the irradiated portion of the sample can be determined from the attenuation coefficient taken from a calibration curve (relating attentuation to the volume fraction of solids). In determining the calibration curve it is important to use a similar experimental system as in the research sample.

This method has been used to analyze the pH effect of particle-particle interactions during the sedimentation of α-Al$_2$O$_3$ suspensions. More recent experiments have determined the settling behavior of submicron polydispersed alumina powders stabilized by fatty acids of varying molecular weight, dispersed in an organic solvent (decalin) [120]. A typical series of volume fraction profiles, obtained with oleic acid dispersant after various aging periods, is shown in Fig. 24. In this study, the interparticle interaction energies were also calculated from colloidal theory using the thickness of the adsorbed fatty acid layer. This enabled the experimental volume fraction profiles to be compared with theoretical predictions derived from mod-

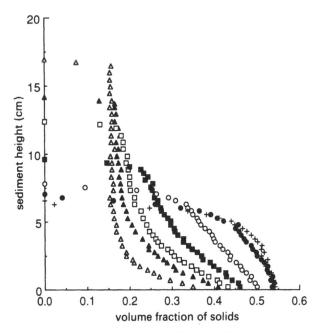

FIG. 24 Volume fraction of solids vs. sediment height for an alumina suspension with oleic acid added. Measurements performed after (Δ) 5, (▲) 11, (□) 29, (○) 57, (●) 126, and (+) 172 days of settling. (From Ref. 120.)

els. These types of experiments are useful for studying the permeability of the aggregated particle network in concentrated ceramic powder systems.

REFERENCES

1. D. R. Ulrich, *Chem. Chem. Eng. News*, 1 Jan. 1990, 28.
2. L. M. Sheppart, *Ceram. Bull. 68*(5): 980 (1989).
3. H. Rhodes, *J. Am. Ceram. Soc. 64*(1): 19–22 (1981).
4. I. Aksay and C. H. Shilling, in *Ultrastructure Processing of Ceramic, Glasses and Composites* (L. L. Hench and D. R. Ulrich, eds.). John Wiley, New York, 1984, p. 439.
5. G. D. Parfitt, *Dispersion of Powders in Liquids.* John Wiley, New York, 1979.
6. M. von Smoluchowski, *Phys. Z. 27*: 585 (1916); *Z. Phys. Chem. 92*: 129 (1917).
7. I. A. Aksay and C. H. Schilling, *Advances in Ceramics*, Vol. 9. American Ceramic Society, 1984, p. 85.
8. F. F. Lange, Powder processing science for structural reliability, *J. Am. Ceram. Soc. 72*(1): 3–15 (1989).

9. M. Persson, Slip casting and pressing of ceramics based on colloidal processing techniques. Ph.D. thesis, Department of Inorganic Chemistry, University of Göteborg, 1989.
10. A. Roosen and H. Hausner, *Adv. Ceram. Materials 3*(2): 131 (1988).
11. W. H. Rhodes and S. Natansohn, *Ceramic Bull. 68*(10): 1804 (1989).
12. E. A. Barringer and H. K. Bowen, *J. Am. Ceram. Soc. 65*(12): C-199 (1982).
13. K. Kendall, *Powder Tech. 58*: 151 (1989).
14. Manufacturing literature for α-SiC Lonko products.
15. R. J. Pugh, M. Kizling, and L. Bergström, in *Ceramic powder processing science*, Proc. 2d Int. Conf. (Berchtesgaden) (H. Hauser, G. L. Messing, and S. Hirano, eds.). Deutsche Keramische Gesellschaft, 1989.
16. E. Carlström, Defect minimization of silicon carbide, silicon nitride and alumina ceramics. Ph.D. thesis, Department of Inorganic Chemistry, University of Göteborg, 1989.
17. L. Bergström and R. J. Pugh, *J. Amer. Chem. Soc. 72*(1): 103 (1989).
18. S. G. Malghan and L. Lum, in *Ceramic Transactions*, Vol. 12, *Ceramic Powder Science III* (G. L. Messing, S. Hirano, and H. Hauser, eds.). American Ceramic Society, 1990.
19. P. K. Whitman and D. L. Feke, *Adv. Ceram. Mater. 1*(4): 366 (1986).
20. M. Volante, B. Fubini, E. Giamello, and V. Bolis, *J. Material Sci. Lett. 8*: 1076 (1989).
21. R. C. Piller, K. P. Balkwill, A. Briggs, and R. W. Davidge, *Brit. Ceram. Proc. 37*: 191 (1986).
22. S. Chander and R. Hogg, *Minerals and Metallurgical Processing,* August 1988, p. 158.
23. R. Rameshand and P. Somasundaran, *J. Colloid Interface Sci. 139*: 29 (1990).
24. H. G. Bruil and J. J. van Aartsen, *Colloid Polymer Sci. 252*: 32-38 (1974).
25. J. L. Yordan and R. H. Yoon, *Interfacial Phenomena in Biotechnology and Materials* (Y. A. Attia, ed.). Elsevier, 1988, p. 333.
26. E. Carlström, M. Persson, E. Bostedt, A. Kristoffersson, and R. Carlström, *Ceram. Trans. Silicon Carbide 87*, Vol. 2. American Ceramic Society, Columbus, Ohio, 1989, pp. 175-185.
27. U.S. Patent 2966719 (1961).
28. J. J. Thomas, *Amer. Cer. Soc. Bull. 42*: 480 (1963).
29. L. J. Warren, in *Principles of Mineral Flotation* (M. H. Jones and J. T. Woodcock, eds.). Australasian Institute of Mining and Metallurgy, 1984.
30a. E. J. W. Verwey and J. Th. G. Oveerbeek, *Theory of the Stability of Lyophobic Colloids.* Elsevier, Amsterdam, 1948.
30b. J. N. Israelachvili, *Intermolecular and Surface Forces.* Academic Press, London, 1983.
31. Gouy, J. *Phys. Radium 9*: 457 (1910).
32. E. A. Barrington and H. Kent Bowen, *Langmuir 1*: 420 (1985).
33. R. J. Hunter, *Foundations Colloid Science*, Vol. 1. Oxford 1987.
34. W. Stum and J. J. Morgan, *Aquatic Chemistry.* John Wiley, New York, 1981.
35. M. J. Crimp and R. C. Piller, *Brit. Ceram. Proc. 46*: 199 (1989).

36. D. E. Yates, S. Levine, and T. W. Healy, *J. Chem. Soc., Faraday Trans. 1, 70*: 1807 (1974).
37. D. Chan, J. W. Perran, L. R. White, and T. W. Healy, *J. Chem. Soc., Faraday Trans. 1, 71*: 1046 (1975).
38. R. O. James, *Advances in Ceramics*, Vol. 1, *Ceramic Powder Science*. American Ceramic Society, 1987, p. 349.
39. P. W. Schindler and W. Stumm, in *Aquatic Surface Chemistry* (W. Stumm, ed.). Wiley-Interscience, New York, 1987, p. 83.
40. A. J. Babchin, R. S. Chow, and Sawatzky, *Adv. Coll. Int. Sci. 30*: 111 (1989).
41. B. V. Velamakanni and F. Lange, Effect of interparticle potentials and sedimentation on packing. Paper presented at the 91st Annual Meeting of the American Ceramic Society, Indianapolis, April 23–27, 1989.
42. L. Giest, L. Vandenberghen, E. Nicolas, and A. Fraboni, *J. Electrochem. Soc. 113*: 1025 (1966).
43. Y. Bérubé and P. L. de Bruyn, *J. Colloid Int. Sci. 28*: 92 (1968).
44. W. Stumm, C. P. Huang, and S. R. Jenkins, *28*: 92 (1968); *Croat. Chem. Acta 42*: 223 (1970).
45. R. W. Gurney, *Ionic Processes in Solution*. Dover, New York, 1953.
46. F. Dumont, J. Warlus, and A. Watillon, *J. Colloid Int. Sci. 138*: 543 (1990).
47. H. M. Jang and D. W. Fuerstenau, *Colloids and Surfaces 21*: 235 (1986).
48. G. A. Parks and P. L. de Bruyn, *J. Phys. Chem. 66*: 967 (1962).
49. G. A. Parks, *Chem. Rev. 65*: 177 (1965); *ACS Adv. Chem. Ser. 67*: 121 (1967).
50. R. H. Yoon, T. Salman, and G. Donnay, *J. Colloid Int. Sci. 70*: 483 (1979).
51. G. D. Parfitt and C. H. Rochester, eds., *Adsorption from Solution at the Solid/Liquid Interface*. Academic Press, New York, 1983.
52. M. C. Anderson and A. J. Rubins, eds., *Adsorption of Inorganics at Solid/Liquid Interface*. Ann Arbor Science, 1981.
53. P. H. Tewari, ed., *Adsorption from Aqueous Solution*. Plenum Press, New York, 1981.
54. R. H. Ottewill, C. H. Rochester, and A. L. Smith, eds., *Adsorption from Solution*. Academic Press, New York, 1983.
55. S. Levine and A. L. Smith, *Discuss. Faraday Soc. 52*: 290 (1971).
56. R. O. James and T. W. Healy, *J. Colloid Interface Sci. 40*: 65 (1972).
57. R. O. James and T. W. Healy, *J. Colloid Interface Sci. 40*: 65 (1988).
58. R. J. Pugh and L. Bergström. *J. Colloid Int. Sci. 124*: 570 (1988).
59. T. M. Shaw and B. A. Pethica, *J. Amer. Ceram. Soc. 69*(2): 88 (1986).
60. A. M. Gaudin and D. W. Fuerstenau, *Trans. A.I.M.E. 202*: 958 (1955).
61. P. Somasundaran and D. W. Fuerstenau, *J. Phys. Chem. 70*: 90 (1966).
62. I. Lin and P. Somasundaran, *J. Colloid Int. Sci. 37*: 731 (1971).
63. S. G. Dick, D. W. Fuerstenau, and T. W. Healy, *J. Colloid Int. Sci. 37*: 595 (1971).
64. B. Vincent, *Adv. Colloid Int. Sci. 4*: 193 (1974).
65. J. Cesarano III, I. A. Aksay, and A. Blier, *J. Am. Ceram. Soc. 71*(4): 250 (1988).
66. J. Cesarano III and I. A. Aksay, *J. Am. Ceram. Soc. 71*(12): 1062 (1988).
67. J. E. Gebhardt and D. W. Fuerstenau, *Colloids and Surfaces 7*(3): 221 (1983).

68. A. Foissy, A. El Attar, and J. M. Lamarche, *J. Colloid Int. Sci. 96*: 275 (1983).
69. N. Shashidhar, J. R. Varner, and R. A. Condrate, *Ceramic Powder Science III* (G. L. Messing, S. Hirano, and H. Hausner, eds.). American Ceramic Society, 1990, p. 443.
70. M. Persson, A. Forsgren, E. Carlström, L. Käll, B. Kronberg, R. Pompe, and R. Carlsson, in *High Tech Ceramics* (P. Vincenzini, ed.). Elsevier Science, Amsterdam, 1987.
71. M. Persson, L. Hermansson, and R. Carlsson, *Ceramic Powders* (P. Vincenzini, ed.). 1983, pp. 735-742.
72. R. Hogg, T. W. Healy, and D. W. Fuerstenau, *Trans. Faraday Soc. 62*: 1638 (1966).
73. R. G. Horn and D. T. Smith, *J. Non-Crystalline Solids 120*: 72 (1990).
74. R. G. Horn, D. R. Clarke, and M. T. Clarkson, *J. Mater. Res. 3*(3): 413 (1988).
75. J. N. Israelachvili and G. E. Adams, *J. Chem. Soc., Faraday Trans. 1, 74*: 975 (1978).
76. E. A. Barrington and H. Kent Bowen, *Langmuir 1*: 420 (1985).
77. B. V. Derjaguin and N. V. Churaev, *Colloids Surfaces 41*: 223 (1989).
78. R. Kjellander and S. Marcelja, *J. Phys. Chem. 90*: 1230 (1986).
79. Y. Hirata et al., *Mat. Res. Soc. Sym. Proc.*, Vol. 155, p. 343.
80. M. J. Crimp and R. C. Piller, *Brit. Ceramic Proc. 45*: 199 (1989).
81. E. Bostedt, M. Persson, and R. Carlsson, Proceedings of the First European Ceramic Society Conference, Maastricht, 1989.
82a. R. J. Pugh and J. A. Kitchener, *J. Colloid Int. Sci. 35*: 656 (1971).
82b. J. D. F. Ramsay and B. O. Booth, *J. Chem. Soc., Faraday Trans. 1, 79*: 173 (1983).
83. R. Laible and K. Haman, *Adv. Colloid Int. Sci. 13*: 65 (1980).
84. M. Green et al., *Advances in Ceramics*, Vol. 21, *Ceramic Powder Science*. American Ceramic Society, p. 449.
85. M. Aizawa, Y. Nosaka, and H. Miyama, *J. Colloid Int. Sci. 139*: 324 (1990).
86. J. Gregory, Effect of polymers on colloidal stability, NATO Advanced Study Series, Vol. E27, 1978, p. 103.
87. D. H. Napper, in *Colloidal Dispersions* (J. W. Goodwin, ed.). Royal Society Chemical Special Publications, London, 1982, p. 99.
88. B. Vincent, *Adv. Colloid Int. Sci. 4*, 193 (1974). Th. F. Tadros, *The Effect of Polymers on Dispersion Properties*. Academic Press, London, 1982.
89. D. H. Napper, *J. Colloid Int. Sci. 58*: 390 (1977); *32*: 106 (1970).
90. T. Sato and R. Ruch, *Stabilization of Colloidal Dispersions*. Marcel Dekker, New York, 1980.
91. D. H. Napper, *Polymeric Stabilization of Colloidal Dispersions*. Academic Press, London, 1983.
92. J. H. Schenkel and J. A. Kitchener, *Trans. Faraday Soc. 56*: 161 (1960).
93. W. B. Russel, D. A. Savelle, and W. R. Schowalter, *Colloidal Dispersions*. Cambridge Univ. Press, Cambridge, 1989.
94. P. L. Luckham, *Powder Tech. 58*: 75-91 (1989).
95. P. L. Luckham, *Adv. Colloid Int. Sci. 34*: 191 (1991).

96. J. Klein, *J. Chem. Soc., Faraday Trans. 1, 49*: 99 (1983).

97. R. J. R. Cairns and R. H. Ottewill, *J. Colloid Int. Sci. 56*: 45 (1976).

98. T. K. Yin, I. A. Aksay, and B. E. Eichinger, *Ceramic Powder Science*, Vol. 1 (G. L. Messing, E. R. Fuller, Jr., and H. Hauser, eds.). American Ceramic Society Publications, 1988, p. 654.

99. J. J. Lannutti, C. H. Schilling, and I. A. Aksay, *Mat. Res. Sympos. Proc.*, Vol. 155, Material Res. Soc., 1989, p. 155.

100. B. V. Velamakanni, J. C. Chang, F. F. Lange, and D. S. Pearson, *Langmuir 6*: 1323 (1990).

101. F. F. Lange, B. V. Velamakanni, J. C. Chang, and D. S. Pearsson, *Colloidal Powder Processing for Structural Reliability*, p. 57.

102. A. Blier, *J. Phys. Chem. 87*(18): 3493 (1983).

103. A. Blier, *J. Am. Cerma. Soc. 66*: C79 (1983).

104. R. E. Mistler, D. J. Shanefield, and R. B. Runk, in *Ceramic Processing Before Firing* (G. Y. Onoda and L. L. Hench, eds.). John Wiley, New York, 1978, p. 411.

105. F. M. Fowkes, *Surfactants in Computer Science* (K. Mittal, ed.). Plenum Press, New York, 1987, p. 3.

106. F. M. Fowkes, *Advances in Ceramics*, Vol. 21, *Ceramic Powder Science*. American Ceramic Society, 1987, p. 412.

107. J. L. van der Minne and P. H. J. Hermanie, *J. Colloid Sci. 7*: 600 (1952); *8*: 38 (1953).

108. R. H. Ottewill, A. A. Rennie, and A. Schofield, *Progr. Colloid Polym. Sci. 81*: 1 (1990).

109. E. J. Verwey and J. H. de Boer, *Rec. Trav. Chim. 57*: 383 (1938).

110. H. Koelmans and J. Th. G. Overbeek, *Discuss. Farad. Soc. 18*: 52 (1954).

111. E. M. deLiso and A. Blier, *Interfacial Phen. in BioTech. and Mat. Process.* (Y. A. Attia, B. M. Mougdil, and S. Chander, eds.). Elsevier, 1988.

112. K. R. Mikeska and W. R. Cannon, *Colloids Surfaces 29*: 305 (1988).

113. R. J. Pugh, *Ceramic Powder Science III, Ceramic Transactions*, Vol. 12, 1990, pp. 375-382.

114. V. Gutmann, *The Donor Acceptor Approach to Molecular Interactions*. Plenum Press, New York, 1978.

115. P. D. Calvert, E. S. Tormey, and R. L. Pober, *Am. Ceram. Soc. Bull. 65*(4): 669 (1986).

116. M. E. Labib and R. Williams, *Colloid Polymer Sci. 264*: 533 (1986).

117. M. E. Labib and R. Williams, *J. Colloid Int. Sci. 97*: 356 (1984).

118. N. Satoh, S. Bandow, and K. Kimura, *J. Colloid Int. Sci. 131*: 161 (1989).

119. L. M. Sheppard, *Ceramic Bulletin 68*(6): 1187 (1989).

120. L. Bergström, Ph.D. thesis, Royal Institute of Technology, Stockholm, 1992, TRITA-FYK 9203.

121. F. F. Lange, unpublished data.

5

Rheology of Concentrated Suspensions

LENNART BERGSTRÖM   Institute for Surface Chemistry, Stockholm, Sweden

I. INTRODUCTION

Rheological methods are widely used to determine the properties of concentrated ceramic powder suspensions. Rheology can be used as an analysis method, e.g., when determining the optimal amount of dispersant from measurements of viscosity versus the amount of dispersant added. Also, rheological measurements are often used for quality control in order to minimize the batch-to-batch variation before a ceramic suspension is processed further, e.g., spray dried or tape cast.

In more sophisticated usage, the rheological behavior can be used as a direct process parameter, which should be appropriately adjusted to obtain optimal green body properties after forming. This approach requires a more fundamental understanding of the ceramic forming methods in question, e.g., slip casting, pressure filtration, injection molding, tape casting, etc., and of the appropriate rheological behavior associated with each forming method.

Since concentrated ceramic suspensions often consist of particles in the colloidal size range, the interparticle forces have a pronounced effect on the rheological behavior. This chapter describes and discusses, in a mainly qualitative way, the rheological behavior of colloidal suspensions in relation to different types and magnitudes of the interparticle forces. More quantitative reviews can be found elsewhere [1-2]. The examples are taken from studies performed on well-characterized model systems and focus mainly on the steady shear properties. The chapter starts with a general background concerning the definitions and terminology used. Then the different contributions to the interparticle potential and a number of characteristic total interparticle potentials are described. The different types of arrangements of the particles in the suspension then follow, before the main section on steady shear rheology. The chapter ends with two smaller sections on viscoelastic properties and compression rheology.

II. GENERAL BACKGROUND

The rheology of concentrated suspensions is concerned with how these materials respond to an applied stress or strain [3]. In reality, both the stress and the strain are tensors each having nine components. In simple shear, which is the most common way of determining the rheological behavior, the shear stress σ_{xy} can be related to the shear rate $\dot{\gamma}$ by

$$\sigma_{xy} = \sigma = \eta\dot{\gamma} \tag{1}$$

where η is the viscosity.

However, rheological measurements are also performed in other types of flow or stress fields. If a uniaxial extensional flow field is applied to a material, the stress distribution can be described by

$$\sigma_{xx} - \sigma_{yy} = \dot{\varepsilon}\eta_E \tag{2}$$

where $\dot{\varepsilon}$ is the extensional strain rate and η_E is the (uniaxial) extensional viscosity.

Rheological measurements can also be made in compression, but the term compression rheology, applied to concentrated suspensions, refers to the compressive yield properties of the particles interacting through repulsive or attractive interparticle forces. Hence, compression rheology, in this chapter, is not related to the compressibility of the suspension as such but only to the compressibility of the particle network.

In general, a material can be characterized by two types of rheological behavior. The material can be described as a solid showing an elastic behavior if the deformation is fully recovered after removal of the applied stress. The material can be described as a liquid having a viscous response if it flows under a very small stress. Commonly, many materials, and this definitely includes concentrated suspensions, are characterized by both an elastic and a viscous response, so-called viscoelastic behavior. The type of response depends on the time scale of the experiment. If a small strain or stress is applied very rapidly to a viscoelastic material, it will respond elastically. If the stress or strain is applied for a long time, the material will flow and hence show a viscous response.

It is possible to classify a particular system by comparing the relaxation time τ of the system with the time scale of the experimental measurement t_0 and construct a dimensionless number called the Deborah number (De) [4]:

$$De = \frac{\tau}{t_0} \tag{3}$$

When De is very large, the system behaves like a solid, and when De is very small, it flows like a liquid. When De is around unity, the response is viscoelastic.

A. Steady Shear

The different types of viscous response in steady shear can be illustrated by plots of shear stress versus shear rate, or viscosity versus shear rate (Fig. 1). In the simplest case (curve a), so-called Newtonian behavior, the flow curve is a straight line passing through the origin. Equation (1) describes such a Newtonian system, the slope of the curve being equal to the viscosity η. In practice, most concentrated suspensions show a more complicated flow behavior where the viscosity is shear dependent. In such non-Newtonian systems, Eq. (1) can still be used to describe the flow behavior if a shear dependent viscosity or an apparent viscosity is used. For example, a shear thickening (also called dilatant) system is characterized by an increasing

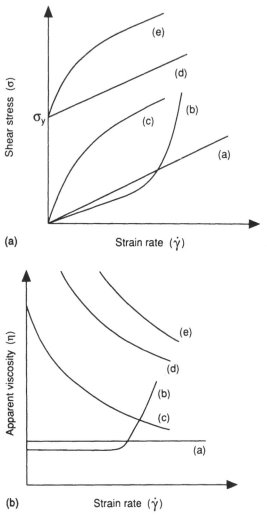

FIG. 1 Classification of rheological behavior in steady shear plotted as (a) σ versus γ̇ or (b) η versus γ̇. Curve a, Newtonian; b, shear thickening, c, shear thinning, d, Bingham plastic, e, nonlinearity plastic.

apparent viscosity with shear rate (curve b). If the viscosity decreases with shear rate, the system is called shear thinning (curve c). If the decrease in viscosity is very large at small shear rates, the system is sometimes called pseudoplastic. Commonly, concentrated suspensions show a plastic behavior with no response until a limiting yield stress σ_y has been exceeded. If the flow curve is linear above σ_y, the system is said to be Bingham plastic (curve d) described by the Bingham model [5]:

$$\sigma = \sigma_y + \eta_{pl}\dot{\gamma} \tag{4}$$

where the plastic viscosity η_{pl} is defined as the slope of the flow curve at $\sigma >$ σ_y. The yield stress σ_y in the Bingham model is sometimes called the Bingham yield stress σ_B. The curve above the yield stress can also be nonlinear (curve e). Such behavior can be described by the Herschel-Buckley model [6]:

$$\sigma = \sigma_y + k_1\dot{\gamma}^n \tag{5}$$

where k_1 and n are constants, or the Casson model [7]:

$$\sigma^{1/2} = \sigma_y^{1/2} + k_2^{1/2}\dot{\gamma}^{1/2} \tag{6}$$

where k_2 is a constant. It is important to realize that fitting the same rheological data to the different models can lead to different yield stresses [8]. The accuracy of the yield stress determined is dependent on the applicability of the model used and should thus be considered a model parameter and not a true material property.

The rheological properties of concentrated suspensions are often time dependent. If the apparent viscosity continuously decreases with time under shear with a subsequent recovery of the viscosity when the flow is stopped, the system is said to be thixotropic [9]. The opposite behavior is called antithixotropy or sometimes rheopexy. Thixotropy should not be confused with shear thinning, which is a time independent characteristic of a system. Systems that show an irreversible decrease in viscosity with shear should be termed shear destructive and not thixotropic.

B. Viscoelasticity

The viscoelastic behavior of concentrated suspensions can be studied by several different procedures [10]. Transient measurements involve two types of measurement: stress relaxation, which involves applying a constant small strain and measuring the time dependence of the stress, and creep measurement, which involves applying a constant stress and measuring the time dependence of the strain.

The most widely used method is probably that of oscillatory measurement [11], also called forced oscillation [4], which belongs to the family of dynamic measurements. Oscillatory measurements consist of subjecting the material to a continuously oscillating strain over a range of frequencies and measuring the peak value of the stress σ_0 and the phase difference between

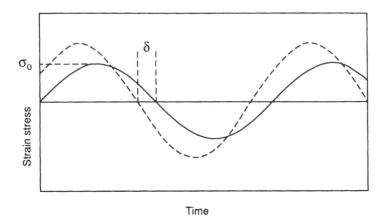

FIG. 2 Schematic illustration of a viscoelastic response to an oscillating strain. The dotted line represents the applied strain, and the full curve represents the stress. δ is a measure of the phase difference.

the stress and strain δ (Fig. 2). A sinusoidal deformation is usually employed. In most cases, an attempt is made to keep the applied strain small enough to be in the linear viscoelastic region.

In the linear viscoelastic region, the mathematical analysis of the data is substantially simplified, since the ratio of stress to strain

$$G^* = \frac{\sigma_0}{\gamma_0} \tag{7}$$

where G^* is called the complex or the dynamic modulus, is independent of the magnitude of the stress or strain. The dynamic modulus can also be expressed in complex form in terms of a storage modulus G' and a loss modulus G'' as

$$G^* = G' + iG'' \tag{8}$$

where i is equal to $\sqrt{-1}$.

The storage modulus G' represents the in-phase stress-to-strain ratio and gives a measure of the elastic properties. The loss modulus G'' represents the out-of-phase stress-to-strain ratio and gives a measure of the viscous properties. It is also possible to define a complex viscosity η^*

$$\eta^* = \frac{G^*}{i\omega} = \eta' - i\eta'' \tag{9}$$

where $\eta' = G''/\omega$, $\eta'' = G'/\omega$, η' is called the dynamic viscosity, and ω is the frequency of the oscillation.

All the rheological parameters, G^*, G', G'', etc., vary with frequency. The general behavior of different types of materials has been described [1,11] and will be discussed later in this chapter. Phenomenalistic models such as the Maxwell, Kelvin, or Berger models can be used to describe the frequency dependence of the rheological parameters [1]. These models are mechanical analogs consisting of combinations of springs representing an elastic, Hookean response and dashpots representing a viscous response.

III. INTERPARTICLE POTENTIALS

There are a number of different contributions to the potential (or force) acting between colloidal particles. The major contributions are the van der Waals potential, the electrostatic double layer potential, and polymeric interactions. These are described thoroughly in Chapter 4. Here a short summary of these interactions is given.

A. The van der Waals Potential

The van der Waals potential is always operating between any kinds of surfaces or molecules. The van der Waals potential has an electrodynamic origin, as it arises from the interactions between atoms and molecules whose orientations are correlated in such a way that they attract each other.

Hamaker [12] calculated the distance dependence of the free energy of macroscopic bodies by performing a pair-wise summation over all the atoms in the bodies. For example, the van der Waals interaction free energy V_A between two spheres of the same size can be expressed as

$$V_A = -\frac{A}{6}\left[\frac{2a^2}{D^2 - 4a^2} + \frac{2a^2}{D^2} + \ln\left(\frac{D^2 - 4a^2}{D^2}\right)\right] \tag{10}$$

where A is the Hamaker constant, which is a characteristic of the material, a is the radius of the sphere, and D is the center-to-center separation between the spheres.

In the original treatment, also called the microscopic approach, the Hamaker constant was calculated from the polarizabilities and number densities of the atoms in the two bodies [12]. However, there are several limitations to this approach. For example, many-body effects are ignored, and the effect of retardation is not included. Lifshitz [13] presented an alternative approach where each body is treated as a continuum with certain dielectric properties. This automatically incorporates many-body effects.

Using this approach, the van der Waals interaction can be thought of as a fluctuation in the electromagnetic field between two macroscopic bodies, modified by the separating media.

A simplified approximation of the Hamaker constant A_{131} has been derived for two identical materials 1 interacting across a medium 3 [14,15]:

$$A_{131} = \frac{3kT}{4} \left(\frac{\varepsilon_1(0) - \varepsilon_3(0)}{\varepsilon_1(0) - \varepsilon_3(0)} \right)^2 + \frac{3h\nu}{32\pi\sqrt{2}} \frac{(n_1^2 - n_3^2)^2}{(n_1^2 - n_3^2)^{3/2}} \tag{11}$$

where $\varepsilon_1(0)$ and $\varepsilon_3(0)$ are the static dielectric constants for the particle and the medium, respectively; n_1 and n_3 are the refractive indices for the particles and the medium, respectively; and ν is the characteristic adsorption frequency in the visible range. If the interacting material and the medium are allowed to have different absorption frequencies, so that $\nu_1 \neq \nu_3$, a more complicated expression for the dispersive contribution is obtained [16]. Furthermore, the value of the Hamaker constant is decreased by retardation effects at relatively large (> 5 nm) surface-to-surface separations [1].

Calculations of nonretarded Hamaker constants for different ceramic materials (Al_2O_3, Si_3N_4, SiC) interacting across various media (air, water, and cyclohexane) showed that these materials, like probably most ceramic materials, have large Hamaker constants [17].

B. The Electrostatic Double Layer Potential

A colloidal particle will in general acquire a charge on the surface when it is immersed in an electrolyte solution. The surface charge plus the diffuse ion layer surrounding the particle constitute the so-called electrical double layer. This is described in more detail in Chapter 4. The thickness of the double layer depends on the ionic strength of the electrolyte, and a convenient measure of the thickness is given by the inverse of the Debye constant κ

$$\kappa^{-1} = \left(\frac{\varepsilon_0 \varepsilon_r kT}{e^2 \Sigma n_i z_i^2} \right)^{1/2} \tag{12}$$

where ε_0 is the dielectric constant for vacuum, ε_r is the dielectric constant of the medium, n_i is the number density of ion i in the medium, z_i is the valency of ion i, and e is the electronic charge.

When two identically charged surfaces approach each other, a repulsive force occurs. This repulsion can be analyzed by considering the ionic pressure generated by the accumulation of ions between the particles [18]. The following analytical expression can be used [18] to calculate the energy of interaction for two identical spheres at small degrees of double layer over-lap $((D - 2a) > \kappa^{-1})$ in symmetrical electrolytes:

$$V_{DL} = \frac{64kTna\beta^2}{\kappa^2} \exp(-\kappa(D - 2a)) \tag{13}$$

where $n = \sum_i n_i$, $\beta = \tanh(e\psi_0/4kT)$, and ψ_0 is the surface potential.

C. Polymeric Potentials

The addition of polymers to a colloidal suspension can cause a variety of effects [19]. Depending on the amount added, the chemical nature, and the molecular weight of the polymer, the added polymer can induce either floc-culation or stabilization of the suspension. In the case of nonadsorbing polymer, an attractive force is created if the polymer chain is evicted from the region between two approaching surfaces. A higher concentration of polymer outside the region between the two surfaces will create an osmotic pressure difference that results in an attractive depletion force. This may give rise to so-called depletion flocculation.

If the polymer adsorbs onto the surface, two different types of effects can be identified. If one polymer molecule can adsorb to more than one parti-cle at the same time, the polymer will hold the particles together. This effect is called bridging flocculation and is usually the result of strong adsorption at low polymer concentration of polymers of high molecular weight. If the polymer instead forms a layer on each particle, this polymer layer can give rise to a repulsive force, so-called steric stabilization.

For steric stabilization to be effective, several requirements, such as strong attachment to the surface, high surface coverage, and a sufficiently thick layer, must be fulfilled [19,20]. The steric interaction is composed mainly of two contributions: an elastic term ΔG_{el} resulting from the loss in configurational entropy of the attached polymer chains when the layer is compressed, and a mixing term ΔG_M arising from the change in segment concentration in the overlap region. It is generally assumed that the ΔG_M contribution dominates the steric stabilization, although Tadros [20] has argued that ΔG_{el} probably plays an important role in the case of thin, dense layers.

Fisher [21] derived the following expression for the free energy of mixing when two polymer-coated spheres with constant radial segment density interact:

$$\Delta G_M = \frac{\pi a k T}{\overline{V}_3} \phi_2^2 \frac{1}{(0.5 - \chi)} (2\Delta + 2a - D)^2 \tag{14}$$

where $\bar{V}_3$ is the partial molar volume of the solvent, ϕ_2 is the volume fraction of polymer, Δ is the thickness of the polymer layer, and χ is the Flory-Huggins interaction parameter.

Although the Fisher theory has been criticized for unrealistic assumptions, it clearly illustrates the importance of the solvency of the solvent. The factor $0.5 - \chi$ determines the sign of ΔG_M. If $\chi < 0.5$, the so-called good solvent condition, ΔG_M is positive and the interaction is repulsive. If $\chi > 0.5$, the so-called bad solvent condition, ΔG_M is negative, resulting in an attractive interaction. When $\chi = 0.5$, the so-called θ condition, ΔG_M is zero and thus does not contribute to any interaction between the two polymer layers. Experimental studies have demonstrated that there is a strong correlation between the critical flocculation point of a sterically stabilized suspension and the θ point (the point where $\chi = 0.5$) of the stabilizing moieties in solution [19].

D. Total Interparticle Potentials

By combining the contributions from the different types of potentials, such as the van der Waals, double layer, and polymeric interactions, an expression for the total interparticle potential (or energy) can be obtained. Four different types of pair-potential curves are illustrated in Fig. 3. A colloidally stable suspension is characterized by a repulsive interaction (positive potential) when two particles approach each other (Fig. 3a). Such a repulsion can be caused by an electrostatic double layer interaction or by a steric interaction in a good solvent condition. If the repulsion varies with distance, as illustrated in Fig. 3a, the term "soft" repulsion is used. The so-called "hard" repulsion illustrated in Fig. 3b constitutes a very important idealization used in modelling. This type of behavior can sometimes be achieved in colloidal model systems.

Figure 3c illustrates a weakly flocculated system with an attractive interaction with a magnitude that is not too large compared to the thermal energy kT. Such an interaction can be created by the adsorption of a thin surfactant layer, by depletion flocculation, or by flocculation in the secondary minimum in the DLVO potential.

Figure 3d illustrates a strongly flocculated system with an attractive interaction several orders of magnitude stronger than the thermal energy kT. Such an interaction can be achieved by strong van der Waals interaction in a nonpolar solvent without any polymers or surfactants added, or in an aqueous suspension where no net charge exists on the particle surface.

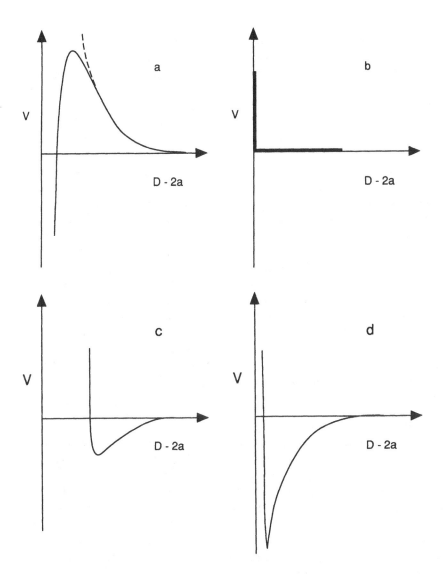

FIG. 3 Total interparticle potentials. Colloidally stable systems: (a) "soft" repulsion; (b) "hard" repulsion. Flocculated systems: (c) weakly flocculated; (d) strongly flocculated.

IV. SUSPENSION STRUCTURE

Colloidal suspensions can show different types of arrangement of the particles, ranging from crystallike structure over disordered liquidlike structures to highly heterogeneous "fractal" structures. The structure of colloidal suspensions can be investigated by different scattering techniques such as light scattering [22], x-ray scattering [23,24], or small-angle neutron scattering [25,26]. Also, direct imaging techniques such as microscopy [27–29] can be used. In a recent study, it was demonstrated that confocal scanning laser microscopy could be used for three-dimensional imaging of the structure in concentrated silica suspensions [30]. As an often less expensive alternative to experimental studies, computer simulations [31] have gained more and more interest and can give important information both about the equilibrium structure and about how the structure of a sheared, concentrated suspension is reflected in rheological behavior [32].

The structure of a colloidal suspension at rest is the result of a balance between the interparticle potential and the Brownian motion of the particles. Monodisperse suspensions with a hard sphere interaction show a disordered, liquidlike, structure (Fig. 4a) with only short-range positional order up to a volume fraction of $\phi = 0.5$, where a coexistence with an ordered, crystallike, structure (Fig. 4b) with long-range positional order occurs [33]. Thermodynamic calculations [34] are in good agreement with this disorder-order transition found experimentally. Above $\phi = 0.54$ the suspension is fully crystallized. At very high volume fractions, close to random close packing where $\phi = 0.64$, it has been found that the suspensions may not crystallize but appear to have an amorphous structure that can be characterized as a colloidal glass [33].

If the particles interact through a long-range, electrostatic repulsion, the volume fraction for the disorder-order transition varies with electrolyte concentration [35]. At a very low electrolyte concentration (10^{-4}–10^{-5} M), the disorder-order transition occurs already at $\phi < 0.10$. This can be understood by considering the range of the electrostatic repulsion, which is strongly dependent on the ionic strength (Eqs. (12) and (13)). At these very low ionic strengths, the double layer thickness is of the order of the particle size, and this results in a large increase in the effective hard sphere radius.

It should, however, be understood that these types of disorder-order transitions probably never occur in practice in ceramic suspensions. All the experiments described above were performed on monodisperse, spherical particles. Dickenson et al. [36,37] used computer simulations of sediments and showed that no crystalline order will occur in moderately polydisperse or bimodal systems with a difference in particle diameter of only 10–20% of

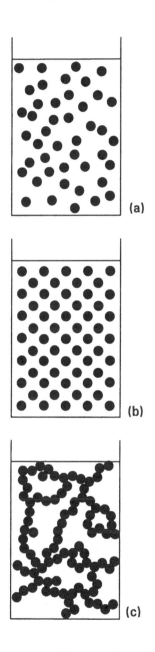

FIG. 4 Schematic illustration of suspension structure: (a) disordered structure (stable suspensions); (b) ordered structure (stable suspension with long-range electrostatic repulsion); (c) disordered, "fractal" structure (flocculated suspensions).

the mean particle size. Instead, the sediments are characterized by a uniform, disordered structure. Most commercially available ceramic powders have a much larger polydispersity than 20%. Their nonspherical shape also favors a disordered structure.

When the particles are subject to attractive interactions, the structure of the suspension at rest will depend on the magnitude of the attraction and the time allowed for equilibration. With an interaction of the order of kT, the hard sphere disordered structure will only be slightly perturbed. A separation into two phases, one concentrated and one dilute, might occur. Such a phase separation can, for example, be induced by temperature variations in sterically stabilized systems [38] or by the addition of nonadsorbing polymers [39,40]. However, phase separation might require a long equilibration time, and hence a weakly flocculated suspension may still be in the one-phase region with a slowly changing metastable structure if the measurements are performed relatively soon after formulation.

When the attractive interactions cause the particles to attach to each other, large, tenuous objects, so-called flocs or aggregates, will be formed. These aggregates have an open, disordered structure (Fig. 4c), and it has been found that the geometry and structure of these aggregates can be successfully described using the concept of fractal geometry [41–43]. A fractal is characterized by a self-similar structure, which means that a fractal is scale invariant between certain system-specific limits. The quantitative measure of how a fractal scale can be expressed by the fractal dimension D_f which is a universal property.

For example, the number of particles in an aggregate N can be expressed as

$$N \sim \left(\frac{R}{a} \right)^{D_f} \tag{15}$$

where R is the radius of the aggregate and a is the radius of the particle [44].

The lower the value of D_f, the more open and ramified is the aggregate structure. The larger the value of D_f, the more dense and space-filling is the structure. Since the fractal dimension D_f is always smaller than the Euclidean dimension d, the density of a floc decreases as the size increases. On the other hand, the density of a nonfractal, close-packed floc ($D_f \to d$) is independent of its size. Meakin [43,45] has reviewed the work on understanding the structure and properties of fractal aggregates using computer simulations. The models that are considered to be most realistic are the so-called cluster-cluster aggregation models where single particles via random walks can form clusters that can stick to other clusters and form large fractal aggregates. In diffusion-limited cluster-cluster aggregation (DLCA), where the sticking probability is one, D_f = 1.78 in three dimensions [46,47].

If many collisions are needed before a particle sticks to another particle or a cluster, i.e., the sticking probability is much less than one, the structure of the aggregate becomes more dense. The so-called reaction-limited cluster-cluster aggregation model (RCLA) describes this case and results in D_f = 2.0 in three dimensions [48]. More recently, simulations have been performed where rearrangements after aggregate formation has been allowed [49,50]. In a rearrangement without any breaking of particle-particle bonds, D_f could be increased to 2.2 [49]. If bond breaking due to thermal motion is allowed, D_f can be increased further [50].

Most of the computer simulations on fractal structures have been related to dilute conditions where single aggregates can grow freely. In a flocculated suspension, there will always be a critical volume fraction ϕ_c, above which the suspension structure is changed from separate flocs to a continuous three-dimensional network. This transition can be thought of as the concentration at which the flocs suddenly grow into each other. The concept of percolation transition can be used to understand this phenomenon. At the percolation threshold, some of the physical properties such as conductivity and network strength change drastically. It has been demonstrated that the percolation threshold changes with the magnitude of the attraction [51,52]. For hard spheres, where there is no attraction, percolation is expected to occur near the volume fraction of random close packing (ϕ = 0.64) [51]. When the attraction is increased, the percolation threshold decreases drastically [51,52].

The rheological properties of a flocculated suspension change at ϕ_c. Below ϕ_c, the suspensions are typically shear thinning and show little or no viscoelasticity. Above ϕ_c, the suspension can develop a yield stress and also a solidlike viscoelasticity [53]. This will be discussed in detail later. There have been several attempts to model the properties of particle networks using a fractal approach [53-56]. Bremer [54] summarized the different models and showed that although they all predict a power-law behavior of the form

$$G \sim \phi^\mu \tag{16}$$

where G is the elastic modulus, the models differ regarding the value of the exponent μ, with μ varying somewhere between 2 and 5. The accuracy of these models will be discussed later. In the following sections on rheology, the shear induced structural changes are discussed in intimate connection with the rheological behavior.

V. STEADY SHEAR RHEOLOGY

A. Hard Sphere Systems

The hard sphere system constitutes the simplest case, where the flow is affected only by hydrodynamic (viscous) interactions and Brownian motion. Although hard sphere systems are not frequently encountered in practice, they represent a very important idealization and starting point when the effect of interparticle potentials on the rheological behavior is evaluated.

Krieger [57] designed a polymeric latex system to behave as a hard sphere system. At relatively high-volume fractions ($\phi > 0.3$), it was found that the suspensions displayed shear thinning behavior with well-defined low and high shear viscosities (Fig. 5). At a constant volume fraction ϕ, it is possible to reduce the flow curves for hard spheres of different particle sizes suspended in continuous media of different Newtonian viscosities η_s by plotting the relative viscosity

$$\eta_r = \frac{\eta}{\eta_s} \tag{17}$$

versus the so-called Peclet number Pe:

$$Pe = \frac{a^2 \dot{\gamma}}{D_0} \tag{18}$$

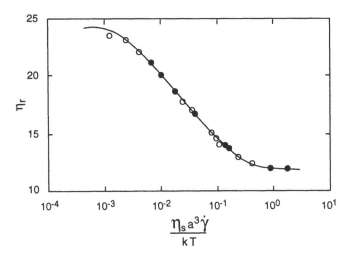

FIG. 5 Steady shear viscosity of polystyrene spheres of various sizes at $\phi = 0.50$, suspended in different fluids; ($\bigcirc$) benzyl alcohol; ($\bullet$) m-cresol, — water. (From Ref. 57.)

where D_0 is the diffusion coefficient for the flow units, which is given as

$$D_0 = \frac{kT}{6\pi\eta_s a} \qquad (19)$$

for spheres at infinite dilution [1]. This type of dimensionless representation is shown in Fig. 5 for polystyrene spheres of various sizes in different media. The success of this scaling can be understood by considering the suspension structure in the different Pe regimes. At low shear rates where $Pe \ll 1$, the suspension structure is not significantly perturbed by the shear, because Brownian motion dominates over the viscous forces. The equilibrium structure is restored more rapidly than it is perturbed. At higher shear rates, the viscous forces affect the suspension structure more, and shear thinning occurs. At very high shear rates, the viscous forces dominate and the plateau in η in this region is a measure of the relative viscosity of a suspension with a completely hydrodynamically controlled structure.

There have been several studies in which the structure of hard sphere suspensions under shear have been investigated [58–61]. A number of these studies [58,59] have shown that shear thinning at higher shear rates is associated with an induced ordering of the suspension structure. Computer simulations also generally show some type of ordering at higher shear rates [32]. However, a recent light scattering study on poly(methylmethacrylate) (PMMA) spheres coated with poly(12-hydroxystearic) acid in an organic solvent showed somewhat conflicting results [60]. Suspensions with liquid-like order at rest did not show any development of a layer structure and only displayed partial string formation at the highest concentration. It was concluded that shear thinning was associated with a distorted liquidlike order. Layer formation under shear was found only when the volume fraction was so high that the structure at rest was at least partly crystalline.

The viscosity of a hard sphere suspension is a function of the volume fraction of particles. Einstein calculated the relative viscosity of a dilute suspension with noninteracting particles [62] as

$$\eta_r = 1 + 2.5\phi \qquad (20)$$

Batchelor [63] extended this approach by also considering two-particle interactions and obtained

$$\eta_r = 1 + 2.5\phi + 6.2\phi^2 \qquad (21)$$

At higher concentrations ($\phi > 0.05$), many-body interactions affect the rheological behavior, and Batchelor's relation no longer applies. Attempts have been made to include parts of the hydrodynamic and Brownian many-body interactions [64,65], but the agreement with experimental data is only qualitative. Instead, semiempirical models such as the Krieger-Dougherty model [66]

$$\eta_r = \left(1 - \frac{\phi}{\phi_m}\right)^{-[\eta]\phi_m} \qquad (22)$$

or the Quemada model [67]

$$\eta_r = \left(1 - \frac{\phi}{\phi_m}\right)^{-2} \qquad (23)$$

are commonly used. ϕ_m is the maximum volume fraction and $[\eta]$ is the intrinsic viscosity, which is 2.5 for spheres.

De Kruif and coworkers [68,69] investigated the steady shear properties of submicron-size sterically stabilized silica spheres in cyclohexane. This system displays a hard sphere behavior. It was found that the low and high shear limiting viscosities, as a function of volume fraction (Fig. 6), could be fitted to both Eq. (22) and Eq. (23), with $\phi_m = 0.63 \pm 0.02$ in the low shear limit and $\phi_m = 0.71 \pm 0.02$ in the high shear limit [69]. The data indicate that a suspension subjected to low shear will cease to flow, i.e., display a yield stress, at a volume fraction of $\phi > 0.63$, which is close to the volume fraction for random close packing. In a recent study performed on the same type of experimental system, Jones et al. [70] investigated the steady shear behavior

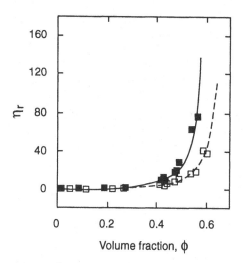

FIG. 6 Relative viscosity as a function of volume fraction for sterically stabilized silica spheres with radius a = 110 nm, suspended in cyclohexane: (■), $\gamma \to 0$, $\phi_m = 0.63$; (□) $\gamma \to \infty$, $\phi_m = 0.71$. (From Ref. 69.)

in the low shear region and found excellent agreement with de Kruif and coworkers [68,69] at $\phi < 0.63$. At $\phi > 0.64$ they detected a dramatic change in the rheological properties, and a yield stress could be measured.

B. Soft Sphere Systems

Most colloidally stable systems are characterized by a more or less strong distance dependence of the magnitude of the interparticle repulsion. In electrostatically stabilized systems, the range of the electrostatic double layer interaction, expressed by the Debye length $1/\kappa$, is strongly dependent on the ionic strength of the solution. The Debye length increases with decreasing ionic strength. Krieger and Eguiluz [71] investigated the influence of the ionic strength on the rheological properties of electrostatically stabilized monodispersed polystyrene latex particles. They found (Fig. 7) that the viscosity increases drastically when the ionic strength is decreased from $I = 1.9 \times 10^{-2}$ down to that of deionized water. The increase in viscosity is much stronger in the low shear limit, where the viscosity actually diverges and goes to infinity when the ionic strength is decreased sufficiently. This indicates that these suspensions will behave as solids at rest and that a critical stress, the yield stress, must be exceeded before the suspension will start to flow. In the high shear limit, the steady shear behavior

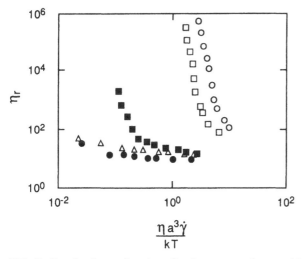

FIG. 7 Steady shear viscosity of polystyrene spheres with $a = 110$ nm at $\phi = 0.40$ suspended in aqueous media at various concentrations of HCl: (O) deionized; (□) 1.9×10^{-4} M; (■) 1.9×10^{-3} M; (●) 1.9×10^{-2} M; (△) 9.4×10^{-2} M. (From Ref. 71.)

is dominated by hydrodynamic interactions. Hence the relative viscosity only changes marginally with ionic strength in this region.

Sterically stabilized suspensions can also display a "soft", long-range interparticle repulsion. The steric repulsion is affected mainly by the surface concentration of the polymer (density of the attached polymer layer), the thickness of the attached layer, and the solvency of the stabilizing moieties (the χ parameter). Croucher and Milkie [72] showed how the steady shear rheology of PMMA particles, sterically stabilized by a relatively thick layer of terminally anchored polydimethylsiloxane and dispersed in n-hexadecane, was affected by temperature. Figure 8 illustrates how the relative viscosity of the suspension increases when the temperature is increased. This effect can be understood by considering how the polymer layer is affected by temperature changes. For polydimethylsiloxane in n-hexadecane, the solvency of the polymer is improved by an increase in temperature. An improvement in solvency, which can also be expressed as a decrease in the χ parameter, means that the mutual repulsion between polymer chains increases. Hence the polymer chains, terminally attached to the surface, will extend further out in the solvent and thus increase the thickness of the polymer layer when the temperature is increased. The hydrodynamic interactions between the particles are affected by an increase in the thickness of the polymer layer and as a result increase the viscosity.

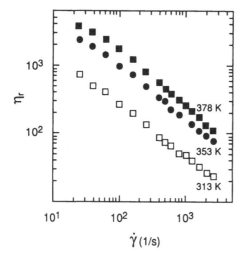

FIG. 8 Steady shear viscosity of sterically stabilized poly(methylmethacrylate) particles with a = 400 nm at ϕ = 0.282 suspended in n-hexadecane. Measurements were performed at three different temperatures. (From Ref. 72.)

It has been found that a sterically stabilized suspension may be treated as a hard sphere system with the use of an effective volume fraction ϕ_{eff}:

$$\phi_{eff} = \phi \left(1 + \frac{\Delta}{a}\right)^3 \tag{24}$$

where Δ is the thickness of the polymer layer [73-75]. The effective volume fraction is commonly estimated by measuring the particle size of the particle with polymer attached or by determining the hydrodynamic volume of the stabilized particles.

However, hard sphere scaling principles and the use of ϕ_{eff} apply only when the ratio of the particle radius to the thickness of the polymer layer a/Δ is large. When the thickness of the polymer layer is increased, the experimental data deviate from the predicted behavior, particularly at high volume fractions [75-77]. The effective volume fraction dependence of sterically stabilized PMMA spheres is shown in Fig. 9 [75]. For the two largest particle sizes, the $\eta_r - \phi$ behavior can be described by the Krieger-Dougherty equation Eq. (22) using $\phi_m = 0.62$. Fitting the data of the smallest particle size with $a/\Delta = 5$ resulted in $\phi_m = 0.96$, which is physically unrealistic for hard spheres. A possible way to explain this deviant behavior is to treat the polymer layer thickness as being both shear and volume fraction dependent. An increase in volume fraction forces the polymer layers to

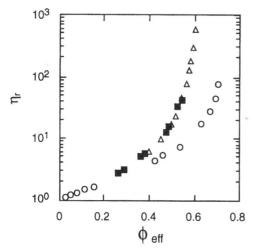

FIG. 9 Relative viscosity in the high shear limit for sterically stabilized poly(methylmethacrylate) particles of different particle sizes suspended in decalin: (O) $a = 42$ nm; (Δ) $a = 240$ nm; ($\blacksquare$) $a = 610$ nm. (Reproduced by permission of the American Institute of Chemical Engineers. © 1989 AIChE, all rights reserved.)

interpenetrate and compress. Also, subjecting the particles to high shear rates causes the polymer layer to deform. Tadros and coworkers [76,78,79] have estimated the thickness of the attached layer by fitting the η_r versus ϕ_{eff} data to the Krieger-Dougherty equation and using Δ as a fitting parameter. Their studies showed that the polymer layer thickness decreases with increasing volume fraction, indicating a substantial compression of the layer.

In principle, it should also be possible to scale electrostatically stabilized systems in a similar fashion. In these cases, the ratio of the particle radius to the Debye length, κ_a, is a convenient estimate of the softness of the repulsion.

Buscall [80] has presented a model in which he treats colloidally stable suspensions with long-range repulsive interactions as hard sphere systems with a shear stress or shear rate dependent volume fraction. An analytical expression for the shear dependency of ϕ_{eff}, based on the softness of the interparticle repulsion, was introduced, and a surprisingly good correlation with the experimental data was found.

C. Polydispersity and Particle Shape Effects

The rheological properties of concentrated suspensions are affected by the particle size distribution of the multimodal mixture and by the shape of the particles. It is well known that the packing of particles can be affected by mixing spheres of two different sizes or using broad continuous particle size distributions. Furnas [81] developed a model for the packing of a bimodal mixture of spheres with a large size difference. His approach has been extended to ternary and multimodal systems [82]. These studies show that the maximum volume fraction ϕ_m can be increased significantly above the value for random close packing of spheres $\phi_m = 0.64$. With the right size ratio it was found that a maximum volume fraction $\phi_m = 0.85$ could be reached with a binary mixture, and even higher volume fractions with ternary or quaternary mixtures [82]. However, it should be remembered that these models only apply to coarse (> 10 µm) particles and that effects such as mixing inhomogeneities [83,84] and inclusions [85] will lower ϕ_m.

The packing of particles also has a pronounced effect on the rheological behavior, particularly at very high concentrations. Several studies have shown how the use of bimodal or continuous broad particle size distributions can lower the viscosity and increase the maximum volume fraction before flow ceases [86–90]. In most of these studies, large (> 5 µm) noncolloidal particles have been used [87–89], where the hydrodynamic interactions determine the rheological behavior. Farris [91] developed a model in which the viscosity of multimodal suspensions was determined from data for unimodal, monodisperse systems. Farris also demonstrated how his model could be used to calculate the composition needed to minimize the viscosity.

Recently, the steady shear rheology of unimodal and multimodal suspensions of colloidal dispersions was compared with the behavior predicted by the Farris model [90]. The colloidal system consisted of submicron polymer particles suspended in water. The particles were electrostatically stabilized and the particle size distribution was rather narrow for most of the particle sizes used.

The relative viscosity of both unimodal and multimodal suspensions were measured, and it was found that all unimodal suspensions at $\phi > 0.6$ showed Bingham's plastic behavior. The suspensions would not flow out and form a smooth surface. Using bimodal suspensions made it possible to decrease the viscosity substantially and to increase the maximum volume fraction. Figure 10 illustrates how the relative viscosity of bimodal suspensions varies

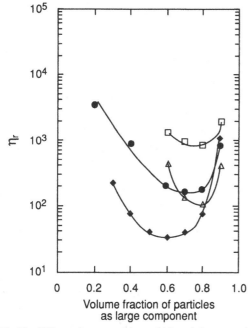

FIG. 10 Effect of composition of bimodal suspensions of colloidally stable polymer particles on the steady shear viscosity measured at $\sigma = 1$ dyn/cm. All the suspensions were at $\phi = 0.65$. The different curves represent ($\square$) mixture of $a = 340$ nm and $a = 135$ nm, size ratio 2.5; ($\bullet$) mixture of $a = 340$ nm and $a = 105$ nm, size ratio 3.2; ($\triangle$) mixture of $a = 475$ nm and $a = 70$ nm, size ratio 6.8; ($\blacklozenge$) theoretical values calculated according to Farris [91]. (From Ref. 90.)

with composition and particle size ratio. It is clearly seen that even a small size ratio can decrease the viscosity substantially. Increasing the size ratio results in a further decrease in viscosity. However, the decrease in viscosity between size ratio 3.2 and 6.8 is far less than expected, and Hoffman [90] attributes this to the increased importance of the repulsive interparticle potential when the absolute particle size is decreased. This can be understood by considering κa, the ratio of particle radius to Debye length (or a/Δ in a sterically stabilized system). With a decrease in κa, the relative volume of the electrical double layer increases and the ϕ_m of the *particles* decreases. Hence the models that consider only hydrodynamic interactions should be used with caution, and the effect of the range and magnitude of the interparticle potentials must be assessed.

The effect of particle shape on the rheological behavior of concentrated suspensions is not well understood. Attempts have been made to include the effect of nonspherical shape and particle orientation in simple models for dilute suspensions [92]. Generally, it has been found that powders with a high aspect ratio (fibers) or irregular shape yield a low ϕ_m and higher viscosities than spherical particles [93]. The higher the aspect ratio, the lower is ϕ_m.

D. Shear Thickening

In a recent review, Barnes [94] stated that concentrated suspensions of nonaggregating solid particles will always show shear thickening if they are subjected to the right conditions. Shear thickening is a phenomenon that needs to be controlled and often minimized in several ceramic processing steps such as the filling of a mold or in general suspension handling such as pumping and pouring.

Hoffman [58,95–97] was one of the first to investigate the shear thickening in detail, although the phenomenon had been observed earlier [98]. Figure 11 illustrates the shear thickening phenomenon in colloidally stable PVC suspensions. These concentrated suspensions show either continuous or discontinuous shear thickening. The severity of the shear thickening increases with increasing particle concentration, and the critical shear rate for the onset of shear thickening decreases with increasing particle concentration. Hoffman [58] attributed the shear thickening to a sort of order-disorder transition of the particle microstructure. He used white light diffraction to probe the suspension structure under shear and found evidence of a hexagonal layered structure at shear rates below the critical. In this shear thinning region, the monodispersed spheres apparently pack into two-dimensional layers with a hexagonal structure in which the layers are

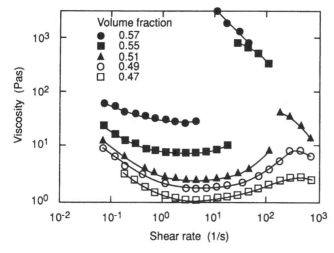

FIG. 11 Shear thickening behavior of suspensions of monodisperse, spherical polyvinyl chloride particles of a = 625 nm in dioctyl phthalate at several volume fractions. (From Ref. 58.)

oriented parallel to the flow direction. At the critical shear rate, the diffraction measurements indicate that the ordered layers break up into a less ordered structure. This less ordered structure dissipates more energy during flow due to particle "jamming," and hence the viscosity increases.

In the last ten years, several experimental and theoretical studies have been performed in order to improve the understanding of the mechanisms behind shear thickening. Boersma et al. [99–101] have investigated the shear thickening phenomena experimentally and have developed a theory to explain the phenomenon based on assumptions similar to those of Hoffman in his early model [95]. Boersma et al. assume that shear thickening occurs when the shear forces overrule the interparticle forces [99]. When a concentrated, colloidally stable monodisperse suspension is subjected to a low shear rate, the repulsive interparticle forces (which can be of electrostatic or steric origin) dominate and keep the particles in a layered structure. With increasing shear rate, the viscous force on the particles increases, and finally a critical shear rate is reached at which the viscous force overrules the repulsive interparticle force. This causes the particles to be moved from their equilibrium position in the two-dimensional layers, and a transition from a layered to a disordered structure takes place.

Boersma et al. [99] introduced analytical expressions for the repulsive electrostatic force and the viscous force acting on the particles and defined a dimensionless number N_d that gives the ratio of the viscous to the repulsive forces:

$$N_d = \frac{6\pi\eta_s a^2\dot{\gamma}}{2\pi\varepsilon_0\varepsilon_r\psi_0^2} \tag{25}$$

This relates to the case of constant surface potential of electrostatically stabilized systems. Shear thickening is predicted to occur when $N_d > 1$. It can clearly be seen that the onset of shear thickening is controlled by parameters such as the viscosity of the medium, the particle size, and the magnitude of the repulsion. Barnes [94] also found in his review on shear thickening that these parameters, together with some other parameters, control the shear thickening behavior.

Boersma et al. [100] demonstrated elegantly that shear thickening is a time dependent phenomenon. This is illustrated by steady shear measurements on a shear thickening suspension at different shear rates (Fig. 12). Measurements performed at a shear rate slightly below the critical shear rate $\dot{\gamma}_{crit}$ show that the low viscosity is essentially time independent (Fig. 12a). At a slightly higher shear rate, the viscosity shows peaks with values up to twenty times higher than the "base" value (Fig. 12b). When the shear rate is increased further, there is a transition to a high viscosity level after a shearing time of a couple of minutes (Fig. 12c). This high viscosity level was reached immediately at a still higher shear rate (not shown). This time dependent behavior can be explained in terms of a fluctuating structural change of the suspension [100]. When the viscosity is at its low "base" level, there is some type of ordered flow. At shear rates very close to $\dot{\gamma}_{crit}$, the ordered flow can become unstable, and a more disordered structure, possibly characterized by cluster formation, is formed that results in a large increase in viscosity. The return to the low viscosity level is then probably associated with the recurrence of ordered flow. Hence the structure fluctuates between an ordered and a disordered state with time. At higher shear rates, slightly above $\dot{\gamma}_{crit}$, the structure is permanently in the disordered state.

Although most researchers agree that the shear thickening phenomenon is associated with an order-disorder transition, the details of this transition are still unclear. Laun and coworkers [102-104] have combined rheological studies and small angle neutron scattering investigations to elucidate the nature of shear induced particle structures of electrostatically stabilized dispersions. They used well-characterized copolymer spheres dispersed in water or glycol. The suspensions showed strong shear thickening at sufficiently high concentrations with the onset of shear thickening preceded by a

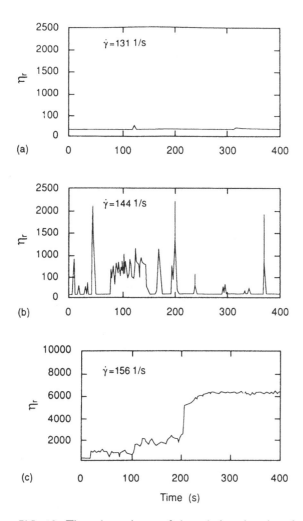

FIG. 12 Time dependence of the relative viscosity of a dispersion of electrostatically stabilized monodisperse, spherical polystyrene particles of a = 1300 nm in water at ϕ = 0.584. The measurements were performed at three different shear rates: (a) $\dot{\gamma}$ = 131 s^{-1}; (b) $\dot{\gamma}$ = 144 s^{-1}; (c) $\dot{\gamma}$ = 156 s^{-1}. (From Ref. 100.)

shear thinning region. Laun et al. [104] investigated two suspensions that showed similar rheological behavior but differed greatly in their shear induced structures. In the shear thinning region, one suspension demonstrated a long-range order with a hexagonal structure, while the other suspension displayed only an anisotropic structure with no long-range order. Furthermore, no major changes in the suspension structure could be detected in the shear thickening region in either of the suspensions. Hence it must be concluded that the association of shear thickening with a transition from an ordered layered structure to a disordered structure is probably more complicated than that originally proposed by Hoffman [58].

There is also some debate regarding the mechanisms underlying the shear thickening phenomenon. Most researchers have assumed that a balance between the viscous forces and the repulsive interparticle forces can explain and predict shear thickening [95-97,99-101], but computer simulations have indicated that these forces do not play a major role [105]. Instead, shear thickening is associated with the diffusion coefficient (relaxation time) of the suspension. In essence, the force balance model requires that the interparticle potential must be repulsive and soft, while the computer simulations have shown that hard sphere systems also should display shear thickening. Rheological studies on the hard-sphere-like system consisting of sterically stabilized silica particles have generally not revealed any shear thickening phenomenon [68-70]. Marshall and Zukoski [106], on the other hand, detected a shear thickening transition using the same type of system at $\phi > 0.58$. There are, however, indications that there was a slight surface charge present on their particles causing a deviation from hard sphere behavior. More experiments with hard sphere systems at high shear rates are probably needed to resolve this question. Finally, it should be noted that shear thickening is strongly dependent on the particle size distribution. The shear thickening phenomenon is most pronounced for monodisperse systems and becomes less severe using a polydisperse system. Figure 13 illustrates the difference between a monodisperse and a polydisperse system [99], where it is clearly seen that the critical shear rate for the onset of shear thickening is higher when a polydisperse system is used. Attempts have been made to attribute this effect of polydispersity to the increase in maximum volume fraction, ϕ_m [99].

E. Flocculated Systems

Characterization and systematization of the rheological behavior of flocculated systems is a difficult task, due mainly to the nonequilibrium nature of the suspension structure. Commonly, flocculated systems dominated by attractive interparticle potentials form disordered, metastable structures with very long relaxation times. This usually results in a history dependent rheological behavior. Since the relaxation time is strongly dependent on the magnitude of the interparticle attraction, it is convenient to distinguish

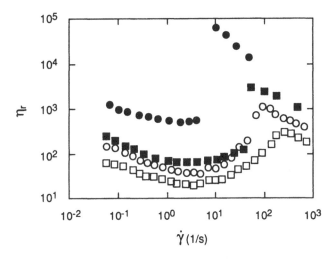

FIG. 13 Shear thickening behavior of polydisperse polyvinyl chloride particles with a number average radius, $\bar{a}$ = 700 nm in dioctyl phthalate, compared to the behavior of monodisperse particles taken from Hoffman (Fig. 11): (□) polydisperse particles, ϕ = 0.53; (○) polydisperse particles, ϕ = 0.57; (■) monodisperse particles, ϕ = 0.53; (●) monodisperse particles, ϕ = 0.57. (Reproduced with permission of the American Institute of Chemical Engineers. © 1990 AIChE, all rights reserved.)

between two types of flocculated systems: weakly flocculated and strongly flocculated. Weakly flocculated systems with $1 < -V/kT < 20$ are able to attain a reproducible rest state after shear in a reasonably short time. In strongly flocculated systems with $-V/kT > 20$, the diffusion of particles is completely eliminated, and these systems do not attain a reproducible rest state after shear. This structural difference between the two types of flocculated systems is manifested in their rheological behavior.

There have been several experimental studies on the steady shear rheology of weakly flocculated systems [107–110]. The attractive interparticle potential has been induced by addition of nonadsorbing polymer [107,108], by flocculation in the secondary minimum in the DLVO potential [109], or by inducing an attractive potential in a sterically stabilized system through temperature changes [110]. The system investigated by Woutersen and de Kruif [110] consisted of sterically stabilized silica particles suspended in the marginal solvent benzene. When the temperature is lowered, the magnitude of the interparticle attraction increases. At high temperatures, the system shows hard sphere behavior. The effect of the magnitude of the attractive interparticle potential on the steady shear properties is illustrated in Fig. 14 [110].

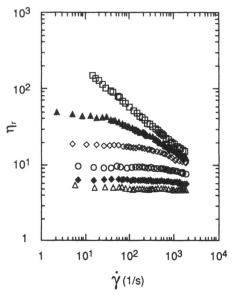

FIG. 14 Relative viscosity of coated silica particles with a = 47 nm suspended in the marginal solvent benzene at ϕ = 0.367. The measurements were performed at temperatures (Δ) 317.3K, (◆) 309.1K, (○) 306.2K, (◇) 304.2K, (▲) 303.2K, (□) 302.2K. (From Ref. 110.)

The steady shear behavior changes drastically with decreasing temperature. In the hard sphere regime (T = 317.3K), the system displays an essentially Newtonian behavior at a low relative viscosity. With an increase in the interparticle attraction (lower temperature) the system becomes more and more severely shear thinning with a strong increase in the relative viscosity at low shear rates. The viscosity at high shear rates is only moderately affected by the increase in interparticle attraction.

Woutersen and de Kruif explained this behavior by assuming that temporary, not permanent, aggregates are formed. The particles are held together by the potential forces while Brownian motion tends to break down the aggregates. The two opposing energies are of the same order of magnitude. Increasing the shear rate (or shear stress) results in a change in the suspension structure, favoring breakdown of the temporary aggregates. At high shear rates, the hydrodynamic interactions dominate the rheological behavior with only a minor influence of the potential interactions. Woutersen's and de Kruif's view of the formation and breakdown of aggregates as a dynamic process is supported by the reversibility of the flow curves. No time dependence was detected when cycling up and down in shear rate or temperature. In addition, a recent small angle neutron scattering study on the same system in the same temperature range with a slightly smaller particle size showed that the amount of permanent clusters is very small and probably a result of the purification of the system [111].

Strongly flocculated systems are in general characterized by an irreversible nonrecoverable rheological behavior when the shear rate (or shear stress) is cycled up and down. Compared to the dynamic process of breakdown and formation of aggregates in weakly flocculated systems, strongly flocculated systems are characterized by irreversibly formed aggregates. The energy from Brownian motion is far too small in magnitude compared to the interparticle attraction to result in a reproducible rest state in a reasonable time. There have been several studies trying to link the changing floc structure during shear with the rheological behavior. Hunter and coworkers [112–115] extensively studied the steady shear behavior of mainly aqueous TiO_2 and PMMA flocculated suspensions. They also developed a model called the elastic floc model trying to link the floc structure and the strength of the interparticle attraction to the rheological behavior. In order to obtain reproducible flow curves, the flocculated suspensions were first subjected to a very high shear rate. With this pretreatment, relatively good agreement between the model predictions and the experimental results was found. They determined the Bingham yield stress σ_B (see Eq. 4) and obtained

$$\sigma_B \sim \phi^2 \tag{26}$$

All strongly flocculated concentrated suspensions yield a severely shear thinning behavior [120,121] similar to the suspension at the lowest temperature in Fig. 14. This behavior can be explained by considering the aggregates as the primary flow units. With increasing shear rate, the viscous forces reduce the size of the aggregates and release liquid immobilized in the aggregates. This leads to a decrease in the effective volume fraction of the aggregates with increasing shear rate, and results in a decreasing viscosity.

Sonntag and Russel [116] investigated the breakup of dilute flocculated suspensions of polystyrene lattices subjected to shear. They used light scattering to determine changes in aggregate size and internal structure using the fractal concept. They found that the average aggregate size, expressed as $\langle R_g^2 \rangle^{3/2}$ varied with the maximum shear stress according to

$$\langle R_g^2 \rangle^{3/2} \sim (\eta_s \gamma_{max})^{-1.1} \tag{27}$$

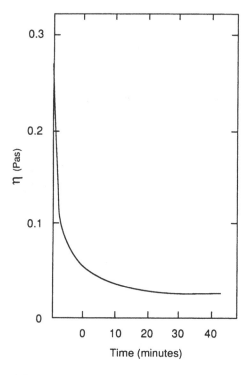

(a)

FIG. 15 Time dependence of the apparent viscosity of a strongly flocculated polystyrene latex suspension with a = 270 nm at ϕ = 0.233. The suspension is sheared at a constant shear stress σ = 19.6 Pa. The figures show the suspension structure (b) initially and (c) following prolonged shearing. (From Ref. 119.)

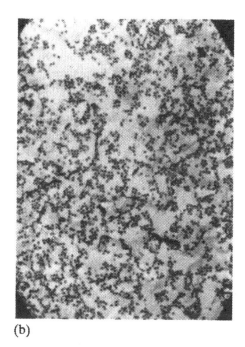

(b)

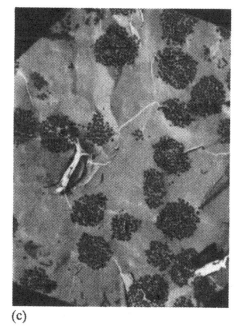

(c)

The determination of the fractal dimension of the aggregates showed that the fractal dimension D_f increased from $D_f = 1.6$ for flocs grown in situ to $D_f = 2.5$ for flocs subjected to continuous shearing. Hence the density of the aggregates increased with shearing. Sonntag and Russel also developed a model for the breakup of an isolated aggregate in a shear flow taking into account the heterogeneous, fractal nature of the aggregates [117]. Recently, Potanin [118] presented a model of the growth and breakup of aggregates in shear flow. He included several different mechanisms that took into account both the decrease in size and the increase in fractal dimension with increasing shear rate.

Several of the characteristics of the steady shear properties of flocculated suspensions are elegantly demonstrated in a recent paper by Mills et al. [119]. Rheological and microscopic studies were combined to elucidate the effect of shear on the morphology of concentrated, strongly flocculated suspensions. Figure 15 illustrates the irreversible decrease in viscosity with time of a salt-coagulated latex suspension. This decrease in viscosity was related to a drastic change in the internal aggregate structure. The minimally sheared suspension has an aggregate structure that is highly ramified and heterogeneous. After prolonged shearing, this fractallike structure has rearranged to yield discrete, relatively tightly packed spherical aggregates.

Since this change in aggregate structure and viscosity is irreversible, the time dependent steady shear behavior should be termed shear degradation and not thixotropy. A thixotropic suspension is characterized by a reversible time dependence of the viscosity. Hence the shear induced changes of the suspension structure have to be reversible, implying that thixotropy mainly relates to weakly flocculated systems.

The qualitative difference between weakly and strongly flocculated systems may also be manifested in the occurrence of a yield stress. Although both types of system display a severe shear thinning behavior, weakly flocculated systems commonly show an apparent Newtonian viscosity at low shear rates (or low shear stress), while a strongly flocculated system typically shows a yield stress. The difference between these two types of behavior is systematically illustrated in Fig. 16, where it can be seen that the presence of a yield stress would result in a straight line with a slope of -1 extending down to zero shear rate in a $\log \eta - \log \dot{\gamma}$ diagram. However, there has been much debate recently as to whether or not a yield stress exists [122].

Barnes and Walters [122] showed that extending the measured shear rate range down to very low shear rates ($\dot{\gamma} = 10^{-4}-10^{-5}$ s^{-1}) gave a low shear Newtonian viscosity in fluids that were originally considered to display a yield stress. Others have disputed these results and modified the yield stress definition somewhat to consider it an engineering reality [123]. This more

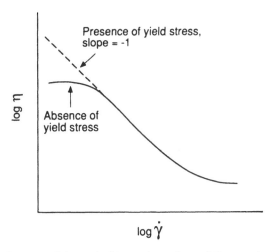

FIG. 16 Schematic illustration of the difference in the viscosity curve in the presence or absence of a yield stress.

pragmatic definition states that the fluid will not flow at a stress $\sigma < \sigma_y$ over a reasonable time scale, long compared with the time scale of the engineering application.

If the viscosity is plotted as a function of a shear stress, the presence of a yield stress becomes easier to detect. Figure 17 shows the steady shear behavior of a weakly and a more strongly flocculated system. The weakly flocculated system consisted of polystyrene latex particles flocculated by the addition of a nonadsorbing polymer, sodium carboxymethylcellulose [124]. The more strongly flocculated system consisted of very fine fumed silica particles ($a = 3.5$ nm) suspended in methyl laurate [125]. The weakly flocculated system displays a shear thinning behavior with an apparent Newtonian viscosity at high and low shear rates. The more strongly flocculated system shows a very large catastrophic drop in viscosity over a narrow stress range. Aerschot and Mewis [125] concluded that for most practical purposes, this critical stress can be considered a yield stress, although the suspension shows a slow creeping flow at lower stresses.

Several studies have attempted to determine the yield stress of suspensions [8,125–128]. Some of them have investigated the effect of prolonged shearing on the yield stress [126], while others have compared different methods to determine the yield stress [8,127–129]. A widely encountered and often ignored problem when measuring yield stress in steady shear is the presence of wall slip. Wall slip must be avoided if reliable yield stress

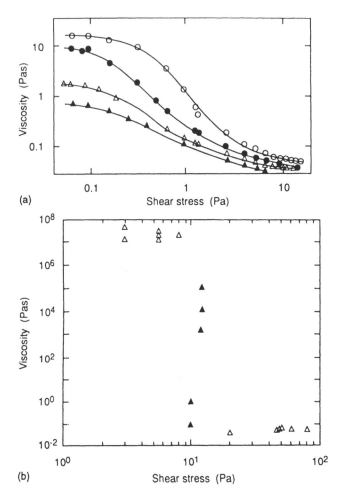

FIG. 17 Viscosity versus shear stress for a weakly flocculated (a) and a strongly flocculated (b) suspension. The weakly flocculated suspension was measured at various volume fractions: (▲) ϕ = 0.25; (△) ϕ = 0.294; (●) ϕ = 0.35; (○) ϕ = 0.427. The strongly flocculated suspension of 2.5% fumed silica in methyl laurate was measured with a stress-controlled (△) or a strain-rate–controlled (▲) rheometer. (From Refs. 124 and 125, respectively.)

data is to be obtained. One method that promises to avoid wall slip problems is the vane technique [8,128,129]. The vane consists of a small number of thin blades arranged at equal angles around a small cylindrical shaft. The vane technique allows the material to yield within the material itself, hence avoiding wall slip problems.

VI. VISCOELASTIC PROPERTIES

Concentrated suspensions commonly display viscoelastic behavior. The viscoelastic properties can preferably be studied by the dynamic rheological measurements (also called oscillatory measurements) described in Section II. Compared to steady shear measurements, oscillatory measurements are made under small deformations, which means that the suspension structure is only slightly perturbed from the structure at rest. Hence oscillatory measurements are suitable for correlating rheological behavior to structural data and interparticle potentials, even for strongly flocculated systems that show irreversible changes if they are subjected to large deformations. From a practical perspective, oscillatory measurements can provide important information regarding the rheological properties of e.g. cast cakes or ceramic slurries that will be subjected to small strains during a forming operation.

When oscillatory measurements are performed, the interpretation is simplified if the measurements are performed in the so-called linear viscoelastic region [11]. In this region, the viscoelastic response is independent of strain, and it can be assumed that no irreversible changes of the structure of the suspension take place. The linear viscoelastic region is usually assessed by performing a test in which the strain is increased and the storage G' and loss G'' moduli are measured. Figure 18 illustrates such a measure-

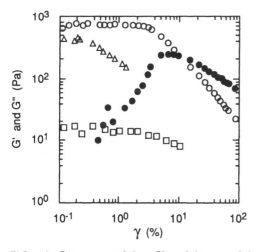

FIG. 18 Storage modulus G' and loss modulus G'' versus strain of a sterically stabilized PMMA suspension in decalin. Open symbols denote G' and closed symbols G''; (□) a = 238 nm, ϕ = 0.595; (△) a = 238 nm, ϕ = 0.624; (○,●) a = 42 nm, ϕ = 0.860. (From Ref. 130.)

ment on a sterically stabilized suspension (PMMA particles suspended in decalin with chemically grafted chains of poly(12-hydroxystearic acid) on the particle surface) [130]. The limit of the linear viscoelastic region can be defined as the deformation at which the storage modulus G' starts to decrease. It can be seen that the linearity limit decreases with increasing particle size and increasing concentration. The nonlinearity can often be observed earlier by an increase in the loss modulus G''. However, the G'' curve often shows a large scatter and is therfore often not shown. In the following discussion, the analysis of the data from the literature is based on the assumption that the measurements were performed in the linear viscoelastic region.

A. Stable Systems

The viscoelastic response of a colloidally stable concentrated suspension is strong when the average distance between the suspended particles is of the same order as the range of the repulsive interparticle potential. Hence the viscoelastic properties originate from the repulsive interparticle potential. Several studies have been performed on the viscoelastic properties of sterically stabilized [74,76–78,130], electrostatically stabilized [131,132], and also hard sphere systems [70,133].

The viscoelastic behavior of a sterically stabilized suspension is illustrated in Fig. 19 [130] (same system as in Fig. 18). The storage modulus increases with increasing frequency ω and increasing volume fraction. With increasing volume fraction, the average distance between the particles decreases, causing a progressively stronger overlap of the polymer layers. This strong overlap results in an increase in the elasticity of the suspension, reflected in an increase in the storage modulus with increasing volume fraction.

The increase in storage modulus with increasing frequency can be understood by considering the relaxation time τ_r of the suspension. At low frequencies, the characteristic experimental time scale, $t_0 \sim 1/\omega$, is longer than τ_r, and the perturbed structure is able to relax during the oscillation. This results in a mainly viscous response (small storage modulus), since the suspension is able to dissipate most of the energy. With increasing frequency, t_0 approaches τ_r, resulting in a viscoelastic behavior where the suspension has both a viscous and an elastic response.

When the frequency is increased further, t_0 becomes smaller than τ_r, and the perturbed suspension structure is unable to relax during oscillation; thus most of the energy is stored, and the suspension shows an elastic response. In this region,

$$G^* \approx G' = G_\infty \tag{28}$$

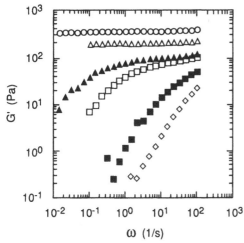

FIG. 19 Storage modulus versus frequency ω of sterically stabilized PMMA suspensions with a = 42 nm in decalin at different effective volume fractions: (◊) φ = 0.638; (■) φ = 0.676; (□) φ = 0.700; (▲) φ = 0.712; (△) φ = 0.742; (○) φ = 0.783. (From Ref. 130.)

since the viscoelastic properties do not change at higher frequencies. In Fig. 19, this is illustrated by the plateau of G' vs. ω for the suspensions at high volume fraction.

Although most suspensions do not have a single but rather a range of relaxation times, it is clear that the relaxation time increases as the volume fraction increases. For the most highly concentrated suspensions in Fig. 19, the response is essentially elastic over the frequency range studied, hence the relaxation time of these suspensions is higher than 100 s ($1/10^{-2}$ s^{-1}).

This example demonstrates how the viscoelastic behavior of a stable, sterically stabilized suspension can be altered by variations in the magnitude of the repulsive interaction through changes in the volume fraction of the particles. In the case of electrostatically stabilized suspensions, the viscoelastic behavior can also be manipulated by changing the ionic strength or the particle size. Figure 20 illustrates how the shear modulus G_0 varies with volume fraction and particle size for a system consisting of charged polystyrene lattices [132]. The shear modulus was measured by shear wave propagation at a frequency of 200 Hz, and the equality $G_0 \approx G_\infty$ probably holds for most of the measurements.

It was found that, with increasing particle size, the shear modulus at a given volume fraction decreases. Also, the onset of a measurable viscoelas-

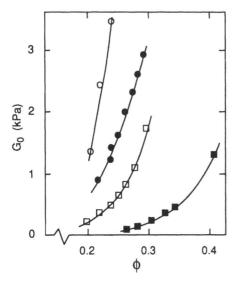

FIG. 20 Shear modulus G_0 as a function of volume fraction of polystyrene suspensions in 10-3 M NaCl electrolyte solutions. The particles have different radii: (○) 26 nm; (●) 34 nm; (□) 39 nm; and (■) 98 nm. (From Ref. 132.)

tic response is shifted to higher and higher volume fractions with increasing particle size. Buscall et al. [132] mention that, for particles with diameters larger than 200 nm, viscoelasticity is only obtained at volume fractions approaching close packing.

The effect of particle size on the viscoelastic properties can be explained in the same manner as the effect of volume fraction. With a decrease in particle size, at a given volume fraction, the average distance between the particles decreases, leading to stronger interactions between the electric double layers. An alternative way of influencing the viscoelastic properties is to change the ionic strength. A low ionic strength results in a long-range electrostatic repulsion, leading to a strong viscoelastic response at low volume fractions.

Tadros and coworkers [76,78,134] have used viscoelastic measurements to investigate the properties of both electrostatically and sterically stabilized suspensions. In one study, aqueous sterically stabilized polystyrene latex dispersions with grafted poly(ethylene oxide) (PEO) chains were investigated [78]. Prestidge and Tadros [78] showed that the dispersion changes from being more viscous ($G'' > G'$) to being more elastic ($G' > G''$) over a narrow volume fraction range (Fig. 21). At $\phi = 0.44$, the suspension shows

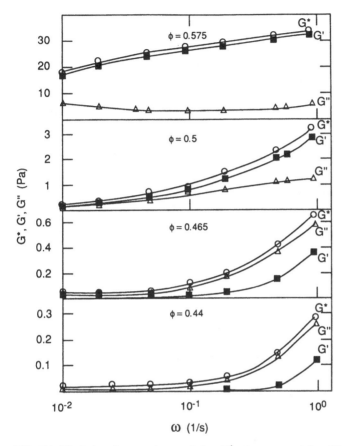

FIG. 21 Variation in complex modulus G^*, storage modulus G', and loss modulus G'' with frequency and volume fraction of a sterically stabilized aqueous polystyrene suspension with $a = 175$ nm having polyethylene oxide chains grafted onto the surface. (From Ref. 78.)

an essentially viscous response. Between $\phi = 0.465$ and $\phi = 0.5$, G' becomes larger than G'', illustrating the shift to an elastic behavior. At the highest volume fraction, $\phi = 0.575$, the response is almost totally elastic.

Prestidge and Tadros related this change in viscoelastic properties to increasing interpenetration and compression of the PEO chains. At very high volume fractions, $\phi > 0.585$, they found an increase of several orders of magnitude in the storage modulus, which they referred to as an elastic "gel" behavior of the suspensions.

B. Flocculated Systems

The viscoelastic properties of flocculated suspensions are strongly dependent on the heterogeneous suspension structure. As already mentioned in Section IV, the suspensions start to show an elastic response at a critical volume fraction ϕ_c, which can be identified as the point at which the particles form a continuous three-dimensional network. Although ϕ_c changes with the magnitude of attraction [51,52], the size of the container and external forces [54], values of ϕ_c = 0.05–0.07 for strong, irreversible flocculation of spherical particles are commonly observed [56]. Hence the very open particle networks formed at these low volume fractions are strong enough to yield an elastic behavior.

The magnitude of the elastic response of flocculated suspensions above ϕ_c depends on several parameters such as suspension structure, magnitude of the interparticle attraction, and particle size, shape, and volume fraction. The volume fraction dependence of the storage or shear modulus has been found to follow a power-law behavior.

$$G, G' \sim \phi^\mu \tag{29}$$

with μ varying between 2 and 5 [53,56,135,136]. This strong dependence on volume fraction illustrates the heterogeneous nature of the particle network. Models based on a homogeneous structure predict a linear relationship between G' and ϕ [137], which is never found in practice.

An increase in the magnitude of the interparticle attraction results in an increase in the storage modulus, provided that the suspension structure remains relatively unaffected. This is illustrated in Fig. 22, where the storage modulus of a flocculated coated silica suspension in hexadecane is shown [136]. Lowering the temperature below 30°C causes the suspension to form a particle network with elastic properties. This is illustrated by the sharp change in the suspension properties from a liquid (viscous) to a solid (elastic) response between 30 and 29°C. Reducing the temperature further results in an increase in the magnitude of the interparticle attraction, which leads to an increase in the storage modulus. The storage or shear modulus usually increases with decreasing particle size at a given volume fraction [53]. This can be understood by considering the increasing number of particle-particle contacts per unit volume with decreasing particle size. Although some general characteristics of the viscoelastic behavior of flocculated systems can be described, much work remains to systematize the properties. In particular, studies where the suspension structure and rheological response of well-characterized systems are studied simultaneously are badly needed.

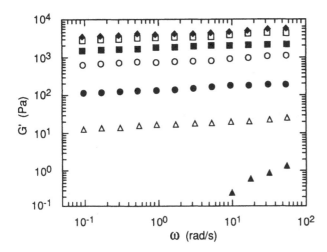

FIG. 22 Storage modulus G' versus frequency ω at temperatures between 20 and 30°C at a strain of 0.01 for a coated silica suspension with $a = 56$ nm in hexadecane with $\phi = 0.182$: ($\blacklozenge$) 20°C; ($\square$) 22°C; ($\blacksquare$) 24°C; ($\bigcirc$) 26°C; ($\bullet$) 28°C; ($\triangle$) 29°C; ($\blacktriangle$) 30°C. (From Ref. 136.)

VII. COMPRESSION RHEOLOGY

In many of the ceramic forming operations, such as slip casting, pressure filtration, and centrifugal casting, the suspensions are subjected to compressive stresses. Although shear rheology provides important information, it is also of great interest to investigate the compressive response of concentrated suspensions. In a number of studies on colloidally stable suspensions, a compression cell was used for this purpose [138–144]. In the early design of this apparatus [138], the suspension was confined between a semipermeable filter and an elastic, impermeable membrane. The pressure was controlled and the volume fraction was measured when steady state was attained. In later modifications of the compression cell, the volume of the sample was controlled and the corresponding pressure was measured [140]. An osmotic pressure can also be applied by allowing the suspension to equilibrate osmotically with a solution of known osmotic pressure [141,143,144]. Compression cell measurements have been performed on both electrostatically [138,140–142] and sterically stabilized suspensions [139,140,143,144].

Callaghan and Ottewill [138] used compression cell measurements to study the compressive and decompressive properties of fine aqueous montmorillonite clay suspensions. They performed measurements at different ionic strengths (10^{-4}–10^{-1} M NaCl) and found systematic variations. The suspensions compressed at low ionic strengths showed a compressive stress response over a much larger concentration range, starting already at 10 wt % clay due to the long-range electrostatic double layer interaction. In addition, a hysteresis effect was found upon decompression, and this was interpreted as indicating a rearrangement of the clay platelets from a disordered structure upon compression to a parallel array upon decompression.

Cairns et al. [139] studied monodispersed PMMA latices with $a \sim 78$ nm suspended in dodecane and stabilized by covalently attached poly(12-hydroxystearic acid) with a thickness $\Delta \sim 9$ nm. They used the compression cell and found that the resistance to compression started to increase around $\phi = 0.55$, and that at $\phi = 0.566$ the compression resistance was very strong (Fig. 23). They interpreted this resistance to compression as being the result of interparticle repulsive forces caused by overlap of the stabilizing polymer chains. Estimates of the average distance between the particles suggested that the polymer layers should start to touch at about $\phi = 0.53$, which is in accordance with the experimental results.

One feature of most colloidally stable suspensions is that the compressive properties are more or less reversible, provided that no major changes in suspension structure occur. However, in the case of flocculated suspensions, the compressive properties are irreversible. In concentrated floccu-

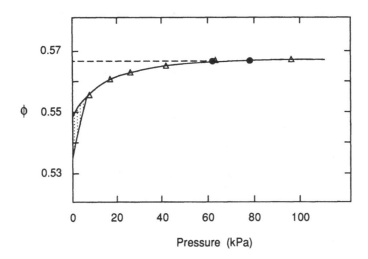

FIG. 23 Volume fraction versus compressive pressure of a sterically stabilized PMMA suspension with $a = 80$ nm in dodecane. Open symbols denote compression and closed symbols decompression. (From Ref. 139.)

lated suspensions, a continuous particle network forms that can support some stress up to a critical value. Once this critical stress, also called the compressive yield stress P_y, is exceeded, the network consolidates to a higher volume fraction with a higher critical stress. In addition to the compression cell, the strength of irreversibly consolidated particle networks can be estimated by two other methods. Buscall [145] has shown how a simple centrifuge method, where the equilibrium sediment height as a function of centrifugal acceleration is measured, can be used to derive the compressive yield stress $P_y(\phi)$, or the compressional modulus $K(\phi)$. A thorough analysis of the method was given in another paper [146]. Another possibility is to calculate $P_y(\phi)$ from the steady state volume fraction profiles of sediments subjected to different force fields [147,148].

Bergström et al. [148] investigated the compressive yield properties of particle networks formed in nonaqueous alumina suspensions with different degrees of flocculation. The magnitude of the attractive interparticle energy at contact was controlled by adsorbing fatty acids of different molecular weights at the alumina/decalin interface. It was found (Fig. 24 and Table 1) that strongly attractive interactions resulted in a particle network that resisted consolidation (compression) and was compressible over a large stress range. The most weakly flocculated particle network (oleic acid added, $V = -11kT$) showed an essentially incompressible behavior above 10 kPa. The compressive yield stress curve for the different particle networks showed a behavior that could be fitted to a modified power law

$$P_y = \frac{\sigma_0 \phi^n}{\phi_m - \phi} \tag{30}$$

suggested by Auzerais et al. [147]. The best fits for the different particle networks are shown in Fig. 24 with the fitting parameters tabulated in Table 1. A striking result from a comparison of the curves is that all systems

TABLE 1   Parameters with the Best Fit to the Stress-Density Results Shown in Fig. 24 Using an Empirical Expression of the Form $\sigma = \dfrac{\sigma_0 \phi^n}{(\phi_m - \phi)}$

Fatty acid	ϕ_m	n	σ_0(Pa)
Propionic	0.54	3.4 ± 0.2	2.8 ± 0.5 × 10^4
Pentanoic	0.595	3.0 ± 0.2	6.0 ± 1.4 × 10^3
Heptanoic	0.615	3.0 ± 0.2	5.2 ± 1.1 × 10^3
Oleic	0.63	3.4 ± 0.2	5.3 ± 1.1 × 10^2

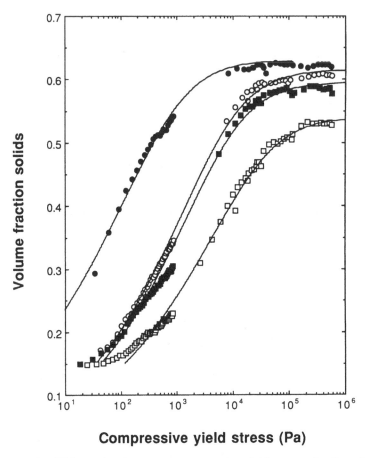

FIG. 24 Volume fraction versus compressive yield stress of a flocculated polydisperse alumina suspension in decalin with various magnitudes of the attractive interparticle energy: (□) V = -54kT; (■) V = -35kT; (○) V = -29kT; (●) V = -11kT. (Reprinted by permission of the American Ceramic Society.)

exhibit an asymptotic approach to a maximum volume fraction ϕ_m. It is apparent that ϕ_m is dependent on the magnitude of the attractive interparticle energy. The maximum volume fraction decreases with an increase in the magnitude of the attractive interparticle energy and varies between $\phi_m = 0.63$ ($V = -11kT$) and $\phi_m = 0.54$ ($V = -54kT$). The most weakly flocculated suspension (oleic acid added, $V = -11kT$) shows a maximum volume fraction close to random close packing of spheres ($\phi = 0.64$).

Buscall et al. [56] investigated the compressive yield properties of particle networks formed from salt-flocculated polystyrene suspensions of various particle sizes. Their data could also be fitted to a power law of the form

$$P_y \sim \phi^n \tag{31}$$

with $n = 4.0 \pm 5$, when the volume fraction was well above the critical volume fraction ϕ_c. They also investigated the particle size dependence of P_y and found that

$$P_y \sim a^m \tag{32}$$

with m between -2 and -3.

The compressive yield properties are also strongly dependent on the suspension structure. Mills et al. [119] showed that extensive preshearing of a flocculated suspension resulted in a much denser floc morphology. This was reflected in a decrease in the compressive yield stress by one order of magnitude.

In general, concentrated ceramic suspensions are subjected to compressive and shear stresses simultaneously in many forming operations. Attempts have been made to study the compaction and shear properties simultaneously on ceramic suspensions with the use of triaxial testers and extrusion tests [149–151]. Soil mechanical principles have been applied to analyze the results. However, most of the studies remain rather inconclusive regarding the effect of interparticle forces due to the use of poorly defined systems.

VIII. SUMMARY

The rheology of concentrated, colloidal suspensions displays a behavior ranging from solidlike, elastic properties to liquidlike, viscous properties depending on the range and magnitude of the particle interactions. There is an intimate link between the particle interactions, suspension structure, and rheological behavior that must be considered for a better understanding of the rheological behavior. This chapter has attempted to provide a qualitative introduction to the rheology and structure of concentrated suspensions with special emphasis on the effect of the nature of the particle interaction, repulsive or attractive, long-range or short-range.

REFERENCES

1. W. B. Russel, D. A. Saville, and W. R. Schowalter, *Colloidal Dispersions.* Cambridge University Press, Cambridge, 1989.

2. W. B. Russel, in *Ceramic Powder Science III* (G. L. Messing, S. -I. Hirano, and H. Hausner, eds.). American Ceramic Society, Westerville, Ohio, 1990, p. 361.
3. H. A. Barnes, J. F. Hutton, and K. Walters, *An Introduction to Rheology*. Elsevier, Amsterdam, 1989.
4. J. W. Goodwin, in *The Structure, Dynamics and Equilibrium Properties of Colloidal Systems* (D. M. Bloor and E. Wyn-Jones, eds.). Kluwer Academic, Amsterdam, 1990, p. 659.
5. E. C. Bingham, *Fluidity and Plasticity*. McGraw-Hill, New York, 1922.
6. H. Herschel and R. Buckley, *Proc. Am. Soc. Test. Mater.* 26(2): 621 (1926).
7. N. Casson, in *Rheology of Disperse systems* (C. C. Mill, ed.). Pergamon Press, London, 1959, p. 84.
8. N. Q. Dzuy and D. V. Boger, *J. Rheol.* 27: 321 (1983).
9. J. Mewis, *J. Non-Newtonian Fluid Mech.* 6: 1 (1979).
10. Th. F. Tadros, *Materials Sci. Forum 25-26*: 87 (1988).
11. G. Marin, in *Rheological Measurements* (A. A. Collyer and D. W. Clegg, eds.). Elsevier, London, 1988, p. 297.
12. H. C. Hamaker, *Physica 4*: 1058 (1937).
13. E. M. Lifshitz, *Sov. Phys. JETP 2*: 73 (1956).
14. D. Tabor and R. Winterton, *Proc. R. Soc. London A312*: 435 (1969).
15. J. N. Israelachvili and D. Tabor, *Prog. Surf. Membr. Sci.* 7: 1 (1973).
16. R. G. Horn and J. N. Israelachvili, *J. Chem. Phys.* 75: 1400 (1981).
17. L. Bergström, Colloidal processing of ceramics, thesis, Royal Institute of Technology, Trita-FYK 9203, 1992.
18. J. N. Israelachvili, *Intermolecular and Surface Forces*. Academic Press, London, 1983.
19. D. H. Napper, *Polymeric Stabilization of Colloidal Dispersions*. Academic Press, London, 1983.
20. Th. F. Tadros, ed., *The Effect of Polymers on Dispersion Properties*. Academic Press, London, 1982.
21. E. W. Fischer, *Kolloid Z. 160*: 120 (1958).
22. A. K. van Helden and A. Vrij, *J. Coll. Interface Sci. 78*: 312 (1980).
23. O. Kratky, *Progr. Colloid Polymer Sci. 77*: 1 (1988).
24. J. Moonen and A. Vrij, *Colloid Polym. Sci. 266*: 1140 (1988).
25. C. G. de Kruif, W. J. Briels, R. P. May, and A. Vrij, *Langmuir 4*: 668 (1988).
26. D. J. Cebula, J. W. Goodwin, G. C. Jeffrey, R. H. Ottewill, A. Parentich, and R. A. Richardson, *Faraday Discuss. Chem. Soc. 76*: 37 (1983).
27. M. Tence, J. P. Chevalier, and R. Jullien, *J. Physique 47*: 1989 (1986).
28. M. Suzuki, Y. Muguruma, M. Hirota, and T. Oshima, *Adv. Powder Tech. 1*: 115 (1990).
29. C. C. Donaldson, J. McMahon, R. F. Stewart, and D. Sutton, *Coll. Surfaces 18*: 373 (1986).
30. A. van Blaaderen, A. Imhof, W. Hage, and A. Vrij, *Langmuir 8*: 1514 (1992).
31. E. Dickenson, in *The Structure, Dynamics and Equilibrium Properties of Colloidal Systems* (D. M. Bloor and E. Wyn-Jones, eds.). Kluwer Academic, Amsterdam, 1990, p. 707.

32. H. A. Barnes, M. F. Edwards, and L. V. Woodcock, *Chem. Eng. Sci. 42*: 591 (1987).
33. P. N. Pusey and W. van Megen, *Nature 320*: 340 (1986).
34. W. B. Russel, *The Dynamics of Colloidal Systems.* University of Wisconsin Press, London, 1987.
35. S. Hachisu, Y. Kobayashi, and A. Kose, *J. Coll. Interface Sci. 42*: 342 (1973).
36. E. Dickenson, S. J. Milne, and M. Patel, *Ind. Eng. Chem. Res. 27*: 1941 (1988).
37. E. Dickenson, S. J. Milne, and M. Patel, *Powder Techn. 59*: 11 (1989).
38. J. W. Jansen, C. G. de Kruif, and A. Vrij, *J. Coll. Interface Sci. 114*: 481 (1986).
39. B. Vincent, J. Edwards, S. Emmett, and R. Croot, *Coll. Surfaces 31*: 267 (1988).
40. A. Kose and S. Hachisu, *J. Coll. Interface Sci. 55*: 487 (1976).
41. B. B. Mandelbrot, *The Fractal Geometry of Nature.* Freeman, New York, 1982.
42. F. Family and D. P. Landau, eds., *Kinetics of Aggregation and Gelation.* North Holland, Amsterdam, 1984.
43. P. Meakin, *Ann. Rev. Phys. Chem. 39*: 237 (1988).
44. D. Weitz and D. Oliveria, *Phys. Rev. Lett. 52*: 1433 (1984).
45. P. Meakin, *Adv. Coll. Interface Sci. 28*: 249 (1988).
46. P. Meakin, *Phys. Rev. Lett. 51*: 1119 (1983).
47. M. Kolb, R. Botet, and R. Jullien, *Phys. Rev. Lett. 51*: 1123 (1983).
48. M. Kolb and R. Jullien, *J. Phys. Lett.* (Paris) *45*: L977 (1984).
49. R. Jullien and P. Meakin., *J. Coll. Interface Sci. 127*: 265 (1989).
50. W. Y. Shi, I. A. Aksay, and R. Kikuchi, *Phys. Rev. A 36*: 5015 (1987).
51. S. A. Safran, I. Webman, and G. S. Grest, *Phys. Rev. A 32*: 506 (1987).
52. A. A. Seaton and E. D. Glandt, *J. Chem. Phys. 86*: 4668 (1987).
53. R. Buscall, P. D. A. Mills, and G. E. Yates, *Coll. Surf. 18*: 341 (1986).
54. L. Bremer, Fractal aggregation in relation to formation and properties of particle gels, thesis, Wageningen, 1992.
55. W. H. Shih, W. Y. Shih, S. I. Kim, J. Liu, and I. A. Aksay, *Phys. Rev. A 42*: 4772 (1990).
56. R. Buscall, P. D. A. Mills, J. W. Goodwin, and D. W. Lawson, *J. Chem. Soc., Faraday Trans. 1, 84*: 4249 (1988).
57. I. M. Krieger, *Adv. Colloid Interface Sci. 3*: 111 (1972).
58. R. L. Hoffman, *Trans. Soc. Rheol. 16*: 155 (1972).
59. B. J. Ackerson and N. A. Clark, *Physica 118A*: 221 (1983).
60. B. J. Ackerson, *J. Rheol. 34*: 553 (1990).
61. N. J. Wagner and W. B. Russel, *Phys. Fluids A, 2*: 491 (1990).
62. A. Einstein, *Investigation on the Theory of Brownian Motion.* Dover, New York, 1956.
63. G. K. Batchelor, *J. Fluid Mech. 83*: 97 (1977).
64. C. W. J. Beenakker, *Physica A, 128*: 48 (1984).
65. W. B. Russel and A. P. Gast, *J. Chem. Phys. 84*: 1815 (1986).
66. I. M. Krieger and T. J. Dougherty, *Trans. Soc. Rheol. 3*: 137 (1959).
67. D. E. Quemada, in *Lecture Notes in Physics Stability of Thermodynamic Systems* (J. Cases-Vasquez and J. Lebon, eds.). Springer Verlag, Berlin, 1982, p. 210.

68. C. G. de Kruif, E. M. F. van Iersel, A. Vrij, and W. B. Russel, *J. Chem. Phys. 83*: 4717 (1985).
69. J. C. van der Werff and C. G. de Kruif, *J. Rheol. 33*: 421 (1989).
70. D. A. R. Jones, B. Leary, and D. Boger, *J. Coll. Interface Sci. 147*: 479 (1991).
71. I. M. Krieger and M. Equiluz, *Trans. Soc. Rheol. 20*: 29 (1976).
72. M. D. Croucher and T. H. Milkie, *Faraday Discuss. Chem. Soc. 76*: 261 (1983).
73. G. N. Choi and I. M. Krieger, *J. Coll. Interface Sci. 113*: 101 (1986).
74. D. A. R. Jones, B. Leary, and D. V. Boger, *J. Coll. Interface Sci. 150*: 84 (1992).
75. J. Mewis, W. J. Frith, T. A. Strivens, and W. B. Russel, *AIChE J. 35*: 415 (1989).
76. W. Liang, Th. F. Tadros, and P. F. Luckham, *J. Coll. Interface Sci. 153*: 131 (1992).
77. H. J. Ploen and J. W. Goodwin, *Faraday Discuss. Chem. Soc. 90*: 77 (1990).
78. C. Prestidge and Th. F. Tadros, *J. Coll. Interface Sci. 124*: 660 (1988).
79. I. T. Kim and P. F. Luckham, *J. Coll. Interface Sci. 144*: 174 (1991).
80. R. Buscall, *J. Chem. Soc., Faraday Trans. 87*: 1365 (1991).
81. C. C. Furnas, *Ind. Eng. Chem. 23*: 1052 (1931).
82. R. K. McGeary, *J. Am. Ceram. Soc. 44*: 513 (1961).
83. G. L. Messing and G. Y. Onoda, Jr., *J. Am. Ceram. Soc. 61*: 1 (1978).
84. G. L. Messing and G. Y. Onoda, Jr., *J. Am. Ceram. Soc. 61*: 363 (1978).
85. F. Zok, F. F. Lange, and J. R. Porter, *J. Am. Ceram. Soc. 74*: 1880 (1991).
86. D. I. Lee, *J. Paint Technol. 42*: 579 (1970).
87. J. S. Chong, E. B. Christiansen, and A. D. Baer, *J. Appl. Polym. Sci. 15*: 2007 (1971).
88. S. C. Tsai, D. Botts, and J. Plouff, *J. Rheol. 36*: 1291 (1992).
89. P. A. Smith and R. A. Haber, *J. Am. Ceram. Soc. 75*: 290 (1992).
90. R. L. Hoffman, *J. Rheol. 36*: 947 (1992).
91. R. J. Farris, *Trans. Soc. Rheol. 12*: 281 (1968).
92. G. B. Jeffrey, *Proc. Royal Soc.* (London), *Ser. A, 102*: 161 (1922).
93. B. Clarke, *Trans. Instn. Chem. Engrs. 45*: 251 (1967).
94. H. A. Barnes, *J. Rheol. 33*: 329 (1989).
95. R. L. Hoffman, *J. Coll. Interface Sci. 46*: 491 (1974).
96. R. L. Hoffman, *Adv. Coll. Interface Sci. 17*: 161 (1982).
97. R. L. Hoffman, In *Theoretical and Applied Rheology* (P. Moldenaers and R. Keunings, eds.). Elsevier, Amsterdam, 1992, p. 607.
98. A. B. Metzner and M. Whitlock, *Trans. Soc. Rheol. 2*: 239 (1958).
99. W. H. Boersma, J. Laven, and H. N. Stein, *AIChE J. 36*: 321 (1990).
100. W. H. Boersma, P. J. M. Baets, J. Laven, and H. N. Stein, *J. Rheol. 35*: 1093 (1991).
101. W. H. Boersma, J. Laven, and H. N. Stein, in *Theoretical and Applied Rheology* (P. Moldenaers and P. Keunings, eds.). Elsevier, Amsterdam, 1992, p. 582.
102. H. M. Laun, *Suppl. Rheol. Acta 26*: 287 (1988).
103. H. M. Laun, R. Bung, and F. Schmidt, *J. Rheol. 35*: 999 (1991).
104. H. M. Laun, R. Bung, S. Hess, W. Loose, O. Hess, K. Hahn, E. Hädicke, R. Hingmann, F. Schmielt, and P. Lindner, *J. Rheol. 36*: 743 (1992).

105. L. V. Woodcock, *Chem. Phys. Lett. 111*: 455 (1984).
106. L. Marshall and C. F. Zukoski IV, *J. Phys. Chem. 94*: 1164 (1990).
107. P. D. Patel and W. B. Russel, *J. Coll. Interface Sci. 131*: 201 (1989).
108. D. Heath and Th. F. Tadros, *Faraday Discuss. Chem. Soc. 76*: 203 (1983).
109. R. Buscall, I. J. McGowan, and C. A. Mumme-Young, *Faraday Discuss. Chem. Soc. 90*: 115 (1990).
110. A. T. J. M. Woutersen and C. G. de Kruif, *J. Chem. Phys. 94*: 5739 (1991).
111. M. H. G. Duits, R. P. May, A. Vrij, and C. G. de Kruif, *Langmuir 7*: 62 (1991).
112. B. A. Firth and R. J. Hunter, *J. Coll. Interface Sci. 57*: 248 (1976).
113. B. A. Firth, *J. Coll. Interface Sci. 57*: 257 (1976).
114. B. A. Firth and R. J. Hunter, *J. Coll. Interface Sci. 57*: 266 (1976).
115. R. J. Hunter, *Adv. Coll. Interface Sci. 17*: 197 (1982).
116. R. C. Sonntag and W. B. Russel, *J. Coll. Interface Sci. 113*: 399 (1986).
117. R. C. Sonntag and W. B. Russel, *J. Coll. Interface Sci. 115*: 378 (1987).
118. A. A. Potanin, *J. Coll. Interface Sci. 145*: 140 (1991).
119. P. D. A. Mills, J. W. Goodwin, and B. V. Groover, *Colloid Polym. Sci. 269*: 949 (1991).
120. S. J. Willey and C. W. Macosko, *J. Rheol. 22*: 525 (1978).
121. B. V. Velamakanni, J. C. Chang, F. F. Lange, and D. S. Pearson, *Langmuir 6*: 1323 (1990).
122. H. A. Barnes and K. Walters, *Rheol. Acta 24*: 323 (1985).
123. J. P. Hartnett and R. Y. Z. Hu, *J. Rheol. 33*: 671 (1989).
124. R. Buscall and I. J. McGowan, *Faraday Discuss. Chem. Soc. 76*: 277 (1983).
125. E. van der Aerschot and J. Mewis, *Coll. Surf. 69*: 15 (1992).
126. R. Buscall, P. D. A. Mills, and G. E. Yates, *Coll. Surf. 18*: 341 (1986).
127. A. E. James, D. J. A. Williams, and P. R. Williams, *Rheol. Acta 26*: 437 (1987).
128. A. S. Yoshimura, R. K. Prudhomme, H. M. Princen, and A. D. Kiss, *J. Rheol. 31*: 699 (1987).
129. M. Keentok, *Rheol. Acta 21*: 325 (1982).
130. W. J. Frith, T. A. Strivens, and J. Mewis, *J. Coll. Interface Sci. 139*: 55 (1990).
131. J. W. Goodwin, T. Gregory, and J. A. Stile, *Adv. Coll. Interface Sci. 17*: 185 (1982).
132. R. Buscall, J. W. Goodwin, M. W. Hawkins, and R. H. Ottewill, *J. Chem. Soc. Faraday Trans. 1, 78*: 2873 (1982).
133. J. C. van der Werff, C. G. de Kruif, C. Blom, and J. Mellema, *Phys. Rev. A, 39*: 795 (1989).
134. Th. F. Tadros, *Langmuir 6*: 28 (1990).
135. R. C. Sonntag and W. B. Russel, *J. Coll. Interface Sci. 116*: 485 (1987).
136. M. Chen and W. B. Russel, *J. Coll. Interface Sci. 141*: 564 (1991).
137. M. van den Tempel, *J. Colloid Sci. 16*: 284 (1961).
138. I. C. Callaghan and R. H. Ottewill, *Disc. Faraday Soc. 57*: 110 (1974).
139. R. J. R. Cairns, R. H. Ottewill, D. W. J. Osmond, and I. Wagstaff, *J. Coll. Interface Sci. 54*: 45 (1976).
140. A. Homola and A. A. Robertson, *J. Coll. Interface Sci. 54*: 286 (1976).
141. S. Rohrsetzer, P. Kovacs, and M. Nagy, *Coll. Polym. Sci. 264*: 812 (1986).

142. J. W. Goodwin, R. H. Ottewill, and A. Parentich, *Coll. Polym. Sci. 268*: 1131 (1990).
143. B. A. Costello, I. T. Kim, and P. F. Luckham, *J. Chem. Soc., Faraday Trans. 86*: 3693 (1990).
144. I. T. Kim and P. F. Luckham, *Coll. Surfaces 68*: 243 (1992).
145. R. Buscall, *Coll. Surfaces 5*: 269 (1982).
146. R. Buscall and L. R. White, *J. Chem. Soc., Faraday Trans. 1, 83*: 873 (1987).
147. F. M. Auzerais, R. Jackson, W. B. Russel, and W. F. Murphy, *J. Fluid Mech. 221*: 613 (1990).
148. L. Bergström, C. H. Schilling, and I. A. Aksay, *J. Am. Ceram. Soc. 75*: 3305 (1992).
149. C. P. Wroth and G. T. Houlsby, in *Ultrastructure Processing of Ceramics, Glasses, and Composites* (L. L. Hench and D. R. Ulrich, eds.). John Wiley, New York, 1984, p. 448.
150. M. A. Janney and G. Y. Onoda Jr., in *Ceramic Powder Science* (G. L. Messing, K. L. Mazdiyasni, J. W. McCauley, and R. A. Haber, eds.). American Ceramic Society, Westerville, Ohio, 1987, p. 615.
151. C. H. Schilling, Plastic shaping of colloidal ceramics, Ph.D. thesis, University of Washington, 1992.

6

Surface Chemistry in Dry Pressing

ELIS CARLSTRÖM   Swedish Ceramic Institute, Gothenburg, Sweden

I. THE PRESSING PROCESS

Forming by pressing consists of granulation of the powder, filling of the mold, pressing to a compact, and ejection of the compact from the mold. Often a final step of removing the organic binder has to be added.

Fine ceramic powders have to be granulated in order to be handled in an industrial pressing process. A finely ground powder sticks to the hopper and the die, resulting in clogging and variations in the powder density of the filled mold. Granulation, for example by spray drying, is a process that forms spherical powder agglomerates (granules) of the fine powder particles. The granules are kept together by addition of an organic binder or by the attractive van der Waals forces between the individual particles.

When the mold is filled, the granules are compacted by the application of pressure. The pressure can be applied by a die (die pressing) or by an isostatic pressure (isostatic pressing). The isostatic pressure is often transmitted by sealing the powder in a flexible container and applying a liquid pressure to the container. During the compaction stage the granules break down or deform and the powder is compacted. Pressures from 50 to 200 MPa are often used. After pressing, the pressed body typically has a compact density of 50–65% of the theoretical density of a pore-free material. Compact density can vary greatly, depending on particle size distribution and shape of the powder.

Prior to granulation, the powders often have to be ground, and several different powder components may have to be mixed. This is often done in a liquid suspension of the powder. A suspension is also used for granulation by spray drying and freeze granulation. In all these steps, surfactants can be added to control dispersion and rheology of the suspensions. Organic binders and lubricants with varying surface activity are also often added. These influence the behavior of the suspension as well as the properties of the granules, and thus the pressing behavior of the powder.

The dispersants, lubricants, and binders also influence the mechanical properties of the pressed body. The pressed body needs sufficient mechanical strength to be handled until it is sintered. Prior to sintering, the organic additives have to be removed. Preferably, they should break down on heating without leaving any residues.

Large, spherical, monosized granules can easily flow into the die and distribute themselves evenly. The granules are not allowed to fracture at this stage, which requires sufficient strength to maintain them. During the actual pressing, the agglomerates have to be broken down. The individual particles have to be able to slide past each other, and the boundaries between the granules should be eliminated at this stage. The forces that hold the granules together must be sufficiently low to allow this.

Lubrication of the particle die wall contact is desired in die pressing. Friction against the die wall obstructs the transmission of the die pressure throughout the pressed body. This produces variation in the density of the pressed body. Lubrication is often provided by organic additives.

When the powder compact is ejected, it expands. This phenomenon is often called springback and is partly attributable to recovery of the elastic deformation of the particle-particle contacts. Additives such as lubricants, dispersants, and binders can also compress elastically during pressing. When the compact is ejected they will expand and contribute to the springback. Large springback can cause cracks and other defects in the pressed body.

If the granule boundaries are not eliminated during pressing, they will remain in the sintered compacts, where they will function as fracture origins and reduce the strength of the final material [1]. If instead the granules break down too quickly, they will cause clogging and uneven filling of the die. The rearrangement of powder is limited during pressing. Uneven filling of the mold results in uneven densities after sintering. The properties of the granules are largely influenced by the colloid chemistry of the system. The degree of dispersion influences such properties of the granules as homogeneity, density, and porosity distribution. These properties are fundamental to the mechanical properties of the granules.

Polymers can be used both as dispersants and as thickeners. As thickeners they increase the viscosity and pseudoplasticity of a liquid medium. The degree of dispersion and the viscosity of the liquid are important factors in controlling the plasticity of a powder-liquid system. Granules often contain residual moisture and can exhibit a plastic behavior. This plasticity is controlled by the addition of dispersants, binders, and lubricants, as well as by the residual moisture.

Lubrication of the powder body against the die wall reduces the friction that obstructs compaction near the die walls. Particle-particle lubrication facilitates flow during pressing. Lubrication is a surface phenomenon and can be enhanced by adsorption. Consequently, the adsorption behavior of the dispersants, binders, and lubricants is important to their behavior during pressing. Adsorption also influences the migration of organic additives that can take place during spray drying or mechanical granulation.

A. Mechanical Granulation

Powders can be granulated from a semidry state in a number of ways. In sieve granulation, the granules are produced by mixing with additives, drying on a pan, and mechanical granulation by forcing the powder through a sieve. The powder is first mixed with the organic additives in a liquid. The mixing can be done semidry in a kneading type of mixer or wet in a ball mill or stirrer. After wet mixing, the powder is often dried in open pans or in a rotary evaporator. The powder cake is then broken up on a screen. The granules pass through the screen and a granulate is collected.

On a larger industrial scale, continuous agglomeration can be achieved by spraying only the amount of moisture needed for agglomeration into a rotary agglomerator. The mixing can also be done batchwise by wetting to a slight excess and then redrying. These methods require intensive mechanical mixing and mechanically forcing the powder through a screen or a sieve.

Sieve granulation is a straightforward method of granulation. It can be done without any special knowledge of the surface chemistry of the powder. The main disadvantage of this method is the risk of metal contamination from the screen. This risk is always present with hard ceramic particles. The resulting granules also have rough shapes and wide size distributions, which may be a disadvantage. The advantages are energy saving (less drying) and no need for dispersants.

In high-performance ceramic processing, the main use of sieve granulation is in the laboratory or for small-scale production. In these cases, the capital cost of a spray drier or the effort required to develop spray drying cannot be justified. In the traditional ceramics field the main method of granulation has been spray drying. Increasing energy costs have lately made the semidry granulation methods more attractive.

B. Spray Drying

Production scale granulation is often done by spray drying [2] (Fig. 1). Prior to spray drying, the powder has to be dispersed in a well-concentrated slurry, usually containing some pressing additives. The pressing additives function as a binder for the granules and often as a lubricant during pressing.

Low viscosity slurry at high powder concentration is necessary to ensure high-density granules. Owing to the high shear forces in the nozzles used in spray drying, a shear thickening slurry cannot be used [3]. While high-density granules are often desired for efficient compaction, too-high-density granules can also be a disadvantage (see below).

The ideal slurry for spray drying is well dispersed and contains a high concentration of solids. This makes it possible to produce high-density granules. It also makes the problem of hollow granules less pronounced. The problems of dispersion are explained in detail elsewhere in this book. In spray drying, dispersion has to take place in the presence of an organic binder/lubricant. It is necessary to ensure that any dispersants and other organic additives are compatible. Competitive adsorption can sometimes cause problems. The binders often increase the viscosity of the liquid. A compromise often has to be made between the viscosity increase of the liq-

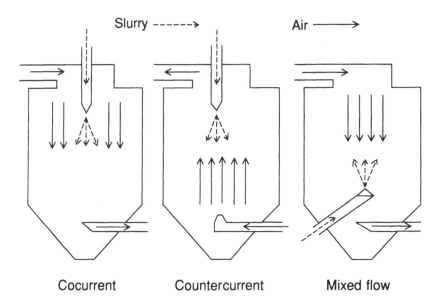

FIG. 1. Common flow patterns of spray drying.

uid and the amount needed to increase the granule strength and/or lubrication.

1. Formation of Hollow Granules

There is a tendency for hollow granules to be formed during spray drying. There are five possible mechanisms that cause formation of large pores in the granules [2]:

1. A low-permeability elastic film forms around the droplet. Reduced evaporation rates result in increased droplet temperatures. The interior moisture vaporizes and causes "ballooning."
2. For soluble salts, the liquid flows to the droplet surface where evaporation and salt crystallization occur. The rate of evaporation exceeds the diffusion rate of salt back into the droplet interior, and internal voids are formed.
3. For slurries of insoluble solids, as liquid flows to the droplet surface under capillary action, particles are carried along. The liquid evaporates, and internal voids in the granule result.
4. The presence of occluded air in the slurry.
5. Too-high drying temperatures that cause boiling inside the droplets.

The third mechanism can, for example, be found in clay-based ceramic systems. In these systems, there is a large particle size difference between the small clay particles and the feldspars and quartz. In high-strength ceramics, this tendency is minor, as the particle size differences are usually smaller.

If all the film-forming mechanisms are absent, a slurry with a high concentration of solids will give dense granules. Particle and salt migration will be minimized, as the amount of water that has to be removed is decreased.

Organic binders used as pressing additives will migrate during drying, especially if they are water soluble. Binder migration might cause hollow granules, due to the mechanisms mentioned above. Binder migration also results in a binder-rich outer layer of the granule. This layer will be stronger and will resist deformation and fracture during pressing. This can give rise to granule structures in the green body, resulting in defects in the sintered body. If most of the moisture is evaporated from the interior of the granule, the binder migration is minimized (Fig. 2).

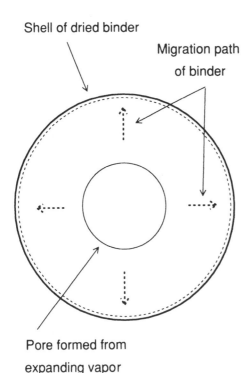

FIG. 2. Migration of binder results in the formation of an outer shell that traps vapor and results in large internal voids.

2. Liquid Transport During Spray Drying

During drying of the granule, three separate phases can be detected (Fig. 3). After a short initial period (A-B), the granule reaches its maximum evaporation rate. The period that follows is called the constant rate drying period (B-C). During this period, the surface is saturated with moisture by liquid flow from the interior of the granule. The temperature of the granule is low and constant during this time. The duration of the constant rate period will be determined by the moisture content, the viscosity of the liquid, and the temperature and humidity of the drying air. When saturation of moisture can no longer be maintained on the surface of the granule, the falling rate period ensues (C-D). The droplet temperature begins to rise as the drying rate decreases. After some time, the entire surface is unsaturated (D), and the drying rate will be controlled by the permeability of the dried solid crust that develops, forming inwards from the surface.

Moisture migration during the constant rate period can be understood by approximating it to liquid flow through a capillary [2]. The pressure drop P across a length L at a linear flow rate dl/dt of the liquid is given by

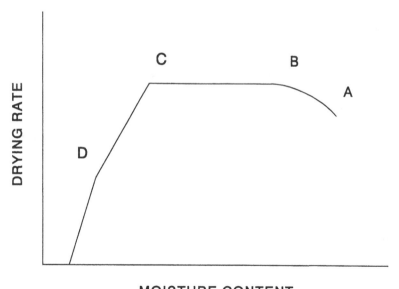

MOISTURE CONTENT

FIG. 3. Evaporation behavior of a droplet injected into the drying chamber of a spray dryer.

$$\frac{d\mathrm{l}}{d\mathrm{t}} = \frac{r^2 P}{8\eta L} \qquad (1)$$

where r is the tube radius and η the liquid viscosity. The pressure drop due to the capillary forces is

$$P = P_0 = \frac{2\gamma \cos \theta}{r} \qquad (2)$$

where γ is the surface tension of the liquid and θ the contact angle. Combining these two equations we get a linear flow rate of

$$\frac{d\mathrm{l}}{d\mathrm{t}} = \frac{r\gamma \cos \theta}{4\eta L} \qquad (3)$$

We find that the liquid flow rate is a function of surface tension, contact angle, liquid viscosity, and radius of the passageway. This is a very rough approximation of the actual situation. The diameter of the passageway is hardly constant and changes with time.

The viscosity of the liquid may be the most easily influenced parameter. However, the viscosity of the liquid that flows through the pores of a drying granule must not be confused with the viscosity of the slip. Adding a polymeric binder that dissolves in the slip liquid increases the liquid viscosity (as well as affecting the slip viscosity). Using a poorly dispersed slip with a high slip viscosity does not affect the liquid viscosity. On the contrary, the drying granule's permeability to liquid flow is strongly influenced by the stability of the slip. The permeability of a drying flocculated slip (prepared using small particles) is significantly higher than that of a well dispersed slip. This has been shown by measurements of the permeability of the cast body during slip casting experiments [4]. Consequently, a well-dispersed highly concentrated slip has the shortest drying rate period and the least tendency of binder migration.

3. Adsorption of Binders

Binder migration during spray drying can be minimized by using an adsorbed pressing aid. As the pressing aid is adsorbed, it does not migrate with the liquid during the constant rate drying period. The adsorbed binder is evenly distributed along the particle surfaces. If it functions as a lubricant, this will improve lubrication.

This mechanism was tested by adsorbing a latex onto alumina particles [5]. A latex is a stable dispersion of polymeric particles in water. This particular latex consisted of negatively charged acrylic particles. It was stable in a wide pH range and retained its charge independent of the pH value owing to the sulfate groups at the surface.

The latex was added to alumina suspension at pH 11. At this pH, both the alumina and the latex particles are negatively charged. The pH was then lowered to pH 4. The latex maintained its negative charge while the alumina reverted to a positive charge at this pH. This resulted in an adsorption of the latex particles on the positively charged alumina particles (Fig. 4).

Two slurries, one with adsorbed latex binder and the other with a conventional polyvinyl type of binder, were spray dried. The spray dried granules were compared in the scanning electron microscope. Samples with the adsorbed latex binder exhibited a dramatic reduction in the number of large pores (often observed as hollow spheres and doughnut-shaped granules) in the granules, as compared with samples with a conventional polyvinyl alcohol type of binder.

The latex binder also enhanced the pressing properties of the granulate. The green densities were higher and the ejection force lower. The

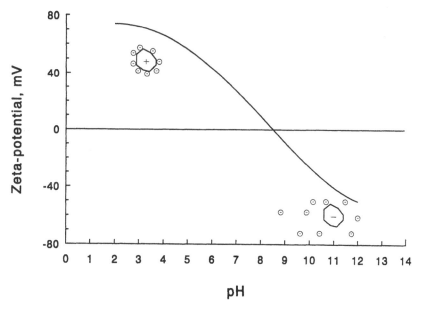

FIG. 4. Adsorption of charged latex particles onto alumina particles.

improved properties of the granulate from using an adsorbed latex could also be observed as an increase in the final sintered density of the ceramic.

C. Freeze Granulation

One alternative to spray drying is freeze granulation. A slurry is sprayed as small droplets into liquid nitrogen. The droplets freeze instantaneously to granules. The frozen granules are then dried, without thawing, by freeze drying. The frozen liquid is removed from the granules by sublimation. This means that no transport of either soluble components or small powder particles takes place. This improves the homogeneity of the granulate, and the formation of hollow granules in spray drying is avoided. In drying from the liquid state, the liquid creates a capillary pressure that makes the granules shrink during drying. This shrinkage is not present in freeze drying. Consequently, freeze dried granules usually have a lower density than the corresponding spray dried granules.

Freeze drying is a common method of drying powders that have been synthesized using wet chemical methods. A freeze dried powder cake can be broken down and used as it is. The granules formed in this way are irregular in shape and size and do not flow freely. Freeze drying can, however, also be used to dry granules that have been prefrozen by spraying into liquid nitrogen. Prochazka [6] produced a silicon carbide material by both freeze granulation and spray drying. By using freeze granulation he improved the homogeneity of the green bodies and reached higher strength values of the sintered bodies. Müller-Zell and Hahn [7] compared spray drying, fluid bed granulation, and freeze drying of silicon carbide powders. Spray drying and freeze drying gave similar green densities and sintered microstructures. Fluid bed granulation gave a lower green density and a rougher microstructure. Spray drying and freeze drying gave the same green density after pressing. However, the freeze granulated powder had a lower granule density (and consequently lower fill density). In other words, the compaction ratio was higher for the freeze granulated powder. Sintered material from the freeze granulated powder had the highest strength and Weibull modulus. A higher compaction ratio can, however, be a disadvantage. In die pressing, the die has to travel a longer path, and this increases the density variation due to wall friction. Freeze granulation has also been used in a few other systems, for example to granulate and add sintering additives to silicon nitride [8].

Freeze granulation makes it possible to use a wide range of pressing additives as sintering additives. This is because there is no problem with transport of soluble binders or sintering additives.

D. Evaluation of Pressing Behavior

To be able to compare the efficiency of pressing additives, we need methods for study pressing efficiency and quality. There are parameters that can be measured on the granule, such as tap density, granule density, and pore size distribution of the granule. The mechanical behavior of single granules under compressive stresses can also be measured. Pressed green bodies can be studied by measurement of strength and observation of microstructure.

1. Die Pressing

Compaction behavior can basically be studied in two ways. One is to press to a fixed pressure and measure the density after ejection of the compact. By repeating this procedure for a number of pressures a diagram that shows compact density against pressure can be plotted [9]. This method requires a number of experiments and may not show what actually happens during the intermediate stages of pressing due to effects of the ejection. Another way is to monitor the displacement of the ram and the force required during pressing. Using this method it is possible to generate continuous compaction curves [10]. To make accurate measurements with this method at higher pressures, it is important to compensate for the compliance of the test apparatus. The compliance of the die and the compact can usually be considered negligible. The compliance of the load cell must be compensated for [11].

When density of a compact is plotted against the logarithm of pressure, a diagram with essentially straight lines is obtained. The first region of the curve has a lower slope. At a certain pressure there is a transition to a straight line with a higher slope. The *break point in the curve* (the intersection of two straight lines fitted to the two straight parts of the curve) is often called the *apparent yield point.* It corresponds to the pressure where the granules start to deform (Fig. 5).

The apparent yield point is a measure of the hardness of the granules. A low or nonexistent break point corresponds to very soft granules.

The slope of the curve has been used to characterize the efficiency of the pressing. A low slope means that more pressure must be used to increase the density. A dense granulate starts at a high density and might have a low slope on the pressing curve and still reach the same density as a less dense granulate for a certain pressure [12].

A second break point is sometimes reported on compaction diagrams. This type of response comes from curves not corrected for elastic compliance. Measurements on samples after ejection at high pressures show that only modest increases in density can be obtained by increasing the pressure in the higher range. Samples pressed above a certain pressure (often called

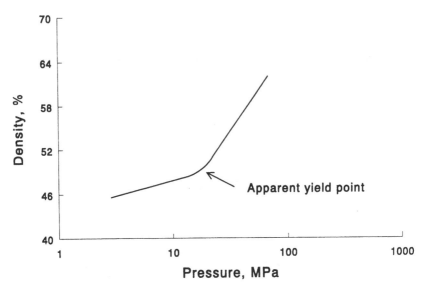

FIG. 5. Compaction response of a granulate.

the joining pressure) show only small increases in compaction. Curves measured with a combination of continuous measurements and measurement after ejection also show a high pressure break point due to the change in method.

2. Isostatic and Triaxial Compaction

No shear (on a macroscopic scale) takes place in isostatic compaction. This makes it more difficult to eliminate all traces of granule boundaries during isostatic pressing and thus to develop granulates and pressing additive formulations for isostatic pressing by testing with die pressing. Compaction curves can be generated for isostatic pressing as well as for die pressing. Continuous isostatic compaction curves can be obtained in the ceramics laboratory by using a mercury porosimeter [13].

Soil mechanics provide further techniques that can be applied to the study of powder pressing [3,14]. Granulates and dry powders can be studied in direct shear tests. Compaction of the powder can be studied in triaxial tests. In a triaxial test, the sample is contained in an elastic membrane. A biaxial force is applied by two pistons. Simultaneously, an isostatic pressure can be applied to the membrane using a liquid. The biaxial and the isostatic stresses can be combined in a number of ways in triaxial testing. By compacting the sample biaxially at a constant isostatic pressure, the internal friction of the powder can be measured. When the isostatic pressure is increased, the porosity decreases. The particles start to interlock and the internal friction increases.

3. Evaluation by Sintering

High density after pressing or low end pressure to achieve a certain density is not the ultimate goal in pressing. To make a strong ceramic the compact has to able to sinter to a high density *and* arrive at this density with as few defects as possible. Some of the conditions cannot be estimated with measurements of density alone. Density gives an average measure of porosity in the compact. However, pores above a certain size do not shrink upon sintering. They remain as possible fracture origins in the finished material. Also, agglomerates that remain in the green body shrink differentially, thus causing cracks and/or leaving pores. The reason for this is the different density of the agglomerate that causes it to shrink either more or less than the matrix around it.

A green microstructure with small variations in the pore size is more important than the absolute value of the porosity. Even so, a high density of the green compact lowers the general pore size and minimizes shrinkage. This is often necessary to enable sintering and eliminate warping problems caused by large shrinkages. A test of the sintering properties of a compact can be used to distinguish the pressing quality of pressed samples. The final test, however, is strength measurement of the samples. Measurement of the strength and the statistical scatter of the strength are the most sensitive tests of defects caused by pressing. In practice, this requires several other variables to be kept constant and is also time consuming. A practical method for developing new pressing additives would be to start with measurements during pressing and on the green body. Only when these are successful is there a need to extend the testing to sintering and strength evaluation.

II. EFFECT OF BINDERS AND LUBRICANTS ON GRANULATION

A. Dispersion and Viscosity

A slurry containing powders, binders, and lubricants has to be well dispersed to be spray dried. The liquid is often water but can also be an organic liquid. Dispersion is described in detail in Chapter 4. When adding a binder we have to consider its effects on viscosity.

B. Viscosity Effects of Organic Additives

Organic binders are often polymers. A long molecule can have a significant influence on viscosity depending on its configuration. In a stretched straight configuration it will increase the viscosity. If the molecule curls it has a much smaller influence on viscosity [15].

Vinyl acrylic polymers have flexible backbones (Table 1). Owing to the possibilities of rotation of the carbon-carbon bond, they can curl. Carbohydrate binders have a less flexible structure (Table 2).

For organic binders to be water soluble, they have to contain polar groups. Even when the polymers contain polar groups, they often develop intermolecular bonds that make them insoluble. This can be avoided by attaching side groups to the polymer. These often bulky side groups inhibit the intermolecular bonds and make solvation possible. A common example of this is poly(vinyl alcohol). Without side groups it is insoluble in water. In practice, approximately 12 to 20% of the –OH side groups are replaced by acetate groups to make it soluble in water. Cellulose is modified in the same manner by incorporating hydroxyethyl, methyl, carboxymethyl, or hydroxypropyl side groups.

The rheology of a polymer in water is controlled by its configuration. High shear tends to align the polymer chains with the shear direction, thus reducing the viscosity. If this happens, a pseudoplastic behavior can be observed, i.e., viscosity decreases with increasing shear rate. If the shear is stopped, it will take some time before a random distribution of the polymer chains reappears. Thus organic binder solutions can show a time dependent viscosity behavior (thixotropy).

Furthermore, a solution of a binder can gel. Gels can develop during heating or cooling of a solution (thermal gelation) or through addition of a chemical gelling agent (chemical gelation). Bonding between the polymer molecules is responsible for gelation. The thermal types of gelation are often reversible, while chemical gelation is often irreversible.

Binders like poly(vinyl alcohol), poly(ethylene oxide), and poly(vinylpyrrolidone) are nonionic. Poly(ethylene imine) is cationic, and acrylics and alginates are anionic. The electric charge behavior of the powder in water will determine the adsorption of cationic and anionic binders in the same manner as when similar polymers are used as dispersants. Nonionic polymers also can adsorb on the powder, but this interaction is more difficult to predict (see the previous chapter).

The binders can also be classified according to their effect on viscosity of a water solution. This effect is, of course, dependent on concentration. The effect on viscosity can be observed as the slope of the viscosity vs. the con-

TABLE 1 Formulas for Some Synthetic Water-Soluble Binders

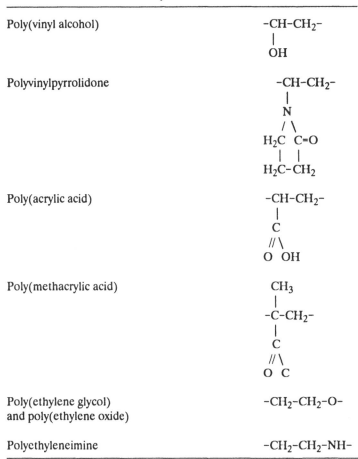

Poly(vinyl alcohol)	$-CH-CH_2-$ $\mid$ OH
Polyvinylpyrrolidone	$-CH-CH_2-$ $\mid$ N $/ \ \backslash$ $H_2C \ \ C=O$ $\mid \ \ \mid$ H_2C-CH_2
Poly(acrylic acid)	$-CH-CH_2-$ $\mid$ C $// \backslash$ $O \ \ OH$
Poly(methacrylic acid)	CH_3 $\mid$ $-C-CH_2-$ $\mid$ C $// \backslash$ $O \ \ C$
Poly(ethylene glycol) and poly(ethylene oxide)	$-CH_2-CH_2-O-$
Polyethyleneimine	$-CH_2-CH_2-NH-$

centration curve of the polymers. Examples of low viscosity polymers in this respect are gum arabic, lignosulphonates, and dextrins. Low-to-medium viscosity polymers include poly(vinyl alcohol)s, poly(vinylpyroolidone)s, poly(ethylene oxide)s, acrylics, and poly(ethylene imine)s. Medium-to-high viscosity polymers include celluloses and alginates.

1. A Practical Example

Several of the above-mentioned effects can be observed in practice when binders are added to a spray drying slurry. Hahn and Müller-Zell [16]

TABLE 2 Some Water-Soluble Carbohydrate-Derived Binders

Binder	Side groups
Methyl cellulose	$-CH_2-O-CH_3$
Hydroxypropylmethyl cellulose	$-CH_2-O-CH_2-CH-CH_3$ $\qquad\qquad\quad \|$ $\qquad\qquad\quad OH$
Hydroxyethyl cellulose	$-CH_2-O-C_2H_4-O-C_2H_4-OH$ $-CH_2-O-C_2H_4-OH$
Sodium carboxymethylcellulose	$\qquad\qquad\qquad\quad O$ $\qquad\qquad\qquad\quad \|\|$ $-CH_2-O-CH_2-C$ $\qquad\qquad\qquad\quad \backslash$ $\qquad\qquad\qquad\qquad ONa$
Starches and dextrins	$-CH_2-OH$
Sodium alginate	$\quad O$ $\quad \|\|$ $-C$ $\quad \backslash$ $\qquad ONa$
Ammonium alginate	$\quad O$ $\quad \|\|$ $-C$ $\quad \backslash$ $\qquad ONH_3$

tested 45 binders for spray drying of a porcelain slurry and had to reject two-thirds of them. They describe the reasons for this as "uncompensatable changes of the rheological properties (excessive viscosity, thixotropy, rheopexy), insufficient solid content, coagulation, foaming ..." They could not use acrylics, alginates, or starches at all. Certain poly(vinyl alcohol)s, poly(acrylnitrile)s, poly(ethylene oxide)s, and dextrins were usable, as were celluloses of all types. They found that the celluloses increased the viscosity of the slurries, and thus the solid contents of these slurries had to be reduced. Poly(vinyl alcohol), poly(acrylnitrile), poly(ethylene oxide), and dextrin did not influence viscosity.

The results above can probably be explained from knowledge of the binders used. There are too few details in the report, but some conclusions can be drawn. Acrylics and alginates are anionic polymers. The rest of the polymers are nonionic. A porcelain slurry usually contains quartz (with a low isoelectric point) and often alumina (with a high isoelectric point). The slurry is described as ready to spray before the addition of the binder, which means that a dispersant has been added. For a porcelain slurry this is usually sodium carbonate, sodium silicate, or an acrylic surfactant. Addition of an anionic polymer to a slurry like this can be expected to cause problems such as bridging flocculation and selective and competing adsorption. A nonionic polymer might compete in adsorption with the dispersant, but the electrostatic adsorption of the surfactant will probably be stronger in a water system. Modified cellulose has the rigid backbone of cellulose and has a higher viscosity in a pure water solution. Therefore it can be expected to give a greater increase in the viscosity of the slurry.

2. Emulsions and Latexes

Organic binders that are insoluble in water can be added to water slurries as emulsions or latexes. Nonpolar substances such as waxes can be dispersed in water by adding an emulsifying agent. Wax emulsions are commonly used for binder applications. A latex usually consists of a stable concentrated polymer dispersion in water. If no heteroflocculation occurs when mixing the latex with the powder, the latex addition should have very little effect on the viscosity of the slurry.

Emulsions and latexes provide the possibility of adding water-insoluble polymers to water slurries. Water-insoluble polymers are not plasticized by water. That means that the granulate is insensitive to the ambient humidity. The polymers used for this purpose usually have low glass transition temperatures, as they will not be further plasticized.

C. Adsorption of Binders

Frisch, Thiele, and Kirsch [17] studied the effects of surface active additives on pressing. By adding a surfactant to a porcelain slurry they improved the pressing properties and the green strength. The porcelain slurry contained clays and feldspar but no silica or alumina, and the added surfactant was tetraethylene pentamine (TEPA). When part of the water was substituted by TEPA, the pressure passage, i.e., the pressure difference between the ram and the middle of the die, decreased. Adsorption of TEPA is explained by hydrogen bonding between the amine groups and the silanol groups on the kaolin. The silanol groups are shielded and the kaolin becomes hydrophobic, repelling the water. The multipolar amine can form bridges between two kaolin particles, adsorbing on both of them. In this way TEPA increases the green strength of compacts. It could be demonstrated by observing the strength at a constant bulk density that the strength increase was not only an effect of increased granule density.

III. EFFECT OF ADDITIVES DURING PRESSING

A. Plasticity

When organic binders are added during granulation, they should make the granules plastic rather than brittle. Water-soluble binders such as poly(vinyl alcohol) are plasticized by water. Completely dry granules with pure poly(vinyl alcohol) are brittle. With increasing amounts of residual moisture they are plasticized. Brittle granules reach lower densities during pressing. The remaining granule boundaries found in such materials are fracture origins and reduce the strength significantly.

Poly(vinyl alcohol) is hygroscopic. Moisture works as a plasticizer and lowers the glass transition temperature T_g. Below the glass transition temperature the polymer is a brittle material, while above it, it softens and can be deformed plastically [18]. A granulate with added poly(vinyl alcohol) binder will change properties depending on the moisture level of the ambient air during storage.

DiMilia and Reed [19] studied alumina granulates with added poly(vinyl alcohol) binder. They measured moisture adsorption isotherms for the granulate. Moisture adsorption rises with increasing relative humidity (Fig. 6). When granules containing 2.3 wt % binder were pressed, the powder compaction diagrams showed a change in pressing behavior with increasing humidity. When the binder content was decreased to 0.95 wt %, the change in pressing behavior was minimal. The reason for this is probably that the binder content is insufficient to cover the surfaces of the powder completely (Fig. 7). The apparent yield point is shifted to lower pressures by an increase in humidity. If the apparent yield point is plotted against the H_2O/binder ratio, a sharp transition is observed in the 0.1–0.2 ratio range (Fig. 8). A moisture level of 8–9 wt % is needed to reach a glass transition temperature of 24°C [20].

Poly(vinyl alcohol) can be plasticized by other compounds, such as glycerine, poly(ethylene glycol), and poly(propylene glycol). Poly(ethylene glycol) acts as a plasticizer for poly(vinyl alcohol) by penetrating the PVA molecules and forcing the secondary (hydrogen) bonds between the chains

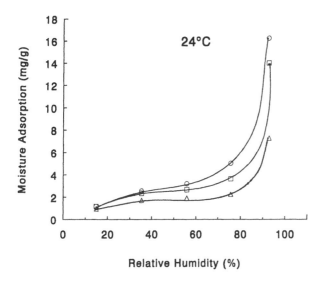

FIG. 6. Adsorption isotherms for spray dried powders (mg H_2O per g sample). Alumina (ALCOA SG A16) with 1 (△), 2.2 (□), and 2.7 (○) wt % poly(vinyl alcohol). [From R. A. DiMilia and J. S. Reed, Dependence of compaction on the glass transition temperature of the binder phase, *Am. Ceram. Soc. Bull.* 62(4): 484–488 (1983), Figure 3. Reprinted by permission of the American Ceramic Society.]

apart. This increases the freedom of movement of the polymer chains and lowers the glass transition temperature.

Nies and Messing [21] studied the influence of poly(ethylene glycol) on poly(vinyl alcohol). The decrease in glass transition temperature is rapid initially, and less pronounced at higher poly(ethylene glycol) concentrations (Fig. 9). The apparent yield point during pressing of an alumina granulate increased with increasing glass transition temperature of the binder (Fig. 10). The green density achieved during pressing was considerably higher for plasticized binder than for binder without addition of poly(ethylene glycol). Increasing the pressing temperature is another way of making the binder softer. The effect of temperature on pressed density was, however, much less than the effect of added plasticizer (Fig. 11).

Moser, Reed, and Varner [22] studied the effect of granule size on the microstructure of green compacts. They found that granules above 106 μm in size gave rise to intergranular porosity in the green compacts. They also measured the green strength of the compacts and found a lower strength and a higher standard deviation in strength (lower Weibull modulus) for the

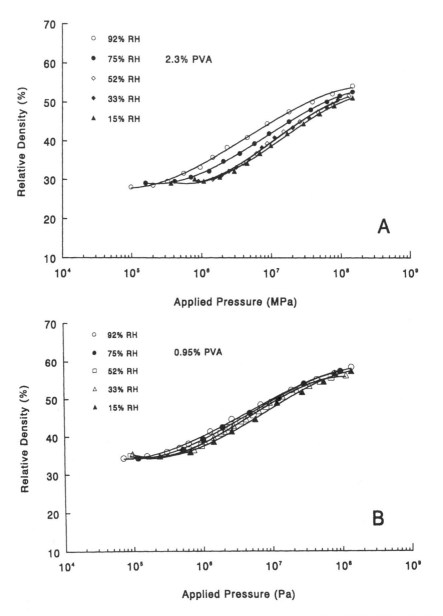

FIG. 7. Compaction diagrams for alumina granules containing (A): 2.3 and (B): 0.95 wt % poly(vinyl alcohol). (From R. A. DiMilia and J. S. Reed, Dependence of compaction on the glass transition temperature of the binder phase, *Am. Ceram. Soc. Bull. 62(4)*: 484–488 (1983), Figure 4. Reprinted by permission of the American Ceramic Society.)

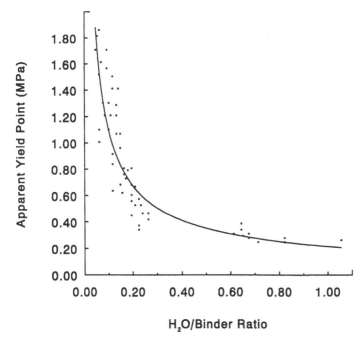

FIG. 8. FIG. 8. Dependence of the apparent yield point of spray dried alumina granulate on the concentration of water in the binder phase. (From R. A. DiMilia and J. S. Reed, Dependence of compaction on the glass transition temperature of the binder phase, *Am. Ceram. Soc. Bull. 62(4)*: 484–488 (1983), Figure 6. Reprinted by permission of the American Ceramic Society.)

larger granules. They attribute this to the finding that finer granules have a higher tendency toward shear deformation. All the granules were well plasticized and equilibrated at the same humidity, to ensure minimal differences in the plastification of the binder (poly(vinyl alcohol) plasticized with poly(ethylene glycol).

The derivative of the pressing curve can be used to describe pressing. A compaction *rate* diagram gives important information about the compaction process. Matsumoto [23] compared compaction of two granulates. One of the granulates contained an acrylic binder and the other a poly(vinyl alcohol) binder. The compaction rate diagram for granulate containing poly(vinyl alcohol) showed two maxima. These maxima correspond to a bimodal granulate size. The smaller granulates dried faster during spray drying and contained less moisture. They fractured at lower pressures, with the larger granules remaining intact. At higher pressures the larger gran-

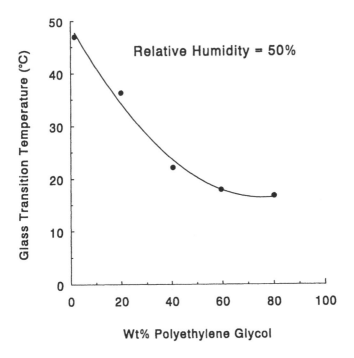

FIG. 9. Effect of plasticizer poly(ethylene glycol) content on the glass transition temperature of poly(vinyl alcohol) at 50% relative humidity. (From C. W. Nies and G. L. Messing, Effect of glass-transition temperature of polyethylene glycol-plasticized polyvinyl alcohol on granule compaction, *J. Am. Ceram. Soc.* 67(4): 301 (1984), Figure 1. Reprinted by permission of the American Ceramic Society.)

ules broke up. The maxima of the compaction rate corresponds to the breakup of the granules. The second granulate contained an acrylic binder. This binder is not plasticized by water. For this granulate, a single maximum could be observed on the compaction rate diagram.

Binders that are insoluble in water, such as latex and emulsion type binders, are not plasticized by water. These binders are insensitive to the humidity of the granulate as well as to the ambient humidity of storage. To make them plastic enough for pressing, they have to be chosen with an inherent low glass transition temperature [24].

B. Granule Density

During spray drying it is important to have a well-dispersed slurry and a high concentration of solids, to avoid excessive migration of binder (see

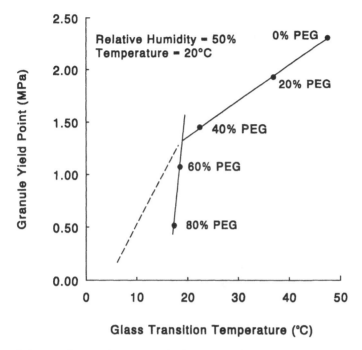

FIG. 10. Change in granule yield point as a function of binder glass transition temperature. (From C. W. Nies and G. L. Messing, Effect of glass-transition temperature of polyethylene glycol–plasticized polyvinyl alcohol on granule compaction, *J. Am. Ceram. Soc. 67*(4): 301 (1984), Figure 2. Reprinted by permission of the American Ceramic Society.)

Section I.B). Binder migration causes large pores and other defects in the granules. A poorly dispersed slurry has high viscosity and generally causes problems during spray drying. A well-dispersed slurry will always yield high-density granules during spray drying (unless binder migration causes large pores in the granules). Even if the slurry is diluted, the drops will shrink during drying to produce a high-density granule.

Zheng and Reed [12] used pressure casting of an alumina slurry followed by grinding and sieving to produce granules for pressing. The slurry was cast in either the dispersed or the flocced state. A 25 wt % slurry containing 2% polyacrylic dispersant was cast in the dispersed state. The pressure casting was performed at 35 MPa. The flocced state was achieved by adding ammonia and $AlCl_3$. The coagulated slurry had a solids concentration of 38 wt %. The dispersed slurry produced a high-density granulate (61.5% of

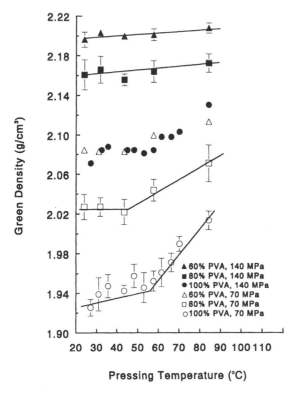

FIG. 11. Variation of green density with pressing temperature. Poly(vinyl alcohol) is used as a binder plasticized with 0, 20, and 40% poly(ethylene glycol). (From C. W. Nies and G. L. Messing, Effect of glass-transition temperature of polyethylene glycol-plasticized polyvinyl alcohol on granule compaction, *J. Am. Ceram. Soc. 67*(4): 301 (1984), Figure 3. Reprinted by permission of the American Ceramic Society.)

theoretical) and the flocced slurry a low-density (46.3% of theoretical) granulate. No binder was added to these two granulates.

The high-density granules were found to compact to a body with higher density during pressing than bodies pressed from low-density granules. Still, the compact with high-density granules did not compact to pressed densities higher than the original granule density. This indicates that the granules were not broken down completely during pressing. The low-density granules compacted to a lower pressed density than the high-density granules. However, the pressed density was significantly higher than the original density of the granules. This indicates that the low-density granules were bro-

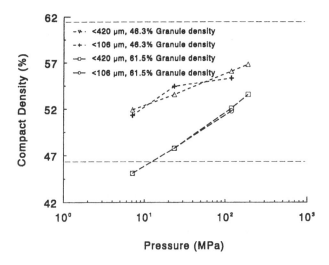

FIG. 12. Density versus pressure for a powder prepared at two granule densities. (From J. Zheng and J. S. Reed, Particle and granule parameters affecting compaction efficiency in dry pressing, *J. Am. Ceram. Soc. 71*(11): C456–C458 (1988), Figure 2. Reprinted by permission of the American Ceramic Society.)

ken down during pressing. The maximum compact pressure was 200 MPa, and the low-density granules exceeded their granule density already at 12 MPa (Fig. 12).

It must be pointed out that the granule density greatly influences the strength of the granulate. The number of contact points between particles increases with granule density. Thus the influence of hardness between two binders or the degree of plasticity of one binder can only be compared correctly at the same granule density.

C. Friction and Lubrication

When surfaces slide over one another, this movement is resisted by adhesion forces on the contact points and mechanical forces produced by irregularities and asperities on the surfaces. If the frictional forces are high, substantial wear of the surfaces can occur and increase the friction even more. To reduce these problems, lubricants can be used.

Tabor [25] states that a lubricant film should possess two main properties: first, it should prevent solid-solid contact, thus reducing wear; second, it also should have a low shear strength, thus ensuring low friction.

There are several types of lubrication, depending on the thickness of the lubricating layer, the contact pressure, and the sliding speed the shear zone is subjected to. In Fig. 13 the transition from hydrodynamic to boundary lubrication with increasing contact pressure P and decreasing sliding speed V is shown. In between, there is a zone called elastohydrodynamic lubrication (E.H.L.). Both hydrodynamic lubrication and E.H.L. are created by a thin liquid film keeping the two surfaces apart. However, for surfaces subjected to high contact pressures at point contacts, there is a strong tendency for a lubricating liquid to escape and leave the contact points unlubricated. For these types of materials, amphipatic molecules that adsorb strongly to the surfaces (such as fatty acids) can lower the friction by boundary lubrication. The molecules are assumed to orient themselves with the polar end toward the solid surface and to form a close-packed monomolecular layer. This layer is expected both to reduce the adhesion at the contact points and, mainly, to reduce wear. It has been shown that fatty acids can reduce the friction of smooth metal surfaces from $\mu = 1$ (clean surface) to $\mu = 0.05-0.1$ (lubricated) [26]. The effects of the carbon chain length of different fatty acids have been investigated [27]. It was found that the magnitude of the

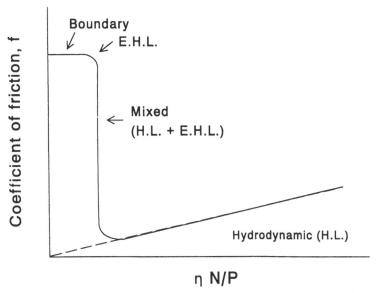

FIG. 13. Schematic diagram showing transition from hydrodynamic to E.H.L. to boundary lubrication. (From D. Tabor, in *Microscopic Aspects of Adhesion and Lubrication* (J. M. Georges, ed.). Elsevier, Amsterdam, 1982, pp. 651–682.)

friction coefficient decreases as the additive chain length increases. Another interesting feature with fatty acids is that the friction suddenly rises when the temperature is increased. This increase in friction is associated with the softening of the lubricant film and is considered to be caused by disorientation of the hydrocarbon chains. These types of lubricant films can take contact pressures up to 10^7 MPa without rupture or extrusion [25].

During pressing, the particles in the powder compact are subjected to different types of forces and associated frictional resistance to flow. During die filling, the friction of the granules against the hopper and the die wall also has to be accounted for. Thompson [28] has presented a model for powder densification with two types of frictional forces. Powder/die wall friction μ_w describes the friction of the powder compact against the die wall, and internal friction μ_i is a measure of the resistance to flow between two powder surfaces. Both the internal and the die wall friction are dependent on particle density, particle shape, and the surface characteristics of the particles (lubrication). The die wall friction is also dependent on the surface roughness of the die wall.

In practice, it is often difficult to distinguish between internal and external friction. A pressing additive can function as an internal or an external lubricant, and these two effects are often difficult to distinguish. During pressing, the flow of the powder is complex as the granules are deformed and compacted at the same time as the individual particles slide past one another. A pressing additive might influence the flow properties of the powder both as an internal lubricant and as a binder that controls the strength and structure of the granules.

Die wall friction can be studied by examining the ejection pressure. Ejection takes place at lower normal stress than compaction, so the results may not be transferable. An accurate method of measuring wall friction has been developed by Samuelson and Bolin [29]. They measured the sliding friction by pressing two powder compacts against each side of a sliding wall. In isostatic pressing there is no die wall friction to account for. Consequently, the shearing that takes place during isostatic pressing is much smaller, which makes it more difficult to break down granules during isostatic pressing.

The pressure during die pressing is transmitted in an irregular fashion depending on the friction against the die wall. High die wall friction results in density gradients in the pressed compact. In the laboratory, a lubricant can be added to the die prior to pressing. In a production situation, this is more difficult. Instead of adding the lubricant to the die wall, it can be added to the granules by tumbling or spraying the granulate with a lubricant. This has the disadvantage of not adding the external lubricant exactly where it belongs. Only if the granules remain intact during pressing (which should be avoided) can this type of lubricant fulfill its purpose. In practice, the internal lubricants or binders added during granulation must function also as external lubricants.

The fast cycles of a production press make application of a lubricant on the die wall during pressing very difficult, but a method of lubricating the die by injecting a fluid lubricant has been developed [30]. The lubricant could be injected against the die wall without interrupting the normal cycle of the press. This process was tried for pressing sponge iron. Normally, sponge iron has to be pressed with the addition of zinc stearate. By injecting a fluid lubricant into the die, the addition of zinc stearate could be avoided, with an increase in green density and no need to burn off the lubricant.

If the pressing additive is to function as a lubricating film, it has to be evenly distributed around the particles. To do this, it is necessary to wet the powder. If the pressing additive wets the powder, the capillary forces will tend to contract it to the contact points between the particles during drying in the granulation process [31]. When the granule is deformed during compaction, new contact points between the particles will be formed. Unless a very high concentration of pressing additive is used, these new contact points will not be coated. To coat the particle evenly, the pressing additive has to adsorb onto the particles. The adsorption forces will ensure that the whole surface is covered and that the additive will not be transported during drying. Experiments with an adsorbing pressing additive showed enhanced pressability as well as lower ejection pressures [5]. The effect on pressability of the adsorbing additive might also arise from the improved green microstructure of the granules (less porosity). The lower ejection pressure is an indication that the die wall friction was reduced by substituting poly(vinyl alcohol) with an adsorbing poly(acrylate) latex.

D. Viscoelastic Behavior

During pressing, the compact acts as a viscoelastic material. Elasticity stems from the contact between the particles. Two particles in point contact will deform elastically at the contact spot. All particle-particle contacts in a certain direction of the body can contribute to this elastic deformation. The elastic modulus for the whole compact will be much lower than the elastic modulus of the individual particles. This is because of the high local contact stresses generated by the much lower bulk stress. If organic binder is added, it can also produce an elastic contribution during pressing. But the binder is normally plasticized and will also deform in a viscous manner. Fragmentation of granules and sliding of particles will cause a viscous (plastic) deformation.

On ejection of the compact, there is a springback effect. The springback is an elastic and sometimes time dependent recovery of the compressed body. If the springback is large and the expansion on ejection is abrupt, the compact can crack. This can be modified by designing the mold so that a gradual expansion results on ejection. The springback can also be influenced by changing the amount or type of pressing additive.

Hsueh [32] measured the elastic properties of a compact during pressing by measuring the relaxation behavior during pressing. The compact is compressed to the maximum pressure. The die is kept in the same place and the pressure then relaxes slowly. The viscoelastic properties can be described using a mathematical model with two springs and a dashpot.

Nyberg et al. [5] found that in a comparison of pressing additive the granulate with the best pressing characteristics also exhibited the largest relaxation after pressing.

IV. GREEN STRENGTH

If the green strength of a ceramic compact is too low, it will be difficult to handle the compact without damaging it. A large amount of the wastage in ceramic production is generated during the handling of green compacts. What is the origin of the strength of a green compact and how can this be controlled for pressed compacts?

The original explanation of green strength was that a mechanical interlocking of rough surfaces produced the binding force for the particles. This has long since been proven wrong, but it still exists as a popular explanation for green strength. Another explanation is that liquid bridges hold together moist powders. A liquid bridge helps keep the particles in a compact together, especially by providing plasticity to help the forming of a green body. Organic binders that form a solid material on drying can also increase the strength of a green body. It is, however, possible to form a green body with a reasonable strength with a dry powder without additives if the particle size is small enough. In such a body the van der Waals attraction between particles in contact is the force that keeps the powder particles in a compact together.

Rumpf [33] has developed a theory of agglomerate strength based on van der Waals attraction. It explains how the strength depends on density of the agglomerates, but it does not account for the large statistical variation in green strength among individual samples, nor does it predict the effects of cracks. Rumpf proposes that the strength of a powder compact is the sum of the bonding forces between the individual particles. This requires that all particles be separated simultaneously when a failure occurs. In this way each particle is subjected to the same tensile stress. In practice, the mechanism of failure for a green compact is by the propagation of a crack.

Kendall, et al. [34] have proposed a theory that accounts for the brittle failure of green compacts. Two spheres in contact are pulled together by attractive van der Waals forces. This results in an elastic deformation at the contact and the formation of a circular contact spot (Fig. 14). Kendall et al. begin by calculating the surface energy required to separate the two spheres. Separating the particles requires that the interfacial energy of the contact area be overcome. Simultaneously, the elastic deformation of the spheres is recovered. This energy has to be subtracted from the attractive energy above. By multiplying by the number of contacts per unit area, the

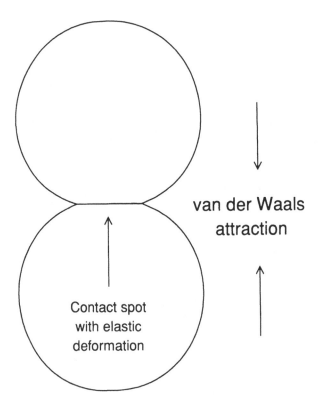

FIG. 14. Two spherical ceramic particles deforming elastically under surface forces to produce a circular contact spot.

fracture energy of a compact can be calculated. This fracture energy is greatly influenced by the solid volume fraction of the compact, roughly to the power of four. When the fracture energy is known, Griffith's criteria for brittle fracture can be applied. In practice, the strength of a green compact is greatly influenced by flaws in the body in the same way as in a sintered ceramic. Kendall et al. could verify their theory of green strength experimentally and measured effects of crack length on green body strength that agreed with the theory.

The number of contacts among particles rises sharply with decreasing particle size. This makes the green strength of fine powders higher. The disadvantage of fine powders is that they resist compaction and require higher pressures for compaction.

V. BINDER REMOVAL

Removal of the binder or lubricant can limit the uses of some organic systems. One of the most commonly used lubricants is zinc stearate. During heating in an inert atmosphere, zinc stearate leaves a residue of zinc oxide. Other metal stearates also leave metal oxide residues.

Stearic acid leaves a small carbon residue when removed by heating in an inert atmosphere. Carbon residues are common problems in the processing of nitride and sialon ceramics. Removal of organic binder has to take place in an inert atmosphere, owing to the oxidation sensitivity of the high surface area silicon nitride that has to be used for sintering. Carbon residues from the binder or lubricant either remain in the final ceramic and function as defects or react with the silicon nitride during sintering, producing other types of defects in the final material.

Alternative lubricants have been developed to avoid contamination from the lubricant, including a lubricant based on bisstearoylamide ($H_{35}C_{17}CONHC_2H_4COC_{17}H_{35}$) [35] developed for lubrication of metal powders.

Thermal degradation of a polymer is a complex reaction that can involve both depolymerization and chain scission. Chain scission is promoted by free radicals. The reactions take place during an induction phase, where radicals are produced, and a propagation phase, where the radicals react with the polymer, producing new radicals in the process. At the end of the reaction there is a termination phase where the radicals are consumed. Parallel to the chain scission, repolymerization also can take place, and side chains and rings are formed. This last type of reaction often leads to the formation of molecules that leave a carbon residue.

Oxygen often takes part in the formation of radicals. The partial pressure of oxygen influences the degradation path. One example of this is polystyrene that decomposes and leaves a carbon residue at lower temperatures in atmospheres with high oxygen partial pressures [36]. This means that a bulk material of polystyrene forms a carbon residue on the surface at the same time as it decomposes cleanly from inside the bulk. The bulk reaction also depends on oxygen, but the need for diffusion into the bulk makes the partial oxygen pressure lower, which results in a different reaction path.

When a binder is removed from a ceramic powder, we also have to consider the catalytic effect of the ceramic. Alumina is known to have a catalytic effect on many organic reactions. Because of this, other types of reactions will take place on a powder surface than in an organic bulk material.

For the reasons mentioned above, it is recommended in practice to test for problems with binder residue. These tests must be performed under the actual conditions that will be used for binder removal: the same time and temperature program, partial pressure of oxygen, and type of powder.

The analysis methods have to be chosen according to the type of ceramic powder that is used. Carbides and nitrides tend to oxidize slightly, making gravimetric methods unreliable for exact measurements of carbon residue. Elemental carbon analysis can be used with precision on nitrides, but not on carbides.

VI. APPLICATIONS

The ceramic systems that have been studied with regard to pressing are mostly alumina and porcelain systems. Most of the references cited in this chapter deal with pressing of alumina powder.

Systems such as silicon nitride, silicon carbide, and zirconia are important as high performance ceramics. Unfortunately, very little systematic research on pressing has been published on these materials.

Bhattacharya and Chaklader [37] studied the pressing of silicon nitride. They found that an increase in pressed density could be achieved by dispersing the powder properly and adding poly(ethylene glycol).

Dietrich et al. [38] pressed sinterable SiC powder. They found that both particle size distribution and the shape of the particles influenced the springback and the green strength of compacts pressed without pressing additives. The powder with the lowest springback also had the highest green strength. This powder had a wide particle size distribution and rounded grains.

Nyberg et al. [39] found that particle shape influenced the pressing properties of silicon nitride. A powder with lower rugosity had the best pressing properties. When a lubricant (oleic acid) was added, a larger amount of lubricant was needed to enhance the pressing properties of the powder with higher rugosity.

REFERENCES

1. S. J. Lukasiewicz and J. S. Reed, *Am. Ceram. Soc. Bull.* 57(9): 798–801, 805 (1978).
2. S. J. Lukasiewicz, *J. Am. Ceram. Soc.* 72(4): 617–624 (1989).
3. J. S. Reed, in *Introduction to the Principles of Ceramics Processing.* Wiley, New York, 1988, p. 24.
4. I. Aksay and C. C. Schilling, in *Ultrastructure Processing of Ceramics, Glasses and Composites* (L. L. Hench and D. R. Ulrich, eds.). Wiley, New York, 1984, pp. 439–447.
5. B. Nyberg, E. Carlström, M. Persson, and R. Carlsson, in *Proceedings of the 2d International Conference on Ceramic Powder Processing Science* (H. Hausner, G. L. Messing, and S. Hirano, eds.). Deutche Keramische Gesellschaft, Köln, 1989, pp. 573–580.
6. S. Prochazka, in *Investigation of Ceramics for High Temperature Turbine Components*, U.S. N.T.I.S. AD/A Report No. 005 830 (1975).
7. A. Müller-Zell and Ch. Hahn, in *Ceramic Materials and Components for Engines* (W. Bunk and H. Hausner, eds.). Deutsche Keramische Gesellschaft, 1986, pp. 339–345.
8. German Patent DE 3711191 A1 to Hoechst (1988).
9. S. J. Lukasiewicz and J. S. Reed, *Am. Ceram. Soc. Bull.* 57(9): 798–801 (1978).
10. G. L. Messing, C. J. Markhoff, and L. G. McCoy, *Am. Ceram. Soc. Bull.* 61(8): 857–860 (1982).
11. R. L. K. Matsumoto, *J. Am. Ceram. Soc.* 69(10): C246–C247 (1986).
12. J. Zheng and J. S. Reed, *J. Am. Ceram. Soc.* 71(11): C456–C458 (1988).
13. B. Nyberg and E. Carlström, unpublished work.
14. O. Alm, *Scan. J. Metallurgy 12*: 302–311 (1983).
15. G. Y. Onoda, Jr., in *Ceramic Processing Before Firing* (G. Y. Onoda and L. L. Hench, eds.). John Wiley, New York, 1978, pp. 235–251.
16. C. Hahn and A. Müller-Zell, in *High Tech Ceramics* (P. Vicenzini, ed.). Elsevier, Amsterdam, 1987, pp. 595–599.
17. B. Frisch, W.-R. Thiele, and R. Kirsch, cfi/Ber. *DKG 58*(4-5): 251–262 (1981).
18. R. G. Frey and J. W. Halloran, *J. Am. Ceram. Soc.* 67(3): 199 (1984).
19. R. A. DiMilia and J. S. Reed, *Am. Ceram. Soc. Bull.* 62(4): 484–488 (1983).
20. R. K. Tubbs and T. K. Wu, in *Properties and Application of Polyvinyl Alcohol*, S. C. I. Monograph No. 30, Soc. Cem. Ind., London, 1968, pp. 167–188.
21. C. W. Nies and G. L. Messing, *J. Am. Ceram. Soc.* 67(4): 301 (1984).

22. B. D. Mosser III, J. S. Reed, and J. R. Varner, in *Ceramic Transactions, Vol. 1: Ceramic Powder Science IIB* (G. L. Messing, E. R. Fuller, Jr., and J. Hausner, eds.). American Ceramic Society, Westerville, Ohio, 1987, pp. 767–775.

23. R. L. K. Matsumoto, *J. Am. Ceram. Soc. 73*(2): 465–468 (1990).

24. N. R. Gurak, P. L. Josty, and R. J. Thompson, *Am. Ceram. Soc. Bull. 66*(10): 1495–1497 (1987).

25. D. Tabor, in *Microscopic Aspects of Adhesion and Lubrication* (J. M. Georges, ed.). Elsevier, Amsterdam, 1982, pp. 651–682.

26. F. P. Bowden and D. Tabor, *Friction and Lubrication of Solids.* Clarendon Press, Oxford, Part I, 1950; Part II, 1964.

27. S. Jahanmir, *Wear 102*: 331–349 (1985).

28. R. A. Thomson, *Ceram. Bull. 60*: 237–243 (1981).

29. P. Samuelson and B. Bolin, *Scand. J. Metallurgy 12*: 315–322.

30. B. A. James, *Powder Metallurgy 30*(4): 273–280 (1987).

31. G. Y. Onoda, *J. Am. Ceram. Soc. 58*(5–6): 236–239 (1976).

32. C.-H. Hsueh, *J. Am. Ceram. Soc. 69*(3): C48–C49 (1986).

33. H. Rumpf and H. Schubert, in *Ceramic Processing Before Firing* (G. Y. Onoda and L. L. Hench, eds.). John Wiley, New York, 1978, pp. 358–376.

34. K. Kendall, N. McN. Alford, and J. D. Birchall, in *Special Ceramics 8, British Ceramic Proceedings, No. 37* (S. P. Howlett and D. Taylor, eds.). Institute of Ceramics, Stoke on Trent, 1986, pp. 255–265.

35. *Metal Powder Review 41*(11): 853 (1986).

36. R. S. Lehrle, R. E. Peakman, and J. C. Robb, *Eur. Polym. J. 18*: 517–529 (1982).

37. S. K. Bhattacharya and A. C. D. Chaklader, *Ceramics International 9*(2): 49–52 (1983).

38. R. Dietrich, A. Kerber, and G. Rünzi, in *Ceramic Materials and Components for Engines* (V. J. Tennery, ed.). American Ceramic Society, Westerville, Ohio, 1988, pp. 76–85.

39. B. Nyberg, R. Lundberg, E. Carlström, M. Ernstsson, and L. Bergström, in *Ceramic Transactions, Vol. 12: Ceramic Processing Science III* (G. L. Messing, S. Hirano, and H. Hausner, eds.). American Ceramic Society, Westerville, Ohio, 1990, pp. 845–853.

7

Surface and Colloid Chemistry in Ceramic Casting Operations

MICHAEL PERSSON* Swedish Ceramic Institute, Gothenburg, Sweden

I. INTRODUCTION

Slip casting, pressure casting, centrifugal casting, and tape casting are colloidal processing techniques for ceramics. Colloidal processing is the key to improving the actual properties of different types of ceramic materials toward their theoretically estimated optimal properties. This is especially important in the field of engineering ceramics, where strength can be increased by a factor of three to four by applying proper colloidal processing.

---

*Current affiliation: Eka Nobel AB, Bohus, Sweden

Colloidal processing not only includes the evaluation of an efficient organic processing agent for a particular forming method. Colloidal processing means that the processing agents chosen, which are often required to be surface active, are used to control the interparticle forces that give a system an appropriate rheology during consolidation. The goal is to obtain green bodies, free from unwanted organic or inorganic inclusions, in which the particles have been homogeneously packed and the remaining pores are small and narrowly distributed.

This chapter focuses on the surface and colloidal chemistry aspects of ceramic casting, especially what are believed to be the most important issues now and for the future. It does not give a complete review of the two subjects. The casting methods are compared with other forming methods regarding their potential for optimum powder consolidation. Water systems are the main systems discussed for slip casting, pressure casting, and centrifugal casting, and organic solvent systems for tape casting.

Slip casting has mostly been practiced in the industry in the area of traditional ceramics where sanitary ware and dinnerware in the triaxial systems, i.e., china and ball clay, feldspar, and quartz, have been produced. Since this chapter mostly concentrates on high-performance ceramics (i.e., SiC, Si_3N_4, Al_2O_3, ZrO_2, etc.), experience from the traditional ceramic area is only reviewed when necessary for a complete understanding of the subject.

II. SLIP CASTING AND PRESSURE CASTING

Slip casting and pressure casting are forming methods in which particles in a suspension are consolidated by a filtration process. In slip casting, the liquid is removed with suction pressure caused by the capillary forces in a porous mold; see Fig. 1. Plaster of Paris is almost always used as the mold material. Plaster molds have a suction pressure in the range of 0.1–0.2 MPa [1,2]. In the literature [2], the expression "colloidal filtration" is sometimes used instead of slip casting. This is to emphasize that the forming process is really a filtration of colloidal particles and not just a casting process. To enhance the casting rate, pressure can be applied to the suspension. In pressure casting, plastic porous molds have been developed that increase productivity and improve the tolerances of the pressure cast components. The plastic molds can be used for at least 10,000 cycles, which is 100 times more than for plaster molds [3,4].

A good slip casting slurry is a well-stabilized highly concentrated particle suspension. The stability against flocculation gives a dense, incompressible consolidated layer, and the high solids content reduces the amount of liquid sucked away by the mold [5]. In slip casting of traditional ceramics, the slips

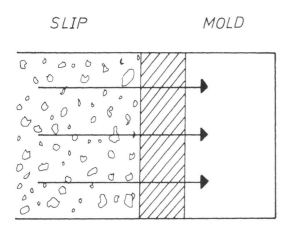

FIG. 1 Schematic picture of the slip casting process.

used do not have the highest stability or the lowest viscosity achievable. The slips have a slightly higher viscosity and some degree of instability to flocculation, in order to reduce the filtration resistance and increase the casting rate [6].

The decreased stability of the slip results in an agglomeration of the colloidal part of the slip, and three-dimensional particle networks can be formed if the particle concentration is high enough. The partially agglomerated slip gives larger filtration channels, which are the main reason for the reduced filtration resistance. Except for producing cast bodies faster, the partial flocculation gives increased moisture retention and an even distribution of moisture through the cast, which favors plasticity and trimming [6,7]. In Fig. 2, Reed [7] exemplifies three different levels of stabilization and their effects on the casting performances.

A. Theory

Models describing the filtration process in slip casting and pressure casting can be found in the literature. Adcock and McDowall [8] described filtration through a consolidated layer. Aksay and Schilling [2] also included the filtration resistance of the porous mold. Tiller and Tsai [1] made the expressions more general by including compressible consolidated layers. Tiller and Hsyung [9] developed the theory further on external and internal cylindrical surfaces. Recently, Bridger and Massuda [10] have reviewed the principles of slip casting.

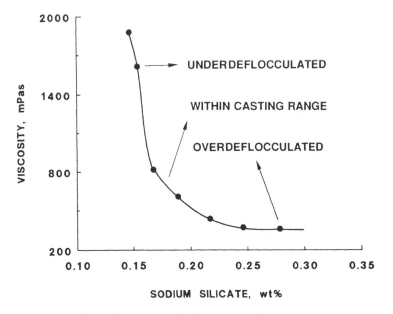

FIG. 2 Casting performances for various stability levels of a whiteware slip. (From Ref. 7.)

Since consolidated layers made from fully deflocculated slips are incompressible [11,12], only the expression for the filtration process of incompressible layers is shown here. Aksay and Schilling [2] expressed the kinetics for the growth of the consolidated layer as

$$P - P_0 = \frac{x^2}{t} \left(\frac{\eta n \alpha_c}{2} + \frac{\eta n^2 \alpha_m}{2\varepsilon_m} \right) \tag{1}$$

$P - P_0$ is the total pressure drop over the cast layer plus the pressure drop in the saturated part of the mold surface. In slip casting, $P - P_0$ is caused by a suction pressure of 0.1 to 0.2 MPa. In pressure casting, the applied pressure P is often in the range of 2 to 4 MPa, and P_0 is the atmospheric pressure of 0.1 MPa. Also, x represents the linear growth of the cast layer from the mold surface ($x = 0$); t is the time, and η is the filtrate viscosity; α_c and α_m are the average specific porous medium resistances for the consolidated layer and the mold, respectively; ε_m is the volume fraction of voids in the mold.

In Eq. (1), n is defined as

$$n = \frac{(1 - \phi_s - \varepsilon_c)}{\phi_s} \tag{2}$$

where ϕ_s is the volume fraction of solids in the slips and ε_c is the void fraction in the cast layer. Here n is critical for the filtration kinetics, and a minimization of n increases the consolidation rate.

Rearrangement of Eq. (1) leads to

$$v = \frac{x^2}{t} = \frac{2\Delta P}{n\eta \left(\alpha_c + \dfrac{n\alpha_m}{\varepsilon_m} \right)} \tag{3}$$

where $v = x^2/t$ is the consolidation rate and ΔP the total pressure drop.

In Eq. (3) it is obvious how the consolidation rate v can be increased. Using pressure casting instead of slip casting in incompressible systems leads to a drastic increase in the casting rate. As mentioned above, minimization of n increases v; n can, of course, be varied only within a limited range. For example, starting with a moderate to bad alumina slip casting slurry with $\phi_s = 0.3$ and $\varepsilon_c = 0.5$ results in $n = (1 - 0.3 - 0.5)/0.3 = 0.67$; $\phi_s = 0.3$ and $\varepsilon_c = 0.5$ corresponds to 30 vol % (63 wt %) of solids in the slurry and a green density of 50 vol %. With further developments of this system, values for ϕ_s and ε_c can easily change to 0.5 (i.e., 79.9 wt %) and 0.4 (60% in green density), respectively. The value of n has now changed to $(1 - 0.5 - 0.4)/0.5 = 0.2$. The ratio between the two n values is 3.33, and this means that improvement of a slip casting slurry can provide better microstructure and material properties and more than triple the casting rate.

The filtrate viscosity η can influence the consolidation rate in at least two different ways. An overdose of organic processing agents is not only a waste of money, it also increases filtrate viscosity and hence decreases the casting rate. In industrial pressure casting production, the slips are often heated to increase the casting rate owing to reduced filtrate viscosity.

Finally, Eq. (3) clearly shows the influence of the filtration resistance α on the casting rate. In slip casting with plaster molds, the mold filtration resistance influences the casting rate. Tiller and Tsai [1] showed that by optimizing the pore sizes of the mold the pressure drop over the cast layer, and thereby also the casting rate, is maximized.

In modern pressure casting equipment, the porous plastic molds used may have pores 40 times larger than the pores in the plaster molds. Hence the plastic porous molds contribute to a high casting rate by making the mold filtration resistance negligible.

Aksay and Schilling [2] calculated the filtration resistance for various consolidated layers having different sizes of monosized particles and differ-

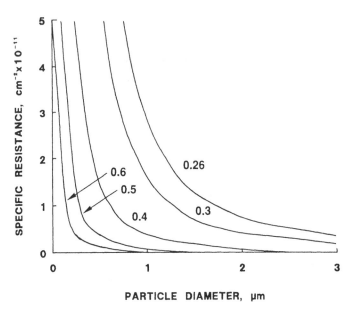

FIG. 3 Specific resistance of consolidated layers (α_c) formed by packing monosized particles to different void fractions. Curves for five different void fractions are shown (ε_c: 0.6, 0.5, 0.4, 0.3, 0.26). (From Ref. 2.)

ent packing densities; see Fig. 3. Figure 3 shows that to increase the filtration rate, the particle size should be increased and/or the packing densities should be diminished, which may contradict the requirements for sinterability and firing shrinkage. Figure 3 also clearly shows that particle sizes close to 0.1 μm (see Section V regarding "the lower limit" of an ideal powder) yield a very high filtration resistance even if the filtration is performed at low packing densities of the layer.

Aksay and Schilling [2] found that by using submicron particles and highly concentrated slips (> 35 vol %), the filtration resistance in the plaster mold becomes negligible compared with the filtration resistance in the consolidated layer ($\alpha_c \gg \alpha_m$).

A question is whether the solid content in a slurry as such influences the green density. It has generally been observed that the green density in a cast body decreases with a decreasing solid content in the casting slurry [13,14]. The main reason for the lower green density is that in a diluted suspension there is more time for the particles to segregate. By maximizing the solid content, the segregation is diminished and the smallest particles can fill the voids between the larger particles [14].

Hampton et al. [15] showed that a 50/50 mixture of two alumina powders with mean particle sizes of 4.2 and 0.4 µm, when slip cast at various solids loadings (0.37–0.47 volume units), has a high, constant green density. Slip casting of only the fine alumina powder shows a tendency of decreasing green density with increased solids loading. On the other hand, the coarse alumina slip with its shear thickening behavior was found to increase slightly the green density with solids loading.

Velamakanni and Lange [16] studied particle segregation during pressure filtration. Two alumina powders with median sizes of 0.5 µm and 1.3 µm were mixed in various proportions and filter-pressed. The fine and coarse powders gave, when consolidated as stable slips, the same green density, 62%. The green density increased to 67% when 40% of the fine powder was mixed with 60% of the coarse powder; 67% in green density was only obtained when solids loading of 50 vol % was used. Slips with solids loadings below 50 vol % lowered the green density, and at 20 vol % the green density was decreased to 63.5%. The density decrease was caused by particle segregation. At high solids content, > 50 vol %, the sedimentation velocity was reduced to 3% of the velocity obtained by applying Stokes' law to a single particle. Hence high solids content reduces particle segregation owing to hindered sedimentation.

The use of flocculated slurries decreases the filtration resistance, owing to the formation of a more porous particle network, and depending on the compressibility of the flocs the pressure gradient causes great problems with particle packing gradients [17]. Density gradients cause extensive crack formation during drying and sintering.

Fennelly and Reed [12,18] investigated the mechanics of pressure casting of alumina. Three different degrees of slip stability were achieved by the addition of varying amounts of a polyacrylate polyelectrolyte. The least stable slip had a casting rate at a pressure of 0.34 MPa comparable to the casting rate of the most stable slip but at a pressure of 2.76 MPa. The specific cast layer resistances of the stable slips were constant with applied pressure, above a pressure of 0.69 MPa, indicating incompressibility. The unstable slip showed a specific cast resistance that increased with applied pressure and a nonlinear relation of the casting rate as a function of applied pressure. A specific cast layer resistance that increases with applied pressure is due to the compressibility of flocculated particles in the slips. Compressible flocs cause the nonlinearity in the casting rate when plotted as a function of applied pressure, since the higher the pressure the denser the outer cast layer, and this results in a filtration resistance that varies with casting pressure.

In the initial stage of slip and pressure casting, a first layer of particles is deposited on the mold surface at approximately zero in effective pressure. This is due to the mechanical pressure from the slip being only infinitesimally greater than the hydraulic pressure. For each layer deposited, the pressure gradient increases. The pressure on the particles in the cast layer is at maximum for the particles adjacent to the mold surface, whereas the growth and the continuous deposition of particles occur without compressive forces. Tiller and Tsai [1] showed that compressible cast layers have a very steep pressure drop close to the mold surface, resulting in a dense outer layer, while the rest, approximately 80% of the cast layer, is unconsolidated with a high porosity. They calculated the local porosity against depth in the cast layer for four different degrees of stability (slip A, B, C, and D) in slips; see Fig. 4.

The most stable slip (D) shows a uniform, low porosity throughout the cast layer. By introducing some degree of instability in the slips (B and C), the cast layer porosity is certainly increased (i.e., reduced filtration resistance), but the density varies slightly and is decreased particularly in the

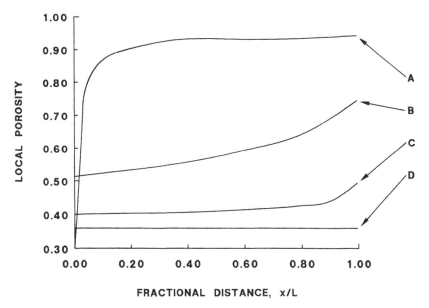

FIG. 4 Variation of porosity in cast layers. (From Ref. 1.)

final cast layers deposited. The flocculated slip (A) shows very low density throughout the cast except for very high density gradient in the first layers deposited. The work of Tiller and Tsai [1] shows the risk of enhanced slip or pressure casting rate introduced by some degree of instability in the slip. This is perhaps not as important in pressure casting of traditional ceramics as for pressure casting of high-performance ceramics, where no clays are present to impart plasticity and where the demand for uniform density is very high.

During the formation of incompressible cast layers as described in Eq. (3), the pressure decreases linearly from the slip cast layer interface to the mold surface. The incompressibility also results in constant porosity throughout the cast layer. Compressibility causes density gradients, as mentioned above, and this results in varying cast layer filtration resistance. When pressure casting of more or less flocculated slips is performed, the expected increase of the filtration rate may diminish owing to increased filtration resistance caused by compressed cast layers.

Lange and Miller [11] noticed difficulties when demolding bodies from pressure cast dispersed or flocculated slips. Bodies from pressure cast dispersed slips show a dilatant behavior on demolding, whereas bodies made of flocculated slips show slow crack extension. Both difficulties could be avoided by dewatering the cast bodies prior to demolding, which is the case in traditional slip casting, where the green body is partly dried in order to shrink it and enhance demolding. However, in modern pressure casting applications, where porous plastic molds are used, the consolidated bodies are released without complete dewatering, i.e., the body has the pores filled with water. The dilatant behavior of wet pressure cast high-performance ceramic components is one of the major problems to be solved in order to be able to mass-produce components using pressure casting. Industrial pressure casting has almost only been applied to traditional ceramics, and no problems have been reported during demolding concerning dilatancy in the wet bodies or slow crack extension. The firmness of pressure cast traditional ceramic bodies, according to Hogg [3], is equal to the firmness of the bodies slip cast in plaster molds followed by drying for two or three hours.

The dilatant demolded high-performance (i.e., non-clay system) ceramic component is stiff and rigid when exposed to high shear rates. When the shear rates are lowered, the body is fluid and deformable, which is not acceptable when high tolerances are required. The dilatant behavior probably also depends on particle morphology, i.e., to what extent the particles tend to lock each other into position. Binders that reveal a shear sensitive rheology in water solutions, especially a shear thinning behavior, can diminish the problems with dilatant bodies. Of course, additions of soluble polymers will decrease the casting rate, depending on the increased viscosity, and will thereby reduce the benefit with pressure casting, through the reduced casting rate. Still, modern pressure casting with plastic porous molds has the advantages of producing more reproducible components with molds that are nonsoluble and noncorrodible and have a repeatability of kinetics, even if the casting rate is reduced by necessary processing aids. This speaks very much in favor of slip casting in plaster molds.

The drawbacks with the large density gradients in pressure cast flocculated slips were counteracted by an isopressing procedure described by Lange and Miller [11]. The isopressing is a combination of pressure casting and cold isostatic pressing performed in the same mold. Using flocculated slips, the handling of multicomponent slips where particles of quite different densities and sizes can be present are enhanced, because particle segregation does not occur in flocs, at least when submicron powders are used. During the first stage of isopressing, i.e., a normal pressure casting step, a density gradient is developed; in the second stage, the cold isostatic pressing part, a uniform pressure is built up by the flexible mold, and components with uniform densities can be produced.

B. Processing Agents in Slip Casting

In order to obtain a suitable consolidation rheology, handling strength, etc., in slip casting, pressure casting, and centrifugal casting, the interparticle forces have to be controlled by dispersants, and green strength has to be controlled by long chain molecules. There is very little literature on processing agents addressing only pressure casting or centrifugal casting, as these methods have had a relatively short period of use. There is, however, much more literature describing processing agents for slip casting and colloidal stability in general, and it is also useful for the other casting methods. The review below presents processing agents for slip casting but may be of interest for all casting methods. Processing agents for tape casting are reviewed in Section IV.A.

Slip casting can be performed in water as well as in organic solvents. Organic solvents are used if the powder properties deteriorate in water. Polyelectrolytes are widely used in industrial applications as dispersants for ceramic powders in water. Polyelectrolytes, when adsorbed, have the potential of creating a large energy barrier preventing flocculation of colliding particles, owing to a combined effect of both electrostatic and steric stabilization. In order to understand the adsorption mechanisms and to be able to optimize slip rheology, a thorough characterization of the powder surface charge characteristics must be performed; see the previous chapters.

The polyelectrolyte polyacrylic acid (PAA) is frequently used in the ceramic industry and is sold under various trade names. PAA adsorbs on oxide particles and stabilizes suspensions of alumina [19–21], rutile, and hematite [22,23]. Anionic polyacrylic acid molecules adsorb with a high affinity type of adsorption on positively charged particles. Gebhardt and Fuerstenau [22] showed that PAA adsorbed strongly onto positively charged hematite and rutile particles from pH 4 up to their different p.z.c. values (6.3 for rutile and 8.3 for hematite). At the p.z.c., the net charge of the particles changes from positive to negative, and the particles begin to repel the anionic polymers. It was also found that PAA did not adsorb on silica that has a p.z.c. value of 2.5 [22]. The mechanism for adsorption at low pH values is believed to be due to both electrostatic forces and hydrogen bonding.

Cesarano et al. [19] presented findings similar to those of Gebhardt and Fuerstenau [22] for the system polymethacrylic acid (PMAA) adsorbing on alumina particles. The PMAA molecules adsorb strongly on alumina particles in the pH range of 3.5 to 8.7. Several authors [19–23] have found that the adsorbed amount of PAA or PMAA increased when the pH value decreased. This is probably due to adsorption of thicker layers of the polyelectrolyte. The degree of ionization of the anionic polyelectrolytes decreases with pH, and hence the polymer chain becomes more and more flexible and permits the existence of more trains and loops when adsorbed at a solid-liquid interface.

Cesarano et al. [19] used the adsorption results to make a "stability map" for the system PMAA (M_w: 15,000) and alumina; see Fig. 5. The solid curve in Fig. 5 indicates the optimal addition of PMAA to alumina to create stable slips.

While polyacrylic acids have a strong affinity, especially to positively charged ceramic particles, the cationic polyelectrolyte polyethyleneimine (PEI) adsorbs on negatively charged ceramic powders (SiO_2, Si, SiC, etc.). Polyethyleneimines have mainly been evaluated for their flocculation capacity of aqueous silica suspensions [24–28], but there is also evidence in ceramic processing that PEIs can stabilize suspensions [29]. In solutions, PEIs are highly branched molecules with spherical shape. At pH > 10, PEIs have a low degree of dissociation. Lowering the pH causes the molecules to become increasingly positively charged, which causes expansion owing to internal repulsion between charged sites. The adsorption isotherms for the PEI silica system are of the high affinity type [24,25].

Lignosulphonates (LS) are polyelectrolytes used both in traditional ceramic applications and in processing of high-performance ceramics. Le Bell et al. [30] studied the stabilization of kaolin particles by LS. The LS molecules in solution are reported to be almost spherical with chains of

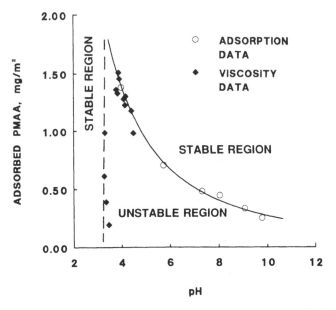

FIG. 5 Stability map for the system PMAA/Al$_2$O$_3$. (From Ref. 20.)

sulphonate and hydroxyl groups at the surface. The LS molecules are negatively charged throughout the pH range of interest in slip casting, i.e.,
approximately pH = 3–10. The LS was adsorbed on the positive sites of the
clay particles at low pH values. At high pH values the LS stabilized the suspension without being adsorbed on the kaolin particles. According to le
Bell et al., the mechanism for stabilization is not a steric stabilization owing
to the lack of adsorption. The highly charged LS molecules are thought to
increase the repulsion between approaching particles, and depletion stabilization can be a relevant explanation.

Slip casting with silicon nitride slips stabilized with LS as the dispersant
has been reported [31–33]. The main effect of LS was the reduction of the
viscosity level at all shear rates and a more Newtonian behavior of the slips,
i.e., less pseudoplastic slips.

Binders such as starches, derivates of cellulose, alginates, polyvinylalco-
hols (PVA), polyethyleneglycols, etc., are used to increase the strength of
the green bodies. The binders are added to the slurry prior to the slip casting operation. The main function of the binders is to increase the strength
of the dried green bodies. The green strength can be increased several
times by the binder after drying. PVA is one of the most frequently used

binders for ceramics processed in water systems, mainly in slip casting but also in dry pressing. Latex binders have been investigated for pressing [34,35] but have also been used in slip casting [34] and tape casting [34,36]. Latex suspensions usually have very low viscosity ($<$ 200 mPas) in relation to the concentration ($>$ 50 vol % polymer). The benefit of latex as a binder is evident after drying above the film forming temperature. The latex will spread onto the particle surfaces during drying, and the polymer in the latex particles will cross-link and form strong bonds. Binder migration is low as compared with water-soluble binders [35].

When slip cast components adhere strongly to the plaster mold, release agents have to be used. Dilute solutions of sodium or ammonium alginates are used as release agents. The molds are impregnated with the solution for one or two minutes and the solution is then sucked into the pores by capillary force. The alginate coagulates with Ca ions to form calcium alginate at the mold surface. The Ca alginate is precipitated as a thin permeable film at the mold surface. This film decreases the casting rate but facilitates the release of the cast component. The film also prevents plaster particles pulled out during mold release from adhering to the green body, which can otherwise cause defects during sintering, when the plaster particles decompose. After drying of the green body, the very thin Ca alginate film is easily removed. The disadvantage of alginates as release agents is that they consume Ca ions and thereby increase the corrosion of the molds.

C. Slip Casting in Plaster Molds

Slip casting of high-performance ceramics is usually restricted to filtration of stable and low viscosity slips. Application of weakly flocculated slips is limited by the requirement of a dense and homogeneous cast layer. The compressible cakes that usually result from flocculation can result in density variations in the cast layer that cause warping and cracking during drying and sintering.

The pore size and particle size distribution influence the filtration resistance and the particle packing characteristics (particle packing gradients) during slip casting of both clay systems [37] and nonclay systems [15,38]. The finest particle fraction can be transported through the cast and deposited close to the mold surface, resulting in uneven packing density [39]. Hampton et al. [15] showed that there are two regions in the cast layer, one close to the mold surface with an enrichment of fine particles and the other close to the slip with a deficit of fine particles. Both regions increase in volume during slip casting, but their growth is governed by different proportionality constants.

Entrapped air is one reason for defects in slip casting. Although the air bubbles give a relatively smooth and harmless shape to the formed pores, they may be large, > 100 µm, and must therefore be removed. It is easiest to treat a slip beaker with gentle stirring under reduced atmospheric pressure. It should be possible gently to stir the slip with paddles that continuously lift a small volume of slip from the bottom up to the slip surface where the entrapped air is released. This behavior is to be recommended when larger volumes of slips are going to be de-aired. Otherwise there will be a long mean distance for the air bubbles to reach the slip surface, and the de-airing will be too time-consuming.

Some powders slip cast in water are known to produce gas, for example silicon metal powders [13,40,41] and silicon nitride powders. The gas formation with silicon powders is mainly due to the hydrolyzation of the particle surfaces, and hydrogen gas is produced. The gas problems can be diminished by conditioning the slips. The experience of the author is that different batches of the same grade of silicon powder can show gassing to different extents, and long-term conditioning does not always completely finish the gas formation. Making reliable gas bubble free slips may only be possible if a nonpolar solvent is used [42]. In the case of silicon nitride, the gas formation is more easily reduced by conditioning in water. The reason for the gas formation may be the hydrolyzation of residual free silicon, but it may also be incomplete wetting of agglomerated and voluminous silicon nitride powders. The degree of agglomeration of the silicon nitride powders can be quantified by the very large specific surface area and the concurrent large particle (agglomerate) size [43,44].

To avoid foaming during degassing or during dispersing of powders, a defoaming agent has to be added. There are several different types on the market, of which alcohols (n-octanol) and silicone emulsions are the most frequently used. Alcohols are efficient in destabilizing foams during addition but have the risk of stabilizing the foam later in the process and creating a more serious problem.

III. CENTRIFUGAL CASTING

Interest has recently been shown in centrifugal casting because of the reduced risk, as compared with other colloid forming methods, of stress gradients in the formed body [45]. The stresses during forming act on every single particle, and during centrifuging the particles move downwards, or outwards, whereas the liquid flows upwards, or inwards. This means that compared with slip casting and pressure casting, the total volume of liquid in the dispersion is not transported through the cast body. Centrifugal cast bodies will therefore show fewer defects such as filtration channels.

Beylier et al. [45] investigated centrifugal casting of electrostatic stabilized alumina dispersions. The alumina powder used was a submicron powder with a relatively narrow particle size distribution (0.2–0.5 µm). Their experiments showed that the solids content of the dispersion very much influenced the sediment buildup rate: The higher the solids content the lower the sedimentation rate. Despite this, solids contents from 10 to 50 vol % of alumina resulted in green densities > 60% after 2 h with a centrifugal acceleration of at least 2500 g. Relatively complex shaped components, i.e., spools and screws, were cast. The components were easily dried without crack formation or nonuniform shrinkage.

Chang et al. [46] studied centrifugal casting of stabilized, flocculated, and coagulated dispersions of both alumina and alumina-zirconia mixtures. Stabilization was accomplished by acid addition to achieve electrostatic stability at pH 2 or 4. Flocculation was caused by increasing the pH to 9, i.e., the iep for alumina. Coagulation resulted when an indifferent salt was added to the stabilized dispersions. The indifferent salt was added until the attractive interparticle forces dominated the repulsive forces. The particle network formed in the coagulated system was weaker than the network formed in the flocculated system. The authors argued that for pressures > 0.5 MPa, i.e., within the range for pressure casting, these weakly coagulated particle structures were pressure independent, in contrast to flocculated structures.

In contrast to the study of Beylier et al. [45], Chang et al. [46] showed that particle/agglomerate size segregation and material segregation occurred when stabilized systems were cast centrifugally. Segregation of particles and agglomerates was not observed by Beylier et al. [45], probably due to better dispersion of the powder and a higher solids content used for the complex shaped components.

The coagulated system tested by Chang et al. [46] had higher green densities (58%) than the stabilized system (57.3%) and considerably higher green densities than the flocculated system (42.4%). The coagulated system was promising in the sense that it had the same level of green density as the stabilized dispersion and showed no particle size segregation or material segregation. The coagulated system was believed, according to the DLVO theory, to have a weak attractive minimum and a short-range repulsive potential. The system acts as a lubricant and makes it possible for the particles to rearrange when they collide during consolidation.

Bergström et al. [47] studied centrifugal casting of alumina suspensions in an organic solvent. The repulsive interparticle forces were controlled by adsorbing fatty acids. By minimizing the steric barrier with oleic acid, weak alumina flocs were consolidated and uniform bodies with high densities were formed.

IV. TAPE CASTING

Tape casting is a forming method for producing thin and flat ceramics mainly for the electronics industry. Ceramic sheets with thicknesses between 0.025 and 1.0 mm can be produced with tolerances of ± 10% or better [48-50]. In precision tape casting, sheets with thicknesses of 0.025-0.25 mm and with tolerances of ± 0.008 mm can be produced [49]. The sheets are used as substrates, capacitors, or sensors. The most widely examined and most frequently published systems are Al_2O_3 for substracts and $BaTiO_3$ for capacitors. In Fig. 6, the principle of the process is illustrated. A well dispersed, stabilized slip is poured into a container that is in close contact with a moving plastic sheet. The slip is transported by the plastic sheet from the container through a narrow gap that gives the slip exact and uniform thickness. The narrow gap consists of one or two knives. This gives the slip an exact thickness [49]. The final sintered thickness, despite the exact thickness of the slip layer, is dependent on slip stability, solids content, the rheology of the slip, the velocity of the plastic sheet, particle packing, green density, and shrinkage during sintering.

Aqueous tape casting has been used [36,49], but nonaqueous systems are more common [36,51-53]. Tape casting differs from slip casting mainly in the way of removing the liquid part of the slip. In slip casting most of the liquid is absorbed by the mold while the component consolidates. The liquid in the consolidated body is expelled by evaporation after mold release. In tape casting, all the liquid is removed by evaporation.

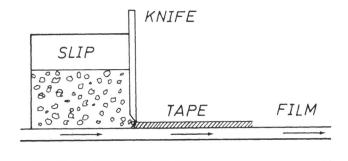

FIG. 6 Schematic picture of the tape casting process.

The evaporation begins as soon as the slip has entered the carrier film. The drying is performed by blowing filtered air in a counterflow toward the cast sheet. In this way the driest part of the continuously produced sheet will meet the driest air. When the fresh tape casting slip meets the drying air, it is saturated with solvent, which results in slow initial drying. The solvent is transported to the surface of the cast during the first stage of drying by influence of the capillary forces. In the second stage, the drying is governed by a slow diffusion process with the drying rate decreasing with time. If, during the first stage of drying, a too large air flow is used, the solvent will not be transported to the surface fast enough, and a skin might form. The skin is a potential source of defects, as it can trap gas bubbles and reduce the drying rate, since this rigid but permeable layer reduces solvent diffusion [50,54].

There are special requirements for green strength in tape casting, as compared, for instance, with slip casting. The cast sheets must be able to be stored on reels and punched without cracking. To create sufficient handling strength by a relatively large amount, up to 30 vol % of binders, including necessary plasticizers to make the binder flexible, are added. The ceramic sheets with green densities in the range of 55 to 60% have tensile strengths of 1 to 2 MPa [54,55].

A. Processing Agents in Tape Casting

The processing agents for tape casting are treated separately from processing agents for the other casting methods, since in industry tape casting is performed almost solely in nonaqueous media and since tape casting makes extraordinary demands with regard to high green strength.

The organic processing agents used in tape casting can be categorized into the following groups: deflocculants, binders, plasticizers, wetting agents, and defoamers. Only nonaqueous systems are described here. Williams [49] and Nahass et al. [36] have shown some of the few available recipes for water-based tape casting systems.

Extensive lists of organic processing agents for nonaqueous tape casting have been drawn up by Williams [49] and Mistler et al. [50]. In nonaqueous tape casting, azeotropic solvents such as methyl ethylketone/ethanol and trichlorethylene/ethanol are often used. The solvent mixtures should give low viscosity and an extended range of solubility for organic processing agents. Azeotropic mixtures are used to avoid changes in concentration of one of the solvents during processing of the tape cast sheets.

Dispersants for the powders in nonaqueous tape casting should be added prior to the addition of the other more or less surface-active processing agents, in order to avoid competitive adsorption and the risk of not reaching an optimum viscosity level. Braun et al. [53] have examined how the order of addition of the processing agents affects the rheology of tape casting slips. When the binder was added prior to the dispersant, an unexpected increase in viscosity occurred. They found that the binder, in this case polyvinylbutyral, could occupy some sites of the powder surfaces if added before the dispersant, a phosphate ester, and thus decrease the amount of adsorbed dispersant.

The standardized tape cast slurry procedures [49,51] consist of a two-step milling (i.e., dispersing) process. The first step involves the mixing and deagglomeration of the powders in the solvents with the aid of the dispersant. In the second step, binders and plasticizers are added and ball milling is applied to homogenize the slip and create proper slip rheology. The binder, including the plasticizers, is added later in the process to avoid competitive adsorption [53] and degeneration of the long chain binder molecules during ball milling. Binder degeneration can increase viscosity in the ball mill, which decreases the deagglomeration efficiency [50]. In Table 1 the composition of typical tape casting slurries is presented [52].

Menhaden fish oil, glyceryl trioleate, and phosphate ester are among the most frequently used dispersants in nonaqueous tape casting [48,56]. Menhaden fish oil has been proven to be an effective dispersant for ceramic powders in organic liquids [50,57]. The main disadvantage of this product is that it is a natural one, with the inherent shortcomings of being a mixture of more or less defined products.

Glyceryl trioleate, believed to be the nearest pure analog to Menhaden fish oil, is not as effective as oxidized Menhaden fish oil. Menhaden fish oil acts as a steric stabilizer by the strongly adsorbing carboxylic acids, which attach to the particle surfaces, and the hydrocarbon chains extending out into the liquid. The versatility of fish oil as a dispersant is explained by the high solubility in various more or less polar liquids. The cross-linked polymeric structure includes several sites for adsorption, and the molecules adsorb densely, which improves the dispersion stability [57].

In Fig. 7, adsorption isotherms for the adsorption of fish oil and glycerol trioleate onto alumina in toluene are shown [58]. It can be seen that the adsorption of fish oil is almost three times higher than the adsorption of glycerol trioleate. A thicker and denser layer of steric stabilizer always results in better colloidal stability. Tormey et al. [58] also investigated the effects of varying the amount adsorbed on the sedimentation volume. They found that the sediment densities increased from 25 to 45 vol % with an increase in the adsorbed amount of fish oil from 0.66 to 1.18 mg/m ^{2}.

TABLE 1   Typical Tape Casting Slurries of Al_2O_3 and $BaTiO_3$

| Components (vol %) | Tape casting slurries | | |
	$BaTiO_3$	Al_2O_3	Al_2O_3
Ceramic powder	35	29.65 +1.06 (MgO)	30.52
MEK/ethanol	33.25		57.6
Trichloroethylene		31.6	
Ethanol		22.18	
Phosphate ester	0.75		0.77
Menhaden fish oil		2.01	
Acrylic binder	20		
Polyvinylbutyral		4.73	4.71
Butyl benzyl phthalate	5		
Polyethylene glycol	5	4.51	2.85
Dioctyl phthalate		4.26	
Dibutyl phthalate			3.47
Cyclohexanone	1		

Source: Ref. 52.

Phosphate esters are another type of dispersant that have ethoxylated chains attached to a phosphorous molecule by an ester bond. The ethoxylated chain consists of a hydrophilic part, i.e., the polyoxyethylene group, and a lipophilic part, i.e., an alkyl group. This amphiphilic characteristic of the molecule is the explanation for its ability as a dispersant of ceramic powders in organic liquids.

Chartier et al. [59] investigated the dispersing effect of various phosphate esters in the system Al_2O_3-MEK/ethanol. Using viscosity measurements of highly concentrated slips, they evaluated phosphate esters with different molecular structures. The best results were achieved using a phosphate ester where (1) the alkyl tail is linear, (2) the phosphate ester is mainly a diester, (3) the phosphate ester has a high degree of phosphatization, and (4) the HLB value (the hydrophilic-to-lipophilic balance) is relatively high.

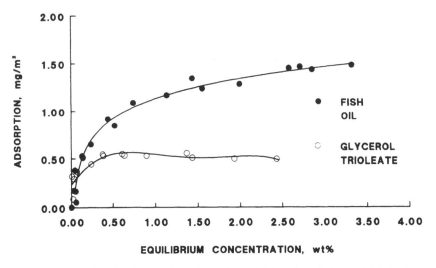

FIG. 7 Adsorption isotherms for adsorption of fish oil and glycerol trioleate on alumina. (From Ref. 58.)

Phosphate esters were also used when tape casting of AlN was developed [55].

There are two main types of binders used in tape casting: polyvinyl butyral and acrylic binders. The binders are used to increase the flexibility of the cast film and therefore have to be used above the glass transition temperature. To decrease the glass transition temperature, which is usually above room temperature, plasticizers are added, often in larger amounts than the binder. Plasticizers for polyvinyl butyral are glycols and phthalates (dioctyl phthalate, dibutyl phthalate). Acrylic binders are plasticized with dimethyl or butyl-benzyl phthalate and polyethylene glycol [52].

Cima et al. [60] found, by studying the binder removal of a polyvinylbutyral/dibutylphthalate system in a tape cast alumina sheet, that capillary forces influenced the distribution of polymer-plasticizer. The first stage of binder removal has similarities with drying in that the more volatile plasticizer can be drawn to the surface through capillary forces.

B. Powder Requirements in Tape Casting

The particle packing during tape casting affects performance concerning reliability and reproducibility of the sintered cast layers. Tormey et al. [58] noticed a tendency toward particle size segregation in a well-stabilized slip and recommended the use of narrow particle size distributions to avoid segregation.

Kaneko et al. [61] also examined the particle segregation issue using two different powders, Al_2O_3 and ZnO, with different densities, 3.98 and 5.66 g/cm^3, respectively. It was found that the particle size difference was the main cause of the segregation. ZnO particles, when slightly smaller than the Al_2O_3 particles, were enriched at the bottom of the cast layer. However, when ZnO particles larger than the Al_2O_3 particles were used, an increased ZnO content was revealed at the top of the cast layer. This was explained by a mechanism wherein the large particles settle to the bottom first and the fine particles fill up the interstices, irrespective of the particle density. The rates of particle settling and segregation are, of course, influenced by particle density. Particle segregation can be diminished by narrow particle size distributions, high solids content, high drying rate, and increased liquid viscosity (for example, through the addition of soluble high molecular processing agents).

Frantz et al. [44] showed that particle size and particle morphology characterization could be improved. By characterization of powders with two different particle size measurement techniques (sedimentation and light scattering) together with specific surface area measurements (BET) it was possible to estimate not only particle size distribution and specific surface area but also properties such as internal particle porosity, surface roughness, and deviation from sphericity. The different calculated values of the properties become more meaningful by examination of the powders under scanning electron microscopy.

V. REQUIREMENTS FOR OPTIMIZED CAST MICROSTRUCTURES

The expression "ideal" powders has been used in the area of high-performance ceramics. Generally, the expression stands for particle sizes between 1 and 0.1 μm, in which the particles should have a narrow size distribution and be equiaxed and nonagglomerated. The upper limit is determined by the requirement of high reactivity (expressed as high specific surface area) to be able to sinter these often highly inert materials. The upper limit is also a matter of decreasing both the flaw size and the risk for abnormal grain growth. The lower limit, 0.1 μm, is a critical particle size below which problems with the various forming techniques accumulate. The problems can appear, for example, as extended difficulties during powder deagglomeration, and decreased maximum solids content in the slips.

At a particle size below 0.1 μm, the preparation of a slip requires increasing amounts of organic surface-active processing aids, owing to the steep increase in surface area. The resulting green density decreases, and problems with crack formation during drying and/or binder removal increase [7,62–65].

Although many of the commercial powders available today are coming closer and closer to satisfying these requirements, they still have to be treated before casting. Lange [66] has summarized the potential risks involved in using submicron powders. Agglomerates are always present in as-received powders and must be removed, in order to enhance densification and avoid defects such as cracklike voids. Impurities in the form of organic material (often with low densities) can cause large pores after firing. Lange also emphasized the importance of obtaining homogenous mixtures of powders when using multicomponent systems.

The goal in every forming process or in the development of forming processes is to achieve defect-free, uniformly packed particles that yield a high, uniform green density and a small, narrow pore size distribution. If this can be accomplished, there is great potential for many ceramic systems to sinter at low temperatures to high densities with a microstructure with small, narrow grains [65,67,68].

Roosen and Bowen [67], for example, showed that for α-Al$_2$O$_3$ shaped with different colloidal forming techniques, the smaller the average pore size in the green bodies, the lower the temperature for maximum shrinkage rate and the higher the final density.

Yeh and Sacks [65] showed that by using an α-Al$_2$O$_3$ powder of mean particle size of approximately 0.1 μm, it was possible to achieve a high green density, 69%, by slip casting. The compacts sintered at a very low temperature, 1150°C, to almost full density, >99.5%. This result can be compared with sintering of α-Al$_2$O$_3$ powder of poorer morphology, or larger and unclassified powders that often require sintering temperatures above 1500°C [69]. Yeh and Sacks also refer to work in the sol-gel area, of which the results, expressed as green densities, sintering temperatures, and final densities, were inferior to their own work using the powder route. The explanation of the very low sintering temperature in the Yeh and Sacks work was the very small mean pore sizes, 10 nm, and the high green density. The well-dispersed and agglomerate-free particle suspension with its small particles is the main explanation for the success.

Green microstructures obtained by different casting operations all have some type of inherent defects. Usually a casting operation creates more uniform particle packing and has a larger potential for reducing the size of agglomerates than dry pressing. On the other hand, forming techniques that are shear induced, i.e., extrusion and injection molding, result in narrower pore size distributions [70,71]. In a viscous processing technique where high shear rates are applied to reduce the agglomerate/aggregate sizes, very promising results with reduced sintering temperature and high strengths have been shown for alumina and titania. For example, viscous processed alumina could be sintered at a temperature of 1200°C and a strength above 1 GPa obtained [68,72]. The residual defects in the sintered microstructures could be related to particle size and dispersability in the viscous processed samples of the powders [73]. This was the case even if two very pure and easily dispersible submicron powders with good particle morphologies were used.

The question is whether the particle packing and green microstructure in the various casting methods can be further developed to achieve the properties obtained by the high shear forming methods. Slip casting of purely electrostatically stabilized slurries has, as described above, been shown to form green compacts that sinter at reduced temperatures. According to Aksay [74], slip casting and tape casting form second-generation pores (pores larger than particle sizes). This is due to the formation of flow channels because of filtration, in the case of slip casting and pressure casting, and because of drying, in the case of tape casting. Cast microstructures have also been shown to improve the sinterability of the green microstructures [14]. Perhaps it would be possible with some degree of steric stabilization to improve the particle packing and green microstructure to make them resemble the particle packing of optimally extruded and injection molded compacts.

Suggestions have been presented for further improvements of green microstructures and their homogeneity in terms of no density gradients, narrow pore size distribution, and even distribution of sintering agents in multicomponent systems. These improvements should result in enhanced sinterability and reduced numbers of defects regarding both number and size. For example, metal alkoxides have been used chemically to bond steric stabilizers [75] or to surface-modify [76,77] powders. This has the advantage that multicomponent systems can be dispersed without having to use different dispersants for the various powders, and the risk of heteroflocculation in the slip is avoided. A properly chosen metal alkoxide can also act as a dopant [77].

Recent developments have shown that slips can be gelled or "frozen" with the particles randomly oriented in the cast. This forming technique has the potential of not only improving the green microstructures of bodies cast with traditional techniques but also replacing traditional casting techniques (and also injection molding) with a new powder consolidation method. A new set of processing agents then has the potential of supporting the forming of both thin ceramic sheets and more complex components. The gelling can be accomplished by polymerization of the continuous media initiated by ultraviolet radiation [78], heating [58], or the addition of initiators. To be able to benefit from this type of gelling system, some requirements have to be fulfilled. The systems must have high solids loading and low viscosity in the ungelled state. The gelling should be performed without altering the particle packing. The gel must have good handling strength, and all organic processing agents must be able to be removed without deforming the powder compact.

Young et al. [79] have successfully gel cast alumina in a water system. The alumina powder was dispersed with ammonium polymethacrylate. To the stabilized slurry a mixture of monomers, acrylamide, and methylenebisacrylamide was added. After homogenization, an initiator, ammonium persulfate, and a catalyst, tetramethyl-ethylenediamine, were added in order to start the polymerization leading to gelling of the slurry. The gelling time could be varied from 5 to 120 min by controlling the amounts of initiator and catalyst as well as the temperature. Alumina slurries with low viscosity, approximately 1800 mPa, at solids content > 60 vol % were gel cast. The drying shrinkage was less than 1%, but the drying had to be performed for several hundreds of hours in a controlled atmosphere with high humidity. The dried bodies contained less than 4 wt % organic binders, which still means that a controlled binder burn-out process is necessary.

VI. SUMMARY

The colloidal processing methods discussed here — slip casting, pressure casting, centrifugal casting, gel casting, and tape casting — all make it possible to deagglomerate, fractionate, and "purify" the as-received powders. The powders are kept in a fluid or semifluid state until the final forming to a component where strong interparticle bondings form. It has been shown that by interparticle force control the consolidated cast microstructures of the ceramic materials can be optimized.

Examples have been given of how to stabilize ceramic powders with dispersants. When developed stabilized casting slips are applied in industrial forming processes, it is important to use the low viscosity obtained to increase solid content (in order to prevent segregation) as well as casting rate.

For both slip casting and pressure casting, introduction of some degree of instability in the slurry clearly increases the production rate, but this must be done without creating compressible systems resulting in density gradients. It is at this moment unclear whether the instability should be achieved by an insufficient amount of dispersant or by adding electrolyte to reduce the repulsive forces. Both methods are probably of interest even for non-clay systems. The increased industrial use of pressure casting with plastic porous molds has increased the need for a more fundamental understanding of how submicron powders can be consolidated even when the mold pore channels are 20 to 40 times larger than the particles. Successful pressure casting of submicron powders is governed by an optimized slurry stability, maximum solids content, suitable slurry rheology, and optimum pressure/time cycle for every material.

As a result of the pressure cast components being released from the molds fully saturated with the filtrate, problems with wet green body distortion have arisen. More research is needed to show if the green body fluidity is counteracted by adjusting the interparticle forces, by organic binders or rheology additives influencing only the filtrate viscosity.

Centrifugal casting and gel casting have recently shown to be promising forming methods. Their greatest potential is probably for delicately shaped components with demands for close tolerances. However, a high, uniform green density must be obtained reproducibly. Centrifugal casting and gel casting may then be interesting alternatives to injection molding and may help to avoid binder removal problems.

For all casting methods, it is of interest to develop further the dispersions of multicomponent systems. Metal alkoxides and sols, used for coating particle surfaces or surface modification, may be another way of receiving interparticle force control in both aqueous and nonaqueous systems, in addition to the use of various organic dispersants.

REFERENCES

1. F. M. Tiller and C.-D. Tsai, *J. Am. Ceram. Soc. 69*(12): 882–887 (1986).
2. I. A. Aksay and C. H. Schilling, in *Advances in Ceramics, Vol. 9: Forming of Ceramics* (J. A. Mangels and G. L. Messing, eds.). American Ceramic Society, Columbus, Ohio, 1984, pp. 85–93.
3. C. S. Hogg, *Ceramics Int. 11*(1): 32–36 (1985).
4. D. Scherzer, cfi/Ber. DKG 66 (1/2), 62–70 (1989).
5. D. W. Richerson, *Modern Ceramic-Engineering—Properties, Processing and Use in Design.* Marcel Dekker, New York, 1982.
6. J. E. Funk, in *Advances in Ceramics, Vol. 9: Forming of Ceramics* (J. A. Mangels and G. L. Messing, eds.). American Ceramic Society, Columbus, Ohio, 1984, pp. 76–84.
7. J. S. Reed, *Introduction to the Principles of Ceramic Processing.* John Wiley, New York, 1988.

8. D. S. Adcock and I. C. McDowall, *J. Am. Ceram. Soc. 40*(10): 355-362 (1957).
9. F. M. Tiller and N. B. Hsyung, *J. Am. Ceram. Soc. 74*(1): 210-218 (1991).
10. K. Bridger and M. Massuda, in *Ceramic Powder Science III* (G. L. Messing, S. Hirano, and H. Hausner, eds.). American Ceramic Society, Westerville, Ohio, 1990, pp. 507-519.
11. F. F. Lange and K. T. Miller, *Am. Ceram. Soc. Bull. 66*(10): 1498-1504 (1987).
12. T. J. Fennelly and J. S. Reed, *J. Am. Ceram. Soc. 55*(8): 381-383 (1972).
13. M. D. Sacks, *Am. Ceram. Soc. Bull. 63*(12): 1510-1515 (1984).
14. Y. Hirata and I. A. Aksay, in *Ceramic Microstructures '86* (J. A. Pask and A. G. Evans, eds.). Plenum Press, New York, 1987, pp. 611-622.
15. J. H. D. Hampton, S. B. Savage, and R. A. L. Drew, *J. Am. Ceram. Soc. 71*(12): 1040-1045 (1988).
16. B. V. Velamakanni and F. F. Lange, *J. Am. Ceram. Soc. 74*(1): 166-172 (1991).
17. S. Appleton, D. A. Hutchins, and M. H. Lewis, *J. Eur. Ceram. Soc. 8*: 339-344 (1991).
18. T. J. Fennelly and J. S. Reed, *J. Am. Ceram. Soc. 55*(5): 264-268 (1972).
19. J. Cesarano III, I. A. Aksay, and A. Bleier, *J. Am. Ceram. Soc. 71*(4): 250-255 (1988).
20. J. Cesarano III and I. A. Aksay, *J. Am. Ceram. Soc. 71*(12): 1062-1067 (1988).
21. A. Bleier and C. G. Westmoreland, in *Interfacial Phenomena in Biotechnology and Materials Processing* (Y. A. Attia, B. M. Moudgil, and S. Chander, eds.). Elsevier, Amsterdam, 1988, pp. 217-236.
22. J. E. Gebhardt and D. W. Fuerstenau, *Colloids Surf. 7*(3): 221-231 (1983).
23. A. Foissy, A. El Attar, and J. M. Lamarche, *J. Coll. Int. Sci. 96*(1): 275-287 (1983).
24. A. Bleier and E. D. Goddard, *Colloids and Surfaces 1*: 407-423 (1980).
25. G. M. Lindquist and R. A. Stratton, *J. Coll. Int. Sci. 55*(1): 45-59 (1976).
26. J. K. Dixon, V. K. La Mer, Li Cassian, S. Messinger, and H. B. Linford, *J. Colloid Interface Sci. 23*: 465-473 (1967).
27. J. C. Kane, V. K. La Mer, and H. B. Linford, *J. Amer. Chem. Soc. 86*: 3450-3453 (1964).
28. H. E. Ries, Jr., and B. L. Meyers, *J. Appl. Polym. Sci. 15*: 2023-2033 (1971).
29. M. Persson, A. Forsgren, E. Carlström, L. Käll, B. Kronberg, R. Pompe, and R. Carlsson, in *High Tech Ceramics* (P. Vincenzini, ed.). Elsevier, Amsterdam, 1987, pp. 623-632.
30. J. C. Le Bell, V. T. Hurskainen, and P. J. Stenius, *J. Coll. Int. Sci. 55*(1): 60-68 (1976).
31. M. Persson, L. Hermansson, and R. Carlsson, in *Ceramic Powders* (P. Vincenzini, ed.). Elsevier, Amsterdam, 1983, pp. 735-742.
32. E. M. Rabinovich, Sh. Leitner, and A. Goldenberg, *J. Mater. Sci. 17*: 323-328 (1982).
33. W. J. A. M. Hartmann, F. K. van Dijen, R. Metselaar, and C. A. M. Siskens, in *Science of Ceramics 13* (P. Odier, F. Cabannes, and B. Cales, eds.). Editions de Physique, Les Ulis Cedex, 1986, p. C1-7983.

34. N. R. Gurak, P. L. Josty, and R. J. Thompson, *Am. Ceram. Soc. Bull.* 66(10): 1495–1497 (1987).
35. B. Nyberg, E. Carlström, M. Persson, and R. Carlsson, in *Proceedings of the Second International Conference: Ceramic Powder Processing Science* (H. Hausner, G. L. Messing, and S. Hirano, eds.), Deutsche Keramische Gesellschaft, Köln, 1989, pp. 573–580.
36. P. Nahass, R. L. Pober, W. E. Rhine, W. L. Robbins, and H. K. Bowen, *J. Am. Ceram. Soc.* 75(9): 2373–2378 (1992).
37. G. W. Phelps, *Mater. Sci. Res. 11*: 57–65 (1978).
38. G. W. Phelps, A. Silwanowicz, and W. Romig, *Am. Ceram. Soc. Bull.* 50(9): 720–722 (1971).
39. J. H. D. Hampton, S. B. Savage, and R. A. L. Drew, *J. Am. Ceram. Soc.* 75(10): 2726–2732 (1992).
40. A. Ezis, in *Ceramics for High Performance Applications* (J. J. Burke, A. E. Gorum, and R. N. Katz, eds.). Brook Hill, Chestnut Hill, Massachusetts, 1974, pp. 207–222.
41. A. Novotny, E. Gugel, and G. Leimer, in *Keramische Komponenten für Fahrzeug-Gasturbinen* (W. Bunk and M. Böhmer, eds.). Springer-Verlag, Berlin, 1978, pp. 161–200.
42. E. Lidén, L. Bergström, M. Persson, and R. Carlsson, *J. Eur. Ceram. Soc. 7*: 361–368 (1991).
43. G. Wötting and G. Ziegler, *Interceram 2*: 32–35 (1986).
44. F. Frantz, I. Burn, and J. Chase, in *Advances in Ceramics, Vol. 11: Processing for Improved Productivity* (K. M. Nair, ed.). American Ceramic Society, Columbus, Ohio, 1984, pp. 131–141.
45. E. Beylier, R. L. Pober, and M. Cima, in *Ceramic Powder Science III* (G. L. Messing, S. Hirano, and H. Hausner, eds.). American Ceramic Society, Westerville, Ohio, 1990, pp. 529–536.
46. J. C. Chang, B. V. Velamakanni, F. F. Lange, and D. S. Pearson, *J. Am. Ceram. Soc.* 74(9): 2201–2204 (1991).
47. L. Bergström, C. H. Schilling, and I. A. Aksay, *J. Am. Ceram. Soc.* 75(12): 3305–3314 (1992).
48. E. P. Hyatt, *Am. Ceram. Soc. Bull.* 65(4): 637–638 (1986).
49. J. C. Williams, in *Treatise on Materials Science and Technology, Vol. 9: Ceramic Fabrication Processes* (F. F. Y. Wang, ed.). Academic Press, New York, 1976, pp. 173–198.
50. R. E. Mistler, D. J. Shanefield, and R. B. Runk, in *Ceramic Processing Before Firing* (G. Y. Onoda and L. L. Hench, eds.). John Wiley, New York, 1978, pp. 411–448.
51. D. J. Shanefield and R. E. Mistler, *Am. Ceram. Soc. Bull.* 53(5): 416–420 (1974).
52. P. Boch and T. Chartier, in *Proceedings of the International Ceramic Conference AUSTCERAM 88, Part 2* (C. C. Sorrell and B. Ben-Nissan, eds.). Trans Tech Publications, Sydney, 1988, pp. 813–819.

53. L. Braun, J. R. Morris, Jr., and W. R. Cannon, *Am. Ceram. Soc. Bull.* 64(5): 727–729 (1985).
54. A. Roosen, in *Ceramic Transactions, Vol. 1: Ceramic Powder Science* (G. L. Messing, E. R. Fuller, Jr., and H. Hausner, eds.). American Ceramic Society, Westerville, Ohio, 1988, pp. 675–692.
55. T. Chartier, E. Streicher, and P. Boch, *J. Eur. Ceram. Soc. 9*: 231–242 (1992).
56. K. Mikeska and W. R. Cannon, in *Advances in Ceramics, Vol. 9: Forming of Ceramics* (J. A. Mangels and G. L. Messing, eds.). American Ceramic Society, Columbus, Ohio, 1984, pp. 164–183.
57. P. D. Calvert, E. S. Tormey, and R. L. Pober, *Am. Ceram. Soc. Bull.* 65(4): 669–672 (1986).
58. E. S. Tormey, R. L. Pober, H. K. Bowen, and P. D. Calvert, in *Advances in Ceramics, Vol. 9: Forming of Ceramics* (J. A. Mangels and G. L. Messing, eds.). American Ceramic Society, Columbus, Ohio, 1984, pp. 140–149.
59. T. Chartier, E. Streicher, and P. Boch, *Am. Ceram. Soc. Bull.* 66(11): 1653–1655 (1987).
60. M. J. Cima, J. A. Lewis, and A. D. Devoe, *J. Am. Ceram. Soc.* 72(7): 1192–1199 (1989).
61. N. Kaneko, W. E. Rhine, and H. K. Bowen, in *Ceramic Transactions, Vol. 1: Ceramic Powder Science II, A* (G. L. Messing, E. R. Fuller, Jr., and H. Hausner, eds.). American Ceramic Society, Westerville, Ohio, 1988, pp. 410–417.
62. E. Barringer, B. Fegley, R. L. Pober, and H. K. Bowen, in *Ultrastructure Processing of Ceramics, Glasses and Composites* (L. L. Hench and D. R. Ulrich, eds.). John Wiley, New York, 1984, pp. 315–333.
63. I. B. Cutler, in *Ceramic Processing Before Firing* (G. Y. Onoda and L. L. Hench, eds.). John Wiley, New York, 1978, pp. 21–29.
64. J. S. Reed, in *Ceramic Transactions: Ceramic Powder Science II, B* (G. L. Messing, E. R. Fuller, Jr., and H. Hausner, eds.). American Ceramic Society, Westerville, Ohio, 1988, pp. 601–610.
65. T.-S. Yeh and M. D. Sacks, *J. Am. Ceram. Soc.* 71(10): 841–844 (1988).
66. F. F. Lange, in *Ceramic Transactions, Vol. 1: Ceramic Powder Science II, B* (G. L. Messing, E. R. Fuller, Jr., and H. Hausner, eds.). American Ceramic Society, Westerville, Ohio, 1988, pp. 1069–1083.
67. A. Roosen and H. K. Bowen, *J. Am. Cer. Soc.* 71(11): 970–977 (1988).
68. N. McN. Alford, J. D. Birchall, and K. Kendall, *Nature 330*(6143): 51–53 (1987).
69. R. Morell, *Handbook of Properties of Technical and Engineering Ceramics, Part 2: Data Reviews, Section I: High-Alumina Ceramics*, National Physical Laboratory, London, 1987.
70. I. A. Aksay, F. F. Lange, and B. I. Davis, *J. Am. Ceram. Soc.* 66(10): C-190-2 (1983).
71. I. A. Aksay, in *Ceramic Transactions: Ceramic Powder Science II, B* (G. L. Messing, E. R. Fuller, Jr., and H. Hausner, eds.). American Ceramic Society, Westerville, Ohio, 1988, pp. 663–674.

72. N. McN. Alford, K. Kendall, W. J. Clegg, and J. D. Birchall, in *British Ceramic Proceedings, No. 39: Engineering with Ceramics 2* (R. Freer, S. Newsam, and G. Syers, eds.). Institute of Ceramics, Stoke-on-Trent, 1987, pp. 237–246.
73. N. McN. Alford, K. Kendall, W. J. Clegg, and J. D. Birchall, *Advanced Ceramic Materials 3*(2): 113–117 (1988).
74. I. A. Aksay, in *Ceramics: Today and Tomorrow*, 20th Anniversary of the Basic Science Division of the Ceramic Society of Japan, Kobe, Japan, 1985, pp. 71–84.
75. M. Green, T. Kramer, M. Parish, J. Fox, R. Lalanandham, W. Rhine, S. Barclay, P. Calvert, and H. K. Bowen, in *Advances in Ceramics, Vol. 21: Ceramic Powder Science* (G. L. Messing, K. S. Mazdiayasni, J. W. McCauley, and R. A. Haber, eds.). American Ceramic Society, Westerville, Ohio, 1987, pp. 449–465.
76. E. Bostedt, M. Persson, and R. Carlsson, in *Euro-Ceramics, Vol. 1: Processing of Ceramics* (G. De With, R. A. Terpstra, and R. Metselaar, eds.). Elsevier, London, 1989, pp. 1.140–1.144.
77. C.-M. Wang and F. L. Riley, *J. Eur. Ceram. Soc. 10*: 83–93 (1992).
78. R. R. Landham, P. Nahass, D. K. Leung, M. Ungureit, W. E. Rhine, H. K. Bowen, and P. D. Calvert, *Am. Ceram. Soc. Bull. 66*(10): 1513–1516 (1987).
79. A. C. Young, O. O. Omatete, M. A. Janney, and P. A. Menchhofer, *J. Am. Ceram. Soc. 74*(3): 612–618 (1991).

8

Interfacial Aspects of Ceramic Injection Molding

J. R. G. EVANS   Department of Materials Technology, Brunel University, Uxbridge, Middlesex, United Kingdom

I. INTRODUCTION

This chapter discusses procedures for the creation of shape and form from assemblies of ceramic particles by injection molding. However, the issues raised are not confined to injection molding. All procedures that have in common the plastic forming of a ceramic suspension in an organic phase and the subsequent removal of the organic fraction prior to sintering are open for exploration. These include injection molding [1,2] extrusion [3], postextrusion forming [4,5], blow molding [6,7], vacuum forming [8], solvent casting (tape casting) [9], and joining before firing [10–13]. Doubtless many others await investigation. In these processes, ceramic particles are brought into their prefiring positions in an organic vehicle that fills the pore space and is present in slight excess to allow particle rotation and displacement in shear or extensional flow. The organic matter is then removed by solvent extraction, capillary flow, or pyrolysis in oxidizing or inert atmosphere, and the ceramic particle assembly is sintered.

Figure 1 shows an injection molded ceramic rotor joined to a metal shaft and a metallic compressor to obtain a turbocharger assembly. Injection molding machines designed for polymer molding are usually employed with some attention to wear resistance on the barrel and screw and with a low

FIG. 1. Ceramic injection molded turbocharger rotor (diameter 55 mm) attached to a metal shaft and compressor. (Courtesy of J. Woodthorpe, T&N Technology, Cawston, Rugby, U.K.)

compression ratio. A schematic diagram of a reciprocating screw machine
is shown in Fig. 2. Figure 3 illustrates schematically some of the other plastic forming operations.

There are three stages in the manufacture of ceramic artifacts by a powder route: (1) preparation of a ceramic powder with controlled properties,
(2) a method for assembling particles into a macroscopic shape, and (3) the
ability to sinter the assembly to full density. Historically, these three
requirements have been developed neither in parallel nor in sequence. This
chapter addresses the issues that arise in stage 2.

Table 1 lists most of the available fabrication routes and the problems
associated with them. Often the demands placed upon powders by the sintering process conflict with the demands placed by the manufacturing route.
Modern powders typically have surface-area-to-volume ratios greater than
10^7 m^{-1} and thus present areas of 10–50 m^2 g^{-1} to the surrounding media.

The driving force for sintering is the excess free energy associated with
the powder surface, which for spherical particles is given by

$$\Delta G_s = \frac{3\gamma M}{r\rho} \tag{1}$$

where ΔG_s is the excess energy in J mol^{-1}, γ is the specific surface energy of
the material, r is the particle radius, ρ is the density, and M is the molecular
weight. For a given ceramic, r is the only parameter, apart from particle
shape, amenable to modification. The distance over which mass transport
occurs to fill the pore space adjacent to each particle is also proportional to
r, and since diffusion distance approximates to $\sqrt{Dt}$, fine powders also
improve the kinetics of sintering.

Even if sintering proceeds to completion and all pores are removed, the
largest grain diameter then approximates to the critical defect size C and

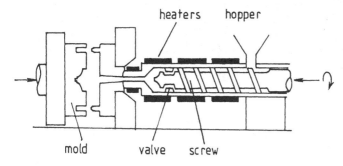

FIG. 2. Schematic diagram of a reciprocating screw injection molding machine.

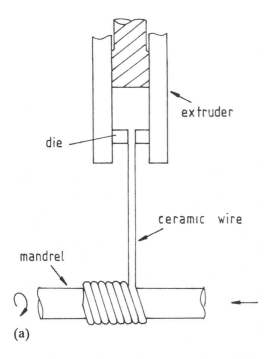

(a)

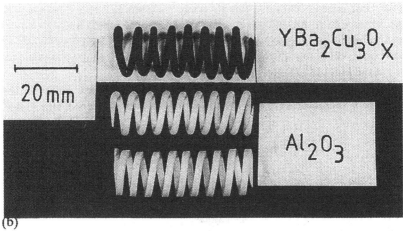

(b)

FIG. 3. Other plastic forming operations for ceramics. Procedures for making (a) ceramic windings; (b) sintered ceramic springs and coils; (c) a ceramic vacuum forming jig; (d) ceramic vacuum formed dome after removal from the binder removal oven; (e) mold for ceramic blow molding; (f) sintered alumina blow molded tubes.

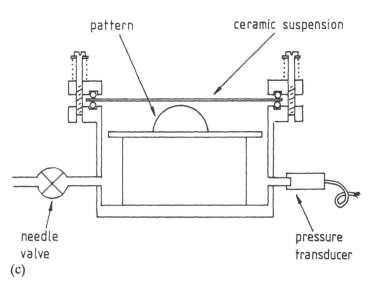

(c)

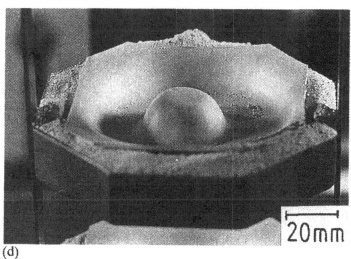

(d)

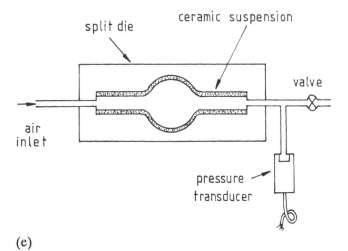

(e)

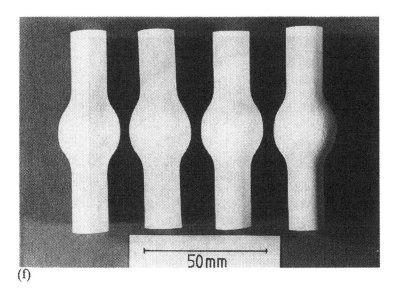

(f)

FIG. 3. (Continued)

TABLE 1 Powder Fabrication Routes

Shaping operation	Advantages	Disadvantages
Die pressing	Rapid, automated	Limited shape, needs agglomerated powders, nonuniform density in large sections
Cold isostatic pressing	Automated	Limited shape, needs agglomerated powders, nonuniform density in large sections
Cold pressing and machining	Complex shapes	Microstructural problems of pressing are carried forward
Hot pressing	High strength, full density	High unit cost, very limited shape
Hot pressing and machining	High strength, full density, complex shape	Very high unit cost
Casting	Complex shape, no binder, low capital cost	Difficult to automate, difficult process control
Pressure casting	Complex shape, automated	Not proven for thick sections
Injection molding	Complex shape, automated	Unsuitable at present for thick sections, high tooling cost, binder to remove
Extrusion	Complex sections, automated	Uniform cross section binder to remove
Tape casting	Automated	Thin flat section only, binder to remove
Vacuum forming	Complex shape in thin-walled body	Binder to remove
Blow molding	Complex shape in thin-walled body	Binder to remove

determines the strenth σ_f according to Griffith's equation, expressed here in its engineering form [14] by

$$\sigma_f = \frac{Z}{y} \frac{K_{1c}}{C^{1/2}} \qquad (2)$$

K_{1c} is the fracture toughness parameter, typically 2–5 MPa m$^{1/2}$ for technical ceramics, while Y is a geometrical constant dependent on the ratio of crack length to sample width and is approximately 1.9 for small cracks [15]. For a halfpenny shaped crack, $Z = \pi/2$. It follows that to achieve strengths of the order of 1 GPa, defects greater than 3–17 µm must be absent.

The driving force for grain growth is the reduction in grain boundary energy, but a metastable position is reached when the condition

$$V\gamma_{gb} \left(\frac{1}{r_1} + \frac{1}{r_2} \right) = 0 \qquad (3)$$

is fulfilled, where V is the molar volume, r_1 and r_2 are the principle radii of curvature of the grain boundary, and γ_{gb} is the grain boundary excess energy per unit area. This situation occurs when each grain is a truncated octahedron and all shared faces between grains are planar [16] and can be approached if all particles are the same initial size. Hence modern developments favor powders that are fine and monodisperse [17]. The condition for minimum grain growth restricts the traditional practice of obtaining efficient packing from multimodal powders. The largest particle diameter must be well below the critical defect size, and the arrangement of the smaller particles into the interstices for efficient packing is then impeded by the strong particle adhesion that characterizes extremely fine particles.

The conflict of interests between sintering and fabrication is most pronounced in the manufacture of ceramics by pressing operations. The powder manufacturer produces a fine powder ideal for a high-strength product. Such powders are then spray dried to confer flow properties by producing agglomerates that may be two orders of magnitude larger than the particles. The pressing operation fails to destroy the agglomerates [18], which leave a succession of defects from large unsinterable pores [19] to cracks [20] of a size comparable to the agglomerates. The resulting strength may thus be one tenth of that which was possible from the powder.

Other approaches to ceramic fabrication have been developed but lack the generality of the powder route. Many glass compositions are now commercially available that can be shaped in the liquid state and subsequently devitrified to produce a fine-grained crystalline solid. This is unlikely to meet the requirements for the extreme operating temperatures proposed for technical ceramics due to residual glass at grain boundaries allowing grain boundary sliding and hence creep at elevated temperatures. There is considerable interest in producing ceramic shapes from polymer precursors [21]. Unfortunately, the volumetric efficiency with which they deposit ceramic residues is very low, resulting in low prefired densities and large shrinkages or the need for successive infiltration.

For these reasons, the main thrust of ceramic technology continues to advance along the particulate route, and since plastic forming processes offer strategic advantages, a consequence is that the interaction between ceramic surfaces and particle conveying fluids must be high on the agenda. Thus flow properties of suspensions, mixing operations, and binder removal are all strongly influenced by interfacial factors.

The surface chemistry of deflocculation has been extensively studied for the case of low-viscosity fluids where surface charge [22] or steric stabilization [23-25] are used to preserve colloid stability, but the corresponding interactions for particles in the melt or where the 'solvent' comprises high molecular weight polymers is far from understood [25].

II. POLYMER-CERAMIC INTERACTIONS

A. Intermolecular and Surface Forces

Adsorption phenomena have an important bearing on a wide range of technologies, notably coatings for corrosion protection, adhesive joining, the fabrication of reinforced polymer composites, the surface treatment of polymer fillers, the preparation of antithrombogenic surfaces, chromatography, the flocculation of suspensions, and the stabilization and manipulation of ceramic suspensions in both low- and high-viscosity liquids.

The forces available in nature for adsorption are mainly of the secondary van der Waals type, which includes London dispersion, dipole-dipole, dipole-induced dipole, and ion-dipole forces; these are surveyed by Moelwyn-Hughes [26]. Although these are referred to as 'weak' forces, the term is strictly relative, for theoretical strength based on dispersion forces alone could more than account for the measured strength of sodium chloride and of thermosetting resins even in the absence of coulombic or covalent interatomic attraction [27]. London dispersion forces [28] provide a universal attraction between matter irrespective of the presence of permanent dipoles. They arise from the attraction between instantaneous synchronized polarizations in neighboring molecules at separations r of less than 20 nm. The energy of interaction of coupled electronic oscillators is proportional to r^{-6} within this region and is given by

$$U = -\frac{3}{4} \frac{\alpha^2 h \nu_c}{r^6} \qquad (4)$$

for identical atoms of polarizability α where $\frac{1}{2}h\nu_0$ corresponds to the quantum mechanical zero point energy. At large separations, the alignment of instantaneous dipoles is hindered by the time taken for electromagnetic wave propagation, and the 'retarded' forces decay with r^{-7}. Tabor and Winterton [29] obtained direct mechnical measurement of retarded and unretarded dispersion forces between crossed, curved mica sheets, and this type of experimental configuration has subsequently been widely used to study the forces between mica surfaces bearing adsorbed polymer layers in solution [30–32]. Such experiments offer a macroscopic route to the experimental study of the forces involved in the stabilization of colloids by adsorbed layers. The difficulty of attaining equilibrium, however, inhibits the investigation of particle interactions in polymer melts.

The presence of permanent dipoles in the polymer provides a higher energy of interaction which, for adsorption on a conducting surface, can be calculated from the interaction between the dipole and its mirror image [33]. Quantum mechanical treatment gives the energy of interaction as

$$U = -\frac{\mu^2}{12r^3} \qquad (5)$$

where μ is the dipole moment and the distance between the dipole center and the conducting plane is $r/2$.

For an ionic substrate, permanent dipoles interact with ions in the surface layers. The energy of interaction is given by [26]

$$U = -\frac{Ze\mu \cos \theta}{r^2} \qquad (6)$$

where θ is the angle subtended by the dipole axis and the line joining the dipole center and the ion. Ze is the electronic charge on the ion. The exceptionally strong polar forces in molecules with $-OH$ or $-NH$ groups give rise to hydrogen bonding with energies of the order 30 kJ mol^{-1} [34]. Such forces may operate where a hydrated ceramic surface makes contact with a hydroxyl-containing organic phase. If the hydrogen bond is treated as a very strong dipole interaction, the energy is given by [28]

$$U = -\frac{2}{3} \frac{\mu_1 \mu_2}{r^6} \frac{1}{kT} \qquad (7)$$

where μ_1 and μ_2 are dipole moments and k is Boltzmann's constant.

Primary forces are uncommon in adsorption phenomena between dissimilar materials, although there are exceptions, notably the siloxane bonds claimed to form between siliceous surfaces and silane coupling agents. In addition, the use of polymers with carboxylic acid groups makes possible and acid-base interaction with basic oxides and hence stronger coulombic adsorption forces. An extreme example is the ionomer cements prepared from polyacrylic acid and divalent metal oxides [35], but full ionic nature is not a necessary requirement for the acid-base interaction. Lewis acids are defined as electron acceptors, and bases as electron donors [36], so that the slight charge asymmetry characteristic of permanent dipoles is sufficient to confer acidic or basic characteristics. Fowkes and coworkers [37,38] therefore relate polymer characteristics to the acidic or basic nature of the ceramic surface in predicting the properties of particle-filled composites.

B. Interfacial Thermodynamics

The work of adhesion between filler and matrix W_A is defined by the Dupré equation in terms of the surface energies γ of phases 1 and 2 and the interfacial energy γ_{12}:

$$W_A = \gamma_1 + \gamma_2 - \gamma_{12} \tag{8}$$

It is thus the negative value of the specific free energy of the adhering system and is equal to the thermodynamic energy needed to separate the surfaces. If positive, it predicts that adhesion is favored. If it becomes negative, for example in the presence of a third (liquid) phase, spontaneous separation is favored.

W_A has been experimentally related to measured adhesive failure energies [39,40]. As discussed below, it is therefore relevant to both the stiffness and more importantly the strength of ceramic-polymer suspensions in the molded state, properties critical for thin sections, notably in solvent casting (tape casting) of electronic substrates and for resistance to fracture under residual stresses in the injection molding process.

The surface energy of polymers is experimentally accessible by contact angle measurements [41–45], and Wu [46] gives a range of values (16–50 mJ m^{-2}). These energies reflect polar and dispersion forces presented to an external medium. The London dispersion force contribution to the surface energy of metals can be estimated, for a close packed solid, from [47]

$$\gamma \approx \frac{A}{24\pi \left(\dfrac{\sigma}{2.5} \right)^2} \tag{9}$$

where A is the Hamaker constant $(2-5 \times 10^{-19}$ J for metals and oxides) and σ is the atomic diameter. This gives $\gamma = 100-200$ mJ m^{-2}, which is much less than the metallic bonding contribution to surface energy, which is $1-2$ J m^{-2}.

Combining the Dupré equation with the Young equation, we obtain

$$\gamma_S = \gamma_L \cos \theta + \gamma_{SL} \tag{10}$$

where θ is the contact angle of the liquid on the solid. Then

$$W_A = \gamma_L (1 + \cos \theta) \tag{11}$$

gives W_A in terms of a fluid property and a measured contact angle. Unfortunately, equilibrium contact angles of viscous fluids of intermediate and high polymers are difficult to obtain, taking several kiloseconds to approach equilibrium [48,49]. Neither do polymer drops on solid surfaces form spherical caps as required for contact angle measurement. Instead, a 'protruding foot' develops at the rim [50].

The thermodynamic work of adhesion between phases 1 and 2 can be expressed in terms of polar and dispersion forces as [51]

$$W_A = 2\sqrt{\gamma_1^d \gamma_2^d} + 2\sqrt{\gamma_1^p \gamma_2^p} \tag{12}$$

where these forces are considered to act independently and to be additive:

$$\gamma_1 = \gamma_1^d + \gamma_1^p \tag{13}$$

Equation (12) rests on the geometric mean relationship for forces between unlike molecules and is regarded as valid for dispersion forces but not always for polar forces [38]. It is argued [38] that the work of adhesion should be expressed in additive terms of dispersion forces, acid-base interactions, covalent forces, and electrostatic forces:

$$W_A = W_A^d + W_A^{ab} + W_A^c + W_A^e \tag{14}$$

where

$$W_A^d = 2\sqrt{\gamma_1^d \gamma_2^d} \tag{15}$$

For acid-base interactions resulting from the interaction of Lewis acids (electron acceptors) and bases (electron donors),

$$W_A^{ab} = \frac{\Delta H^{ab} S}{N} \tag{16}$$

where S is the number of acid-base sites per unit area, ΔH^{ab} is the enthalpy of the acid-base interaction, and N is Avogadro's number. ΔH^{ab} can be found from the theory [52] and experimental data [53] of Drago et al.:

$$\Delta H^{ab} = C_A C_B + E_A E_B \tag{17}$$

where C and E are constants related to the covalent and ionic contributions to bond strength and have units $(J\ mol^{-1})^{1/2}$. Rather than solve Eq. (17) simultaneously, the authors [53] took iodine as a reference acid, putting $E_A = 1$ and $C_A = 1$, and used a series of similar amines as bases in order to obtain C and E for other materials.

Values for a wide range of organic species are tabulated by Drago [53] based on heats of mixing. Values for similar species or homologs can often be derived from Drago's data. As an example, for the interaction between amines and Fe_2O_3, $W_A^{ab} \approx 180$ mJ m^{-2}. Covalent bonds, which are rare in adsorption phenomena, have energies in the region 250–500 kJ mol^{-1} and if present would give rise to $W_A^c \approx 500$ mJ m^{-2}, whereas the electrostatic contribution is typically less than 1 mJ m^{-2}. Several attempts have been made to explain both the mixing and the mechanical properties of mineral-polymer suspensions in terms of acid-base theory, and these are discussed below.

The acidic or basic nature of mineral surfaces can be quantified by calorimetry using test acids or bases with known constants C and E in Eq. (17). Also, infrared spectroscopy can be used to measure the proportion of acid or basic groups whose absorption frequency is shifted by adsorption on the solid. Enthalpy-spectral parameter correlation can thus be used to obtain the enthalpy of mixing and hence values of C and E [53]. A further method is inverse gas chromatography, wherein the column is packed with the mineral powder of interest and the relative elution volumes of acidic or basic vapors are measured [38].

In addition to the adsorption forces at ceramic-polymer interfaces, electrostatic forces arising from charge transfer may arise, and therefore electrokinetic phenomena may be present in ceramic suspensions as with aqueous suspensions [54]. This is explained by adsorption-desorption equilibrium and is especially likely where acid-base interactions are present [55]. If a basic polymer desorbs from an acidic surface after proton transfer has occurred, then particles are left with a negative charge, and a double layer forms that can have a zeta potential of up to −200 mV. It should be remembered, however, that in static solution adsorption-desorption kinetics are generally slow [56].

III. WATER ADSORPTION ON HIGH-ENERGY SURFACES

It would be incorrect to assume that polymers attach directly to a ceramic surface. High-energy surfaces that have been exposed to ambient humidities acquire an adsorbed water layer. The ubiquity of this layer is illustrated in the work of Zisman and coworkers [57,58], who showed from contact angle measurements of organic liquids that after exposure to humidities of 0.6–95% RH, thirteen metals and oxides, as well as glasses, presented a surface energy characteristic of adsorbed water and independent of the underlying metal or oxide. Zettlemoyer and coworkers [59,60] established, by dielectric measurements, that several monolayers of water were hydrogen bonded to surface hydroxyls on Fe_2O_3. On clean metal surfaces it is argued [61] that one monolayer should adsorb, and experiments with very clean surfaces confirm this [62]. It is generally believed that adsorbed water is not fully displaced in the presence of organic fluids even at temperatures of 200°C. Bright and Malpass [63] suggest that adsorbed water layers accounts for the reversibility of adhesion in humid conditions and that the organic phase interacts not directly with an oxide but with a water layer. Such layers may offer a facility for hydrogen bonding with polar polymers but may also screen other interactions with the substrate because of their high dielectric constant.

The structure of surface hydroxyls on ceramic substrates can be studied by near-infrared spectroscopy [64], which distinguishes the free and self-associated surface hydroxyls and those associated with carbonyl groups. This study was particularly interesting in that the adsorbate was polyoctade-cyl methacrylate. The copolymerization of long and short chain methacrylate esters not only allows selective adsorption to control loop size and hence the effective thickness of the adsorbed layer [65,66] but also allows glass transition and the size of decomposition fragments to be adjusted.

The adsorption of water by fine powders and the fact that it is not all displaced during mixing with organic polymers and waxes means that the water must be dispelled during the pyrolytic removal of the binder. If diffusion of water in the organic phase is low, as it is for nonpolar media, this may be a source of defects.

The adsorption area per molecule a_m is given as a function of molecular weight M and liquid density ρ_1 as [67]

$$a_m = f \left(\frac{M}{\rho_L L} \right)^{2/3} \tag{18}$$

where L is Avogadro's number and f is a shape factor taken as 1.019. For water, $a_m = 10.6 \times 10^{-20}$ m^2 per molecule. The weight fraction of water adsorbed on a powder of surface area S (m^2/g) is then

$$A = \frac{nMS}{a_m L} = 0.028 \text{ n.S. wt } \% \tag{19}$$

where n is the number of monolayers. Thus a 36 m^2/g powder may acquire 1 wt % water per monolayer. For a powder of density 4000 kg/m^3, this represents 100 times the volume of the powder as steam at 250°C.

IV. DIMENSIONS OF POLYMER MOLECULES

A polymer molecule with n bonds in the chain, each of length L, tends to configure in the melt or in solution as a random coil. The coil is characterized by two lengths. The average chain end-to-end distance is defined as the root mean square displacement of the ends $\sqrt{\bar{r}^2}$. The radius of gyration of the coil is defined as the root mean square distance of mass elements of the chain from the center of gravity $\sqrt{\bar{s}^2}$. These two distances are related by [68]

$$\sqrt{\bar{s}^2} = \frac{1}{\sqrt{6}} \sqrt{\bar{r}^2} \tag{20}$$

The number of bonds can be found from the degree of polymerization and the number of bonds per monomer unit. For example, vinyl polymers contribute two bonds per unit to the chain. For a freely jointed chain with n links,

$$\sqrt{\bar{r}^2} = L \sqrt{n} \tag{21}$$

The restriction of bond angles together with steric restriction due to side groups and the preference of molecules to adopt low energy conformations means that the mean end-to-end distance of real molecules is enlarged. In general the end-to-end distance can be expressed as

$$\bar{r}^2 = \beta^2 n \tag{22}$$

where β is determined experimentally. For many synthetic organic polymers $\beta \approx 3L$ [69].

In polymer solutions where there is a strong interaction between polymer and solvent (a 'good' solvent), the polymer coil is expanded, while in poor solvents, polymer segments interact with each other, and the dimensions of the coil are reduced. The coil has its unperturbed dimensions in a given solvent at the theta temperature, defined as the temperature at which the excess partial molar free energy of mixing is zero. At much reduced temperatures, the coil contracts and eventually precipitates [70].

V. THE STRUCTURE OF ADSORBED LAYERS

Most studies of polymer adsorption on high-energy surfaces have used solutions, and some extrapolation is needed to understand the situation in the absence of solvent. A common feature of adsorption studies is that the amount of polymer adsorbed from solution per unit area of surface increases with polymer concentration and then reaches a plateau which can, for high molecular weight polymers, extend over the entire experimental range [71].

Theories of polymer adsorption generally ignore the competition between solvent and polymer adsorption [72], although it is a general observation that adsorption from solution is reduced in the presence of a good solvent. Theories of polymer adsorption and structure have been well reviewed [72,73], and the process can be modelled on computer [74]. The fact that measured adsorbed layer thicknesses exceed the value predicted for a flat monomolecular layer was first explained by Jenckel and Rumbach [75] by proposing that molecules protrude from the surface as bristles or loops with only the segments at the ends of loops in an adsorbed state. Such morphology is shown schematically in Fig. 4. This theory was developed by Frisch et al. [76] and by Silberberg [71].

In general, if the adsorption energy is large, then a greater proportion of segments are adsorbed, the loops are smaller, and adsorption is closer to a two-dimensional layer. Conversely, for low adsorption energies, fewer segments are attached, and the loops are larger. In the absence of strong acid-base reactions, the van der Waals attraction generally provides a low adsorption energy, and this is reflected in both a low temperature dependence of adsorption and a strong dependence of adsorption on the type of solvent. Thus according to Silberberg's theory, assuming the chains are sufficiently flexible, if the adsorption energy $E = kT$, 70% of polymer segments

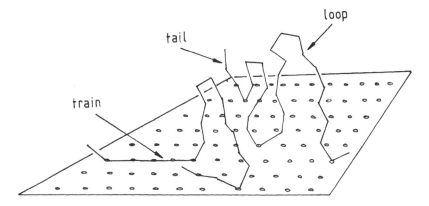

FIG. 4. The structure of adsorbed layers on high-energy surfaces.

are adsorbed, and if $E = 8kT$, the polymer is likely to adsorb in a flat mono-layer configuration.

The structure of the adsorbed layer can be substantially modified by changing the polymer structure. Thus Fontana and coworkers [77] modified polylauryl methacrylate (PLMA) by copolymerizing with 17 mol % N-vinyl-2-pyrrolidone. Unmodified PLMA was attached to SiO_2 by 40% of segments, and the film thickness was only 3 nm. Preferential adsorption of the more polar pyrrolidone groups produced a 20 nm layer and acted as a better dispersant by steric stabilization. The effect could be further quantified by using copolymers of methacrylate with polyglycols [65]. Steinberg [66] obtained similar results for PLMA copolymerized with methylvinyl pyridine.

In general, high molecular weight polymers are preferentially adsorbed from solution [78]. The specific adsorption A has been given by [79]

$$A = KM^a \tag{23}$$

where K is a constant, M is the molecular weight, and $0 < a < 0.3$. However, at low molecular weights ($< 10^4$), end group effects predominate, and at very high molecular weights ($> 10^6$), specific adsorption becomes independent of molecular weight [79].

A number of studies of adsorption from solution indicate that the adsorbed layer thickness of polymers in low adsorption energy situations is similar to the average end-to-end distance of the molecule. Thus Stromberg and coworkers [80] obtained adsorbed layer thicknesses measured by ellipsometry up to 80 nm for polystyrenes in the molecular weight range 76,000 to 3,300,000 on metals (oxides) including chromium, iron, silver, and copper, and on gold. Thickness was proportional to $M^{1/2}$ and compared to the random coil dimensions.

Similarly, Priel and Silberberg [81] measured the adsorption of polystyrene on glass as a hydrodynamic layer thickness deduced from the restriction of capillary viscometer diameters. The adsorbed layer thickness was in the range 5 to 100 nm for molecular weights from 20,800 to 1,700,000 and corresponded to the mean end-to-end distance for the polystyrene molecules.

Experimental measurements of the reduction in flow rate of polymer solutions through sintered borosilicate glass discs also show the hydrodynamic layer thickness for polystyrene, polymethylmethacrylate, and polyvinylacetate to be equivalent to the free coil dimensions [82,83].

VI. THE PREPARATION OF CERAMIC-POLYMER BLENDS

The mixing of powder and organic vehicle involves two stages, although they generally occur together. In dispersive mixing, particles are separated and agglomerates destroyed by successive fracture or by erosion under a high shear stress [2]. In distributive mixing, forced convective flow causes the spatial rearrangement of particles.

The preparation of ceramic suspensions is generally performed in mixing devices with a minimum of two principal moving parts [2], which include the two roll mill, the twin screw extruder, and a range of proprietary double axis lobed mixers.

Three stages are recognized in the dispersion of powders in liquids [84]. Wetting involves the replacement of solid-vapor with solid-liquid interfaces and may affect agglomerate strength. Mechanical disruption involves the fracture of junctions until the required degree of dispersion has occurred. This, for ceramics, is likely to be governed by the maximum permitted residual agglomerate size. Finally, stabilization describes the protection of the suspension against flocculation. Three types of stabilization of relevence to ceramic processing are understood [85]. There are discussed below.

In the mixing of high ceramic content suspensions, particularly in the early stage, before melting of the organic phase and encapsulation of the ceramic particles, there is a problem of abrasive wear, and the lubricant behavior of minor additives is influential in the minimization of contamination.

Previous work has shown that the incorporation of stearic acid into an alumina-polypropylene blend reduced the iron content of the resulting ceramic from 2200 ppm to 120 ppm [86]. It is well known that monomolecular layers of stearic acid on metal surfaces effect a pronounced reduction in coefficient of friction [87]. Measurements on films of calcium stearate of 48 nm thickness between mica surfaces suggest a shear strength of 3–4 MPa [88] for the adsorbed stearate layer.

Philips and Wightman [89] studied the adsorption of n-decyl derivatives onto alumina. These included n-decanoic acid, ethyl octanoate, n-decanol, and n-decylamine. Preferential adsorption of the acid was observed, and the acid assumed a perpendicular adsorbed structure. The overall adsorption of acid was 14×10^{-4} mol g^{-1} compared with 6–8×10^{-4} mol g^{-1} for the other species. This result seems to suggest that carboxylic acids are preferable in this role to esters. Similarly, Wolfrum and Ponjee [90] used infrared spectroscopy to reveal the acid-base interaction between carboxylic acids and a range of mineral surfaces, noting the unreacted carboxylic acid absorption in the 1700 cm^{-1} region and salt absorption in the 1570 cm^{-1} band.

Many claims are made for the benefits of matching the acidic and basic nature of dispersed and continuous phases. Schreiber and coworkers suggest that acid-base interactions can be used to promote dispersion of minerals in molten polymers [91,92]. Choosing polyvinyl chloride and polyethylene as nominally acidic and neutral polymers and $CaCO_3$ surface-modified by plasma polymerization, they argue that enhanced mechanical properties such as yield stress and elongation at break, as well as enhanced dispersion, accompany the selection of materials suitable for acid-base interactions.

Paul and Barlow [93] suggest that the acid-base behavior may also influence the miscibility of polymers. Since organic vehicles for the plastic forming of ceramics are frequently polymer blends containing various oils, plasticizers, and waxes [1], factors affecting their miscibility are significant. The conventional solubility parameter approach [94] is frequently used as a miscibility guide; the solubility parameter δ being given by

$$\delta^2 = \frac{L_v - RT}{V} \tag{24}$$

where L_v is the molar latent heat of vaporization and V is the molar volume. A high solubility parameter is an indication of internal hydrogen bonding, and a low value indicates that intermolecular forces are predominantly of the London dispersion type. It is argued that the acid-base interaction is a better indicator of the miscibility of polymers [95,96].

Dispersive mixing requires that shear stress generated in the melt must be sufficient to overcome interparticle forces. The London dispersion force between spherical particles of diameter d is given by

$$F = \frac{Ad_0}{24a_0^2} \tag{25}$$

where A is the Hamaker constant and a_0 is the separation distance, taken as about 0.2 nm for particles in contact [97]. Thus the force between 1 μm particles is about 0.1 μN.

The force between particles with an included liquid lens arises from the hydrostatic tension within the liquid caused by its curvature, and for spherical particles of radius R and lens radius x, the negative pressure (acting against intermolecular forces in the liquid) is given by

$$P = \gamma \left(\frac{1}{\rho} - \frac{1}{x} \right) \cos\theta \approx \frac{2R\gamma}{x^2} \cos\theta \tag{26}$$

where γ is the liquid surface energy, θ is the contact angle, and ρ is the smallest principal radius of curvature of the liquid surface. Rumpf [98] develops this to give the strength of liquid-bridge agglomerates.

The most serious form of agglomeration often arises in chemically derived powders or where additions have been made from aqueous solution followed by calcination, resulting in solid bridges. Milling is frequently then a prerequisite.

Hydrolytic ceramic powders, such as the nitrides, may form agglomerates in storage. The lower vapor pressure over condensed water at the junction between particles, which itself arises from the liquid-air curvature, encourages capillary condensation. A cementing oxide or hydroxide may be formed by reaction. In this context, it is worth noting that silane coupling agents, and by implication other nonsilicon-containing coupling agents, if added during wet milling, can "waterproof" ceramic powders [99].

Kendall [100] regards the brittle fracture of an agglomerate as being controlled by an internal defect of size c giving a failure stress of

$$\sigma_f = \frac{15.6 V^4 \gamma_c^{5/6} \gamma^{1/6}}{d_0^{1/2} c^{1/2}} \tag{27}$$

where γ_c is the specific interfacial energy between particles, γ is the measured fracture energy, d_0 is the particle diameter, and V is the packing efficiency. Thus as agglomerates are broken in shear flow, the fracture of fragments requires higher stresses.

Mixing devices tend to force the ceramic suspension through small clearances and thus to generate shear stresses that fragment agglomerates. The resulting microstructures are frequently superior to pressed ceramic bodies [101], but fine agglomerated powders are not well dispersed [102].

Calculations of shear stress in mixing are possible for machines of simple geometry such as the twin roll mill, and these show that the applied shear stress must be 5 to 10 times the agglomerate strength to effect dispersion [103].

VII. STABILIZATION OF SUSPENSIONS

The process of reagglomeration of particles dispersed in viscous polymers or waxes is little explored. The Smoluchowski theory of reagglomeration relies on collisions between particles initiated by Brownian motion (perikinetic flocculation). Particle mobility is severely restricted in viscous fluids, but this may not be the case for the low molecular weight waxes sometimes used. Orthokinetic flocculation occurs by particle motion in a velocity gradient. Both theories are surveyed by Ives [104] and assume that all collisions result in aggregation in the absence of surface force repulsion.

Such repulsion may emerge from three sources. Electrostatic stabilization occurs in aqueous media or media with high dielectric constant and is extensively reviewed [23,105,106]. Steric stabilization occurs in aqueous or low molecular weight organic media irrespective of ceramic volume fraction and is promoted by amphipathic dispersants [23-25]. Elastic stabilization results from the compression of adsorbed polymer layers and prevails in high molecular weight organic media [107,108].

Electrostatic charge at a solid surface can arise from (1) the dissolution of ions from the solid so that the concentrations of ions in solution are controlled by the solubility product or (2) the adsorption of ions from solution, including H^+ or OH^- ions. To maintain electrical neutrality, the surface charge is balanced by opposite charges in solution, producing a Helmholtz electrical double layer in the liquid near the interface. This double layer is the source of the four electrokinetic phenomena: electroosmosis, streaming potential, electrophoresis, and sedimentation potential.

Provided ionic concentration in the liquid is controlled so that the isoelectric point of the surface (IEPS) is avoided, a mutual electrostatic repulsion exists between the particles. The suspension will be stabilized against flocculation.

There is great interest in the use of water-based systems for injection molding using cellulose ethers [109,110] or agar [111,112] because of the long tradition of water-based processing in the industry and the perceived ease of binder removal. In general such systems show poor rheological properties unless the polymer fraction is high.

The presence of polymer molecules in water can affect the stabilization in several ways. If the adsorbing polymer can be ionized and carries an opposite charge to that on the solid surface, then electrostatic repulsion may be reduced. Conversely, stabilization is enhanced if the polymer has the same charge as the surface. If the polymer is not ionized, it may still affect the electrostatic stabilization by displacing the layer of counterions to greater distances from the solid surface, increasing the overlap distance and increasing the repulsion between particles [24]. On the other hand, nonionized polymer may also displace adsorbed ions and reduce surface charge leading to loss of stabilization.

Napper [113] is at pains to point out that electrostatic stabilization is less effective in nonaqueous dispersions than in water-based media because of the low dielectric constant. This is particularly true at high solids content because of the extension of the double layers. Steric stabilization, on the other hand, is effective in both types of medium.

Though not named as such, steric stabilization was first noted by Michael Faraday in the stabilization of gold sols by gelatin (113), though it has long been empirically used in the formulation of inks and paints with natural gums. It relies on the adsorption of amphipathic molecules with strongly adsorbing functional groups and free segments that have a strong affinity for the solvent or dispersing medium (Fig. 5). If there is low surface coverage, the adsorbed layer thickness varies with $M^{0.5}$. At close surface packing the adsorbate adopts a chain extended configuration, and the layer thickness varies with M. The overlap of segments from adjacent particles leads to a reduction in the number of ways of arranging chain segments in space and hence in entropy. The resulting free energy increase gives rise to a force of repulsion between particles. This is known as the 'volume restriction' or the 'elastic' effect. In addition there is an 'osmotic' effect above the theta temperature arising from the exclusion of solvent (the continuous phase) in the overlap region.

The term steric stabilization was introduced by Heller and Pugh [114] to distinguish it from electrostatic stabilization. It was recognized at an early stage that inorganic particles could be stabilized at high volume loadings in nonpolar, low-viscosity media (benzene, n-hexane) using adsorbed oleic acid. Electrostatic stabilization is ineffective under these conditions.

The energy of interaction per unit area of two surfaces bearing adsorbed polymer layers at low coverage is given as a function of separation distance D by [25]

$$W_{(D)} \approx 36 \Gamma k T e^{-D/R_g} \tag{28}$$

where Γ is the surface coverage in molecules/m^2 and R_g is the radius of gyration. This is valid in theta solvents and for $2R_g < D < 8R_g$.

For high surface coverage where chains are extended to a length L,

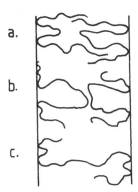

FIG. 5. Schematic representation of elastic stabilization of particles bearing adsorbed polymer films. (a) Interaction of loops and tails; (b) "denting," decrease of conformational entropy without interpenetration; (c) bridging.

$$W_{(D)} \approx \frac{100L}{\pi s^2} kTe^{-\pi D/L} \tag{29}$$

where s is the average distance between polymer adsorption sites. Hesselink [115] calculates the configurational free energy for a tail, loop, and bridge. The latter is a stabilizing molecule that has become attached to two particles. For the tail and loop, free energy rises continuously as particles approach. For the bridge, there is a minimum; the bridge molecule behaves as a spring with particle repulsion at close distances of approach but attraction at large distances. This work is developed [116] to calculate the repulsion due to the decrease in configurational entropy using the Flory-Huggins theory for equal loops, equal tails, and a loop size distribution. For the situation above the theta temperature, the osmotic effect is also treated. Stabilization is enhanced for long adsorbed chains, extreme size distribution of loops, large amount adsorbed, good solvents, low Hamaker constant, and small particle size.

A large number of dispersant copolymers are now available on the market, and they are developed for specific media varying from the strongly polar to the nonpolar [117]. They are also effective with waxes intermediate in molecular weight between solvents and polymers.

The situation in the polymer melt is somewhat less certain. Israelachvili [25] argues that since a polymer molecule in the melt is surrounded by like molecules it will behave as an adsorbate in a theta solvent. A repulsive potential comparable to that given by Eq. (28) should apply. Measurements of forces between mica surfaces in pure polymers do indeed show repulsion extending over distances of $10R_g$ [25]. A complication in the case of polymer melts is that restrictions on mobility may mean that molecular relaxation times are higher than in the bulk and that adsorbed layers may be effectively in the glassy state well above T_g.

Napper [108] also discusses the technologically important situation in a polymer melt where the free energy change due to mixing of polymer and 'solvent' is zero. The maximum size of particles that can be elastically stabilized is dependent on the molecular weight of the adsorbing species. The elastic mechanism does not rely on interaction with solvents and is expected to operate in ceramic-polymer suspensions intended for injection molding.

In addition to the forces of repulsion between particles, the insertion of a different medium between particles changes the Hamaker constant for the system and reduces the van der Waals attraction. Similar materials continue to attract, but the Hamaker constant is given by the combining rule for media of low dielectric constant by [25]

$$A_{132} = (A_{11}^{1/2} - A_{33}^{1/2})(A_{22}^{1/2} - A_{33}^{1/2}) \tag{30}$$

Thus for the system SiO_2-vacuum-SiO_2, $A = 6.3 \times 10^{-20}$ J, but in octane this is reduced to $A = 0.13 \times 10^{-20}$ J.

VIII. FLOW PROPERTIES OF SUSPENSIONS

The primary purpose of the fluid phase in ceramic suspensions is to convey the particles, hence the term 'organic vehicle'. There are four reasons why the vehicle should be as crowded as possible:

1. To reduce slumping during heat treatments to remove the vehicle
2. To reduce shrinkage caused by removal of the organic vehicle
3. To increase the prefired mechanical strength
4. To reduce the firing shrinkage and hence variation in firing shrinkage and distortion of the body

As the ceramic volume fraction of suspensions is increased, there comes a point where processing machinery is overloaded or suffers seizure. Depending on the machine capacity and mold design, it has been suggested that injection molding suspensions should have a viscosity at 100 s^{-1} of 1000–1500 Pas [118].

There is a very large number of expressions linking relative viscosity to volume fraction of ceramic [1], but only two provide a good fit to experimental data at high volume loadings. One of these is due to Eilers [119],

$$\eta_r = \frac{\eta}{\eta_0} = \left(1 + \frac{1.25V}{1 - V/V_{max}} \right)^2 \tag{31}$$

and one to Chong et al. [120],

$$\eta_r = \left(1 + \frac{0.75V/V_{max}}{1 - V/V_{max}} \right)^2 \tag{32}$$

More recently it has been suggested that Chong's equation be made more general to accomodate a wide range of suspensions, by recasting it as [121]

$$\eta_r = \left(\frac{V_{max} - CV}{V_{max} - V} \right)^2 \tag{33}$$

The variable C, equal to 2.5 for the original equation, can be found from a minimum of two relative viscosity measurements on an unknown powder, from which the entire curve can be produced. Figure 6 shows how Eq. (32) does not always fit data for ceramic suspensions and can be made more versatile by modification of C in Eq. (33). The steep dependence on volume

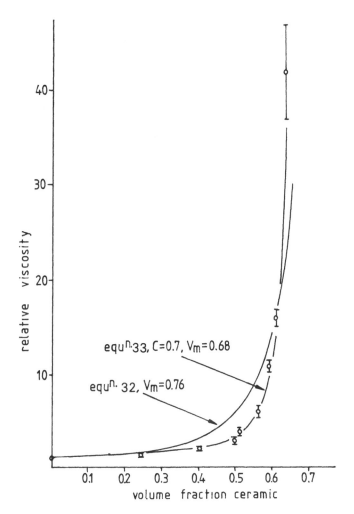

FIG. 6. Relative viscosity curves for an alumina suspension in polypropylene at 200°C and a shear rate of 100 s^{-1} showing the steep dependence of viscosity on ceramic volume fraction. Equation (32) has been used to fit the data in the viscosity range of interest for ceramic injection molding but fails elsewhere for this powder. Equation (33) with C as a disposable parameter fits the data and confers predictability.

fraction of ceramic in the injection molding region is dramatically illustrated. Both equations, and many others besides, contain a function of V_{max} − V as denominator. This represents the free volume fraction of organic vehicle over and above that needed to fill the interstices, the latter quantity being $1 - V_{max}$ (Fig. 7). This free volume fraction is related to the film thickness between particles, which can have dimensions of ~ 10 molecular layers [122].

In the region of the relative viscosity−volume loading curve relevant to plastic forming operations, the slope $d\eta_r/dv$ is steep, and small variations in volume loading (~ 1 vol %) exert a pronounced influence on viscosity. There are some difficulties therefore in defining V. First, there may be significant adsorbed water on fine powders that may not be displaced during mixing and can account for nearly 1 vol % [121]. Second, the volume fraction at room temperature differs significantly from that at the processing temperature [121] because of differential thermal expansion between polymer and ceramic phases. Third, the presence of adsorbed polymer layers may effectively increase V.

Thus Otsubo [123] showed that suspensions of high surface area silica in polyacrylamide solutions present a viscosity lower than the unfilled medium at loadings below 7 wt %, and at higher loadings viscosity increased markedly above that of the medium. The first effect is due to depletion of

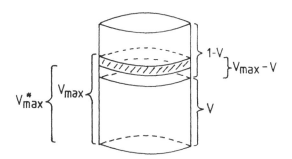

V_{max} from viscosity by extrapolation

V^*_{max} from shrinkage by measurement

$V^*_{max} \approx CPVC$

FIG. 7. Occupation of space in plastic forming operations for ceramics. Ceramic volume fraction V, maximum packing fraction for a given powder derived from viscosity measurements V_{max}, maximum packing fraction after binder removal V^*_{max}.

polyacrylamide in solution due to adsorption on the silica; the latter effect is enhanced because effective particle diameters and hence volume fraction are increased by an adsorbed layer.

A number of authors have used viscosity data to calculate the thickness of adsorbed layers. In order to achieve this, it has to be assumed that the system obeys one of the many relative viscosity–volume loading relationships. Deviation from theory then gives the effective volume fraction of adsorbed matter.

The Einstein equation [124]

$$\eta_r = 1 + 2.5V \tag{34}$$

is valid only at very low concentrations $<$ 5 vol %, and high concentrations are preferable for assessing the deviation. Such equations take two forms [1]: those based on a series,

$$\eta_r = A + BV + CV^2 \tag{35}$$

and those that use a function of $V - V_{max}$ such as Eqs. (30)–(32). The latter contain the maximum packing fraction V_{max}, which can only be obtained by fit to the viscosity data and therefore cannot immediately be used for this purpose [121].

Sato [125] made use of the relationship

$$\eta_r = 1 + 0.67fV + 1.62f^2V^2 \tag{36}$$

where f is a particle shape factor (ratio of elliptical axes), to obtain the film thickness of polyacrylamides on TiO_2 and Fe_2O_3 by deviation of volume fraction from theory. The results correlated with adsorption isotherm thicknesses.

Doroszkowski and coworkers [126] dispersed TiO_2 in a solution of oleic acid in an aliphatic hydrocarbon. The adsorbed layer thickness of oleic acid was allowed for in fitting the viscosity data to a series type equation:

$$\eta_r = 1 + 3V + 23V^2 \tag{37}$$

up to 15 vol % TiO_2. The effect of adsorbed layers on effective volume fraction could then be obtained in a range of copolymer solutions mainly based on methacrylates with long chain esters. Film thicknesses of 6–24 nm were thus obtained by dividing the difference between apparent and actual volumes by the powder surface area.

Minor organic additions to the system exert an influence on viscosity in two ways. In the first case, low molecular weight additions, such as plasticizers, modify the viscosity of the organic vehicle. Thus phthalates have been incorporated in suspensions based on polystyrene [1,4,8], and oils have been incorporated in waxes [1] to this end. Second, coupling agents that adsorb strongly on mineral surfaces have been added to ceramic suspensions to reduce viscosity [99,127].

Lindquist and coworkers [127] explored the effect of amino-functional and epoxy-functional silanes, together with an alkyl-titanate and stearic acid, on the viscosity of silicon nitride-polyolefin systems. Equimolar addition of silane and titanate were most effective in reducing viscosity, and stearic acid was regarded as a lubricant.

Similarly, Zhang and coworkers [99] used alkyl trialkoxy-silanes in conjunction with polypropylene for silicon and silicon nitride suspensions, obtaining a significant reduction in viscosity in the lower shear rate region (100 s^{-1}) of the flow curve.

In low viscosity fluids, coupling agents may act by steric stabilization, provided there is a strong interaction between the fluid phase and the pendent coupling agent group. Thus silanes have been used to obtain steric stabilization of silica and silicon carbide powder and whiskers in hexane [128,129].

In higher molecular weight vehicles, elastic stabilization may provide particle repulsion, and the coupling agent may have the effect of enhancing the adsorption energy of the polymer through either entanglement or chemical coupling. The presence of a coupling agent may also effectively increase the volume fraction of filler.

It is because these several mechanisms can operate that the literature on the effect of coupling agents on viscosity of suspensions is contradictory. While some authors report a reduction in viscosity caused by coupling agents [99,127,130,131] others report an increase [132,133]. Unfortunately, few authors have examined the effect of their coupling agent on the viscosity of the unfilled organic vehicle. Hunt and coworkers [134] obtained a substantial reduction in viscosity of zirconia suspensions in atactic polypropylene with titanate and zirconate coupling agents. However, when the coupling agent was added to polypropylene in the same weight ratio, it also reduced the viscosity of polypropylene in the absence of powder. Glycerol tristearate, with a similar structure but no coupling function, also reduced the viscosity of the suspension. It also reduced the flow behavior index substantially, suggesting lubricant action. Thus some coupling agents may actually be efficient plasticizers in their own right or may form effective plasticizer molecules by oligomer formation.

With alkyl-trialkoxy silanes there is a competition between adsorption and cocondensation that may provide plasticizing oligomers and itself impedes adsorption. Amine catalysts are used to promote adsorption of alkyl silanes [135]. Adsorption is also influenced by the pH of the treating solution in relation to the zeta potential of the substrate. It was noted, for example, that octyl triethoxy silane formed a waterproof barrier on silicon powder but that adsorption on silicon nitride was much slower [99]. The isolectric point of the surface (IEPS) of an oxide in water is a guide to the adsorption of silanes [136]: low IEPS acidic oxides such as silica provide strong adsorption, while high IEPS oxides such as iron, magnesium, or calcium are less effectively coupled.

IX. PHYSICAL PROPERTIES OF SUSPENSIONS

While biaxial stretching operations, such as blow molding [7] and vacuum forming [8], rely on melt strength, which is dependent on polymer molecular weight and degree of dispersion of the powder, injection molding relies on strength in the solid state, for resistance to residual stresses that are inherent in the injection molding process [2,137–142]. The fabrication of thin-walled ceramic bodies such as electronic substrates also relies on the development of substantial mechanical strength in the prefired state, and the binder is often selected for its adhesive properties rather than its pyrolytic characteristics. Most mechanical properties of particle-filled composites are influenced to some extent by interfacial adhesion and by the local influence of the high-energy surface on the segment mobility of organic molecules.

By comparing the storage modulus and loss tangent for a 50 vol % polypropylene with two calcium carbonate fillers whose specific surface areas differed by a factor of 10, it was concluded that a continuous layer of adsorbed polymer was present in the composite with the finer powder [143]. Thus a ceramic injection molded body can be considered a ternary system of ceramic, free organic phase, and an immobile adsorbed organic phase. This may account for the deviation of coefficient of expansion [144] and Poisson's ratio [145] from theory at high volume loadings.

The effect of inorganic fillers on glass transition temperature (T_g) of a composite is also attributed to the restriction of molecular mobility adjacent to the interface. T_g increases with filler loading and with decreasing surface area [146]. Increases of up to 12.5°C have been noted with a range of fillers [147], and the restriction on segmental motion can extend up to 150 nm into the matrix [148]. The adsorption energy may also affect the shift in T_g, for Greenberg [149] observes an increment in T_g for polyacrylic acid only with fillers that can interact with the matrix by acid-base reactions.

The effect of interfacial adhesion on the flexural strength, elastic modulus, and toughness of particle filled systems is analyzed theoretically and experimentally by Sahu and Broutman [150], who used a mold release agent and a coupling agent to decrease and increase adhesion, respectively. Elastic modulus was unaffected by adhesion below about 50 vol % filler, as was impact energy. Flexural and tensile strength on the other hand were significantly influenced by the filler-matrix interface.

The matrix in ceramic injection molding suspensions is not usually selected to achieve high mechanical strength, because pyrolytic and rheological criteria are more highly ranked. For these reasons, low molecular weight additives are almost universal, and direct comparison with the reinforced polymer experience is limited.

An exception is in the use of low molecular weight coupling agents intended to act as 'molecular adhesives' and to provide primary bonding between mineral surfaces and resins. As well as influencing rheology as described in Section VIII, these materials may influence mechanical strength of a particle-filled composite.

Silanes of the general formula

$$\begin{array}{c} X \\ | \\ RO-Si-OR \\ | \\ OR \end{array}$$

are easily hydrolyzed to a silanol by aqueous solution or by adsorbed water:

$$3H_2O + XSi(OR)_3 \rightarrow XSi(OH)_3 + 3ROH$$

whereupon they either hydrogen bond to surface hydroxyls

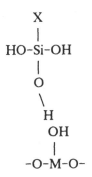

or condense with hydroxyls on a siliceous surface to give a siloxane bond

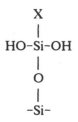

The X group can often be selected to react chemically with the matrix, whereupon improved mechanical strength is assured. Unfortunately, a large proportion of ceramic injection molding blends are based on olefinic matrices, often of low molecular weight, and the opportunity for such chemical reaction is extremely restricted.

Amorphous polymers may be bonded to silane-treated fillers in the absence of chemical reaction by solubility of compatible silane X groups in the polymer [146], but it is believed that syneresis rejects unreactive silane groups from semicrystalline polyolefins. Reactive sites can be introduced by peroxides, and chlorinated polyolefins are also claimed to improve adhesion [151], but both operations have considerable repercussions for ceramic molding. Azide functional silanes are claimed to react with polyolefins [152], while an interpenetrating network is claimed to develop with amino-functional silanes [153,154]. Thus the polar group appears to provide an interpenetrating network with polyolefins where alkoxytri-alkoxy silanes fail: a surprising result and an indication that the concepts of coupling agent interphases should never be oversimplified.

The problems of strengthening by entanglement or cocrystallization are addressed in an interesting experiment by Pukanszky et al. [155]. They explored stearic acid and maleated polypropylene as coupling agents for $CaCO_3$ in polypropylene. The stearic acid, being of low chain length and incompatible with polypropylene, reduced the mechanical strength for all volume loadings. In contrast, maleated polypropylene, compatible and cocrystallizable with the matrix, increased the strength for all volume loading up to 40 vol %.

The mechanical properties of particle-filled composites are also affected by the acidic and basic nature of filler and matrix. Marmo and coworkers [156] measured the relative modulus and relative tensile strength of solvent cast filled polymer films. Generally modulus and tensile strength exceeded theoretical bounds for good adhesion for acidic-basic combinations of filler and matrix. Unfortunately, the system SiO_2–chlorinated polyvinyl chloride also presented excellent adhesion when in fact it is an acidic combination. The experiment is complicated by particle size effects, mixing technique, and the use of solvents. In a range of composites based on ethylene-vinyl acetate copolymers and treated silicas prepared by shear mixing in the melt, Manson and Williams [157] found a general correlation between relative tensile strength, modulus and elongation at break and the extent of the acid-base interaction. Similar conclusions were drawn from a study of zinc and Wollastonite fillers in epoxy, pvc, polystyrene, or polymethylmethacrylate resins [158]. These approaches to the mechanical properties of composite materials have been reviewed by Manson [159].

X. REMOVAL OF THE ORGANIC PHASE

The methods of extraction of organic vehicle can be summarized as

1. Solvent extraction
2. Capillary flow into a porous body
3. Pyrolysis

Solvent extraction is recommended by Wiech [160] but appears not to be extensively used. In general, two-component binder systems are preferred in which one component is soluble [161]. Solvent extraction at low temperatures causes cracking, while high temperatures can result in slumping. It seems that to attain dissolution with minimal swelling, the process should be carried out above the theta temperature. Solubility may also be affected by adsorption. Thus Nahass et al. [162] found that in solvent cast alumina tapes, there was resistance to re-solution of the methacrylate polymer even on heating in a good solvent. This is attributed to chemisorption, and particles are considered to be bridged by the high molecular weight polymer. Particle separation distance was regarded as corresponding to the dimensions of the random coil, and the restricted polymer mobility was thought to be the cause of the low shrinkage on aging of cast ceramic tapes.

Capillary flow into a surrounding finer powder or a porous tile was also suggested by Wiech [163] and relies on a capillary pressure difference between the fluid in the body and the fluid in the surrounding powder.

The usefulness of capillary flow in extracting sufficient binder to reveal open porosity is described by Wei and German for coarse 3 μm heavy metal powders [164]. Since a polyethylene wax was used, minimal restriction to fluid flow from adsorption is to be expected. In the same way, the flow of polyvinylbutyral-dibutylphthalate binders has been directly observed in glass capillaries of diameter 200–300 μm [165]. Since the pore size between four spherical particles of radius r is approximately $0.2\,r$ [166], it follows that ceramic particles in the region 0.1–0.5 μm diameter present pore diameters in the region 20–100 nm and that such distances are of the same order of magnitude as the mean end-to-end distances of intermediate and high polymers. It follows that the segment mobility of organic binders in the pore space between particles is likely to be restricted by adsorption, and although capillary flow may be ideal for coarse powders, the viscosity of polymers in the bulk will bear little resemblance to viscosity within the pore architecture of fine ceramic bodies.

This is a molecular weight restriction over and above that introduced by bulk flow where the zero shear rate viscosity of a polymer is given as a function of molecular weight by [167]

$$\log \eta = 3.4 \log m + Av \qquad (38)$$

where Av is a constant. The method is therefore likely to be restricted to low molecular weight waxes and coarser ceramic powders. With waxes, however, significant fractions of binder can be removed by capillary flow, and in general the residue is uniformly distributed in the ceramic body [168–170].

Pyrolysis of ceramic binder is the favored extraction method, and three mechanisms of weight loss can be identified by their activation energies [171], namely, thermal degradation, oxidative degradation, and evaporation. The oxidative degradation is limited by oxygen diffusion in the suspension and can be modelled by shrinking unreacted core reaction kinetics [172]. Thermal degradation, on the other hand, involves chain scission at all points in the body uniformly and is therefore capable of generating volatile products throughout large sections. Although its activation energy is high, it is still an active reaction mechanism during all forms of pyrolysis. The diffusion of products of decomposition to the surface controls the center concentration and the likelihood of nucleation of bubbles at the center. Thus although mass transport of binders and their degradation products in the liquid or gaseous states in porous bodies have been modelled [173,174], it is the mobility of degradation products in solution in the parent polymer in the composite that controls the incidence of defects at the early stage of binder removal when all pores are full [175,176].

Although theories of diffusion of small molecules in polymer melts are well advanced [177–179], and methods of measurement of diffusivity in polymer melts have been devised [180], the influence of high filler loadings, especially of fine powders, is less explored.

Barrer [181] reviews expressions for electrical and mass transport processes in two-phase media. These relate the diffusion coefficient D in the composite to the diffusion coefficient D_2 in the continuous phase for a volume fraction of ceramic V_1. If diffusion in the dispersed phase is zero, the Maxwell model gives

$$\frac{D}{D_2} = 1 - \frac{3V_1}{V_1 + 2} \tag{39}$$

The Rayleigh model gives

$$\frac{D}{D_2} = \frac{2 - 2V_1 + 1.238V_1^{10/3}}{2 + V_1 + 1.238V_1^{10/3}} \tag{40}$$

and Bruggeman's model gives

$$\frac{D}{D_2} = (1 - V_1)^{3/2} \tag{41}$$

More recently, Bedeaux and Kapral [182] and Sridharan and Cukier [183] present models that reduce to

$$\frac{D}{D_2} = 1 - 1.5V_1 \tag{42}$$

which has the unfortunate prediction that $D/D_2 = 0$ at $V_1 \approx 0.67$.

All the models neglect the effect of adsorbed layers on diffusion coefficient. Since diffusion mechanisms in polymers, like flow mechanisms, involve segmental motion, the restriction on polymer mobility adjacent to high-energy surfaces is likely to impair diffusion rates. It is suggested [184] that these mass transport equations should first be validated for ceramic suspensions using coarse ceramic powders. For fine powders, the effective volume fraction of ceramic V_1 can then be defined as

$$V_1 = V(1 + k \rho_c Sh) \tag{43}$$

where ρ_c is the ceramic density, S is the specific surface area of the powder, h is the mean end-to-end distance of the wax or polymer molecule of the binder, V is the apparent volume fraction of ceramic, and k is a constant ($0 \leq k \leq 1$) that defines the effective fraction of the adsorbed layer for which $D = 0$.

It is sometimes suggested that low molecular weight oils or plasticizers added to ceramic injection molding suspensions emerge early in the pyrolysis stage and provide pore channels for enhanced mass transport [185]. One of the main beneficial effects of plasticizers is likely to be the increase in D_2 afforded by increasing free volume in the molten state.

Some indication of the effects of adsorbed layers on diffusion are obtained from results of water permeability of filled polymers. Stronger interfacial adhesion brought about by acid-base interaction produces lower water permeability [186]. Water sorption studies suggest that if a critical amount of diluent is added to a filled polymer it possesses the same sorptive properties as the unfilled polymer [148]. Kwei assumes that the water does not displace the polymer from mineral surfaces and goes on to propose a thermodynamic model for the decrease in entropy brought about by filler additions.

Water permeability of organic binders in the molten state is important in its own right. High surface area powders can convey a significant amount of water that may not be liberated from its adsorbed state until 200–300°C, whereupon it occupies a substantial volume as steam. Fine powders, even if dried thoroughly, tend to dessicate ambient air within a few minutes, and it would seem that provision for the exodus of water during binder removal must be made, probably by attention to the polarity of the organic vehicle. Nonpolar polymers tend to have very low water permeabilities and may obstruct the exodus of adsorbed water when pyrolyzed in the absence of oxygen. When oxidized, on the other hand, the polarity increases, and diffusion of water may be enhanced. Indeed, water is an oxidation product of polyolefins. The role of silanes as waterproofing agents for powders has already been demonstrated [99], but their influence on binder removal involves many other variables [187].

The decompositions of a range of organic binders were studied by thermogravimetry in finely divided form and as bulk moldings [188], and it was concluded that there was no obvious correlation between the weight loss-temperature profile and the incidence of defects. Thermogravimetry presents only half the story, dealing with the generation of low molecular weight products of decomposition. It does not give unambiguous information on the kinetics of diffusion throughout the molding. Nevertheless, thermogravimetry does indicate the suitability of organic binders for process control [188], and thermogravimetric weight loss control has been used for ceramic injection molded bodies [189]. Furthermore, organic vehicles have been tailored by blending to give steady weight loss with increasing temperature [190], and a simple computer selection technique is available for this purpose [191].

The degradation mechanisms for some polymers are themselves influenced by the presence of high-energy surfaces. The catalytic effects of transition metals on the oxidative degradation of polyolefins is well known [192,193]. A study [194] of the degradation of polyvinylbutyral mixed with 98 wt % of various oxides showed that all the oxides catalyzed the oxidative degradation; what is more surprising is that the thermal degradation was also catalyzed (Figs. 8a and 8b). These results have severe implications for the application of polymer degradation studies to ceramic processing. At least they mean that activation energies and rate constants for degradation

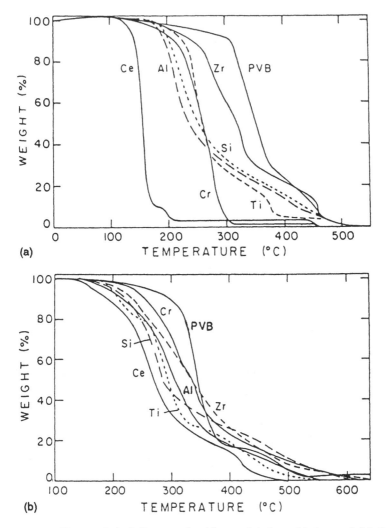

FIG. 8. The catalytic influence of oxides on (a) the oxidative and (b) the thermal degradation of polyvinylbutyral, assessed by thermogravimetry in the presence of 98 wt % powder. (Reproduced from Ref. 194 by kind permission of Chapman and Hall.)

must be measured in the presence of ceramic powder and at worst that theories of polymer extraction will be difficult to apply to powders of different chemical compositions and specific surface areas.

ACKNOWLEDGMENTS

The author is grateful for the wide explorative remit permitted by the Science and Engineering Research Council of the United Kingdom in its support of the Ceramics Fabrication Programme at Brunel University. Mrs. K. Goddard is thanked for typing the manuscript.

REFERENCES

1. M. J. Edirisinghe and J. R. G. Evans, *Int. J. High Tech. Ceramics 2*: 1 (1986).
2. M. J. Edirisinghe and J. R. G. Evans, *Int. J. High Tech. Ceramics 2*: 249 (1986).
3. J. E. Shuetz, *Amer. Ceram. Soc. Bull. 65*: 1556 (1986).
4. J. K. Wright, R. M. Thomson, and J. R. G. Evans, *J. Mater. Sci. 25*: 1833 (1990).
5. N. McN. Alford, T. W. Button, and J. D. Birchall, *Supercond. Sci. Tech. 3*: 1 (1990).
6. K. Kobayashi, M. Furuta, and Y. Maeno, Eur. Pat. 34056, 19 Aug. 1981. Assigned to NGK Insulators, Ltd.
7. P. Hammond and J. R. G. Evans, *J. Mater. Sci. Lett. 10*: 294 (1991).
8. K. M. Haunton, J. K. Wright, and J. R. G. Evans, *Br. Ceram. Trans. J. 89*: 53 (1990).
9. D. W. Richerson, *Modern Ceramic Engineering*, Marcel Dekker, New York, 1982, pp. 206 et seq.
10. K. Katayama, T. Watanabe, K. Matoba, and N. Katoh, SAE Tech. Pap. 861128 (1986).
11. H. Rashid, K. Lindsey, and J. R. G. Evans, *J. Euro. Ceram. Soc. 7*: 165 (1991).
12. H. Rashid and J. R. G. Evans, *Ceram. Int. 17*: 259 (1991).
13. H. Rashid, K. N. Hunt, and J. R. G. Evans, *J. Euro. Ceram. Soc. 8*: 329 (1991).
14. R. W. Davidge, *Mechanical Behaviour of Ceramics*, Cambridge University Press, Cambridge, 1979, pp. 39 et seq.
15. W. F. Brown and J. E. Srawley, ASTM Special Tech. Publ. 410, ASTM, Philadelphia, 1966.
16. R. L. Coble and J. E. Burke, *Prog. Ceram. Sci. 3* (J. E. Burke, ed.). Pergamon, 1963, pp. 197-251.
17. E. A. Barringer and H. K. Bowen, *Comm. Amer. Ceram. Soc. 65*: C199 (1982).
18. K. Kendall, *Proc. Brit. Ceram. Soc. 42*: 81 (1989).
19. F. F. Lange, *J. Amer. Ceram. Soc. 67*: 83 (1984).
20. B. Kellet and F. F. Lange, *J. Amer. Ceram. Soc. 67*: 369 (1984).
21. B. E. Walker, R. W. Rice, P. F. Becher, B. A. Bender, and W. S. Coblenz, *Bull. Amer. Ceram. Soc. 62*: 916 (1983).

22. J. Lyklema, in *The Scientific Basis of Flocculation* (K. J. Ives, ed.). Sijthoff and Noordhoff, Alphen aan den Rijn, Netherlands, 1978, pp. 3–36.
23. D. H. Napper, *Polymeric Stabilization of Colloidal Dispersions*. Academic Press, London, 1983, pp. 18–30.
24. J. Gregory, in *The Scientific Basis of Flocculation* (K. J. Ives, ed.). Sijthoff and Noorhoff, Alphen aan den Rijn, Netherlands, 1978, pp. 101–130.
25. J. N. Israelachvili, *Intermolecular and Surface Forces*, 2d ed. Academic Press, London, 1991, pp. 288–311.
26. E. A. Moelwyn-Hughes, *States of Matter*. Oliver and Boyd, Edinburgh, 1961.
27. J. H. De Boer, *Trans. Faraday Soc. 32*: 10 (1936).
28. F. London, *Trans. Faraday Soc. 33*: 8 (1937).
29. D. Tabor and R. H. S. Winterton, *Proc. Roy. Soc. A. 312*: 435 (1969).
30. J. N. Israelachvili, M. Tirrell, J. Klein, and Y. Almog, *Macromolecules 17*: 204 (1984).
31. M. Tirrell, S. Patel, and G. Hadziioannou, *Proc. Nat. Acad. Sci. USA 84*: 4725 (1987).
32. J. N. Israelachvili, R. K. Tandon, and L. R. White, *J. Coll. Interf. Sci. 78*: 430 (1980).
33. S. J. Cyzak, *Amer. J. Phys. 20*: 440 (1952).
34. J. C. Speakman, *The Hydrogen Bond and Other Intermolecular Forces*. The Chemical Society, London, 1975.
35. A. D. Wilson and S. Crisp, *Organolithic Macromolecular Materials*. Applied Science, 1977, pp. 103–154.
36. P. Sykes, *A Guidebook to Mechanism in Organic Chemistry*, 4th ed. Longman, London, 1975.
37. F. M. Fowkes, *Mat. Sci. Eng. 53*: 125 (1982).
38. F. M. Fowkes, *Rubber Chem. Technol. 57*: 328 (1984).
39. E. H. Andrews and A. J. Kinloch, *Proc. Roy. Soc. London A. 332*: 385 (1973).
40. E. H. Andrews and A. J. Kinloch, *Proc. Roy. Soc. London A. 332*: 401 (1973).
41. W. A. Zisman, *Adv. Chem. Ser. No. 43*, American Chemical Society, Washington, D.C., 1964, p. 1.
42. R. J. Good and L. A. Girifalco, *J. Phys. Chem. 61*: 904 (1957).
43. F. M. Fowkes, *Ind. Eng. Chem. 56*(12): 40 (1964).
44. D. H. Kaelble and K. C. Uy, *J. Adhesion 2*: 50 (1970).
45. M. Sherriff, *J. Adhesion 7*: 257 (1976).
46. S. Wu, in *Polymer Blends*, vol. 1 (D. R. Paul and S. Newman, eds.). Academic Press, London, 1978, pp. 243–293.
47. J. N. Israelachvili, *Intermolecular and Surface Forces*, 2d ed. Academic Press, London, 1991, p. 203.
48. H. Schonhorn, H. L. Frisch, and T. K. Kwei, *J. Appl. Phys. 37*: 4967 (1966).
49. M. Pegoraro, L. DiLandro, R. Trifilo, and A. Penati, *J. Polym. Sci. Polym. Phys. 23*: 2499 (1985).
50. P. G. de Gennes, *Reviews of Modern Physics 57*: 827 (1985).
51. R. J. Good, *J. Coll. Interf. Sci. 59*: 398 (1977).
52. R. S. Drago and B. B. Wayland, *J. Amer. Chem. Soc. 87*: 3571 (1965).

53. R. S. Drago, G. C. Vogel, and T. E. Needham, *J. Amer. Chem. Soc. 93*: 6014 (1971).
54. J. Lyklema, in *The Scientific Basis of Flocculation* (K. J. Ives, ed.). Sijthoff and Noordhoff, Alphen aan den Rijn, Netherlands, 1978, pp. 25–30.
55. F. M. Fowkes, in F. J. Micale, Y. K. Lui, and A. C. Zettlemoyer, *Disc. Farad. Soc. 42*: 238 (1966).
56. C. Pefferkorn, A. Carroy, and R. Varoqui, *J. Polym. Sci. Polym. Phys. 23*: 1997 (1985).
57. M. K. Bernett and W. A. Zisman, *J. Coll. Interf. Sci. 28*: 243 (1968).
58. E. G. Shafrin and W. A. Zisman, *J. Amer. Ceram. Soc. 50*: 478 (1967).
59. E. McCafferty and A. C. Zettlemoyer, *J. Coll. Interf. Sci. 34*: 452 (1970).
60. E. McCafferty, V. Pravdic, and A. C. Zettlemoyer, *Trans. Faraday Soc. 66*: 1720 (1970).
61. C. Kemball, *Proc. Roy. Soc. A. 190*: 117 (1947).
62. F. P. Bowden and W. R. Throssel, *Nature 167*: 601 (1951).
63. K. Bright and B. W. Malpass, *Euro. Polym. J. 4*: 431 (1968).
64. C. E. Miller and T. K. Yin, *J. Mater. Sci. Lett. 8*: 467 (1989).
65. B. J. Fontana, *J. Phys. Chem. 67*: 2360 (1963).
66. G. Steinberg, *J. Phys. Chem. 71*: 292 (1967).
67. S. J. Gregg and K. S. W. Sing, *Adsorption, Surface Area and Porosity*. Academic Press, London, 1982, pp. 62, 238.
68. P. J. Flory, *Polymer Chemistry*. Cornell University Press, Ithaca, New York, 1953, 1971, pp. 399–431.
69. C. Tanford, *Physical Chemistry of Macromolecules*. John Wiley, New York, 1961, pp. 151–170.
70. R. J. Young, *Introduction to Polymers*. Chapman and Hall, London, 1981, pp. 86–122.
71. A. Silberberg, *J. Phys. Chem. 66*: 1872, 1884 (1962).
72. F. Patat, E. Killman, and C. Schliebener, *Rubber Chem. Technol. 39*: 36 (1966).
73. R. R. Stromberg, in *Treatise on Adhesion and Adhesives*, vol. 1 (R. L. Patrick, ed.). Marcel Dekker, New York, 1967, pp. 69–118.
74. R. F. Hoffman and W. C. Forsman, *J. Polym. Sci. A2 8*: 1847 (1970).
75. E. Jenkel and B. Rumbach, *Z. Electrochem. 55*: 612 (1951).
76. H. L. Frisch, R. Simba, and F. R. Eirich, *J. Chem. Phys. 21*: 365 (1953); *J. Phys. Chem. 57*: 584 (1953).
77. B. J. Fontana and J. R. Thomas, *J. Phys. Chem. 65*: 480 (1961).
78. K. Furusawa, K. Yamashita, and K. Konno, *J. Coll. Interf. Sci. 86*: 35 (1982).
79. G. J. Howard and S. J. Woods, *J. Polym. Sci. A2 10*: 1023 (1972).
80. R. R. Stromberg, D. J. Tutas, and E. Passaglia, *J. Phys. Chem. 69*: 3955 (1965).
81. Z. Priel and A. Silberberg, *J. Polym. Sci. Polym. Phys. 16*: 1917 (1978).
82. F. W. Ryland and F. R. Eirich, *J. Polym. Sci. A1 4*: 2401 (1986).
83. F. Rowland, R. Bulas, E. Rothstein, and F. R. Eirich, *Ind. Eng. Chem. 57(9)*: 46 (1965).
84. S. H. Bell and V. T. Crowl, in *Dispersion of Powders in Liquids*, 2d ed. (G. D. Parfitt, ed.). Applied Science, London, 1973, pp. 267–307.

85. K. J. Ives, ed., *The Scientific Basis of Flocculation.* Sijthoff and Noorhoff, Alphen aan den Rijn, Netherlands, 1978, pp. 25–30, 324–329.
86. J. G. Zhang, M. J. Edirisinghe, and J. R. G. Evans, *Industrial Ceramics 9*: 72 (1989).
87. F. P. Bowden and D. Tabor, *The Friction and Lubrication of Solids.* Clarendon Press, Oxford, 1950, pp. 185–188.
88. J. N. Israelachvili and D. Tabor, *Wear 24*: 386 (1973).
89. K. M. Philips and J. P. Wightman, *J. Coll. Interf. Sci. 108*: 495 (1985).
90. S. M. Wolfrum and J. J. Ponjee, *J. Mater. Sci. Lett. 8*: 667 (1989).
91. H. P. Schreiber, M. R. Wertheimer, and M. Lambla, *J. Appl. Polym. Sci. 27*: 2269 (1982).
92. M. Lambla and H. P. Schreiber, *Euro. Polym. J. 16*: 211 (1980).
93. D. R. Paul and J. W. Barlow, *J. Macromol. Sci. Rev. Macromol. Chem. C18*(1): 109 (1980).
94. P. D. Ritchie, *Plasticizers, Stabilizers and Fillers.* Iliffe, London, 1972, pp. 50–65.
95. F. M. Fowkes, *Org. Coatings Plast. Chem. 51*: 522 (1984).
96. F. M. Fowkes, D. O. Tischler, J. A. Wolfe, L. A. Lannigan, C. M. Ademur-John, and M. J. Halliwell, *J. Polym. Sci. Polym. Chem. 22*: 547 (1984).
97. J. Israelachvili, *Intermolecular and Surface Forces*, 2d ed. Academic Press, London, 1991, pp. 176–212.
98. H. Schubert, W. Herrmann, and H. Rumpf, *Powder Technology 11*: 121 (1975).
99. J. G. Zhang, M. J. Edirisinghe, and J. R. G. Evans, *J. Mater. Sci. 23*: 2115 (1988).
100. K. Kendall, *Powder Metall. 31*: 28 (1988).
101. M. J. Edirisinghe and J. R. G. Evans, *Proc. Brit. Ceram. Soc. 38*: 67 (1986).
102. M. J. Edirisinghe, J. G. Zhang, and J. R. G. Evans, *Proc. Brit. Ceram. Soc. 42*: 91 (1989).
103. J. H. Song and J. R. G. Evans, *J. Euro. Ceram. Soc.* in press.
104. K. J. Ives, in *The Scientific Basis of Flocculation* (K. J. Ives, ed.). Sijthoff and Hoordhoff, Alphen aan den Rijn, Netherlands, 1978, pp. 37–61.
105. J. N. Israelachvili, *Intermolecular and Surface Forces*, 2d ed. Academic Press, London, 1991, pp. 213–259.
106. P. Sennett and J. P. Oliver, *Ind. Eng. Chem. 57*(3): 32 (1965).
107. D. H. Napper, *Polymeric Stabilization of Colloidal Dispersions.* Academic Press, London, 1983, pp. 106–107.
108. D. H. Napper, *Polymeric Stabilization of Colloidal Dispersions.* Academic Press, London, 1983, pp. 324–329.
109. R. D. Rivers, U.S. Patent 4113480, Sept. 12, 1978.
110. K. Maeda, Japan Kokkai Tokyo Koho 01249650, Oct. 14, 1989.
111. A. J. Fanelli, R. D. Silvers, W. S. Frei, J. V. Burlew, and G. B. Marsh, *J. Am. Ceram. Soc. 72*: 1833 (1989).
112. A. J. Fanelli and R. D. Silvers, Euro. Patent 246438, Nov. 25, 1987.
113. D. H. Napper, *Polymeric Stabilization of Colloidal Dispersions.* Academic Press, London, 1983, pp. 20–21, 25.
114. W. Heller and T. L. Pugh, *J. Chem. Phys. 22*: 1778 (1954).

115. F. Th. Hesselink, *J. Phys. Chem.* 75: 65 (1971).
116. F. Th. Hesselink, A. Vrij, and J. Th. G. Overbeek, *J. Phys. Chem.* 75: 2094 (1971).
117. J. D. Schofield, in *Recent Developments in the Technology of Surfactants* (M. R. Porter, ed.). Critical Reports on Applied Chemistry, vol. 30. Elsevier, London, 1990, pp. 35-63.
118. M. J. Edirisinghe and J. R. G. Evans, *J. Mater. Sci.* 22: 269 (1987).
119. H. Eilers, *Kolloid Z.* 97: 313 (1941).
120. J. S. Chong, E. B. Christiansen, and A. D. Baer, *J. Appl. Polym. Sci.* 15: 2007 (1971).
121. T. Zhang and J. R. G. Evans, *J. Euro. Ceram. Soc.* 5: 165 (1989).
122. J. K. Wright, M. J. Edirisinghe, J. G. Zhang, and J. R. G. Evans, *J. Am. Ceram. Soc.* 73: 2653 (1990).
123. Y. Otsubo, *J. Coll. Interf. Sci.* 112: 380 (1986).
124. A. Einstein, in *Investigations on the Theory of Brownian Movement* (R. Forth, ed.). Dover, New York, 1956, p. 55.
125. T. Sato, *J. Appl. Polym. Sci.* 15: 1053 (1971).
126. A. Doroszkowski and R. Lambourne, *J. Coll. Interf. Sci.* 26: 214 (1968).
127. K. Lindquist, E. Carlström, M. Persson, and R. Carlsson, *J. Am. Ceram. Soc.* 72: 99 (1989).
128. J. R. Fox, P. C. Kokoropoulos, G. H. Wiseman, and H. K. Bowen, *J. Mater. Sci.* 22: 4528 (1987).
129. J. R. Fox, M. J. Unguriet, and H. K. Bowen, *J. Mater. Sci.* 22: 4532 (1987).
130. S. H. Morrell, *Plast. Rubb. Proc. Appln. 1*: 179 (1981).
131. C. D. Han, C. Sandford, and H. J. You, *Polym. Eng. Sci. 18*: 849 (1978).
132. C. D. Han, H. L. Lu, and J. Mijovic, *S.P.E. Antec 28*: 82 (1982).
133. D. M. Bigg, *Polym. Eng. Sci. 22*: 512 (1982).
134. K. N. Hunt, J. R. G. Evans, and J. Woodthorpe, *Polym. Eng. Sci. 28*: 1572 (1988).
135. R. L. Kaas and J. L. Kardos, *Soc. Plast. Engnrs Tech. Props.* 22: 22 (1976).
136. E. P. Plueddeman, in *Composite Materials*, vol. 6, *Interfaces in Polymer Matrix Composites* (L. J. Broutman and R. H. Krock, eds.). Academic Press, New York, 1974.
137. M. S. Thomas and J. R. G. Evans, *Br. Ceram. Trans. J. 87*: 22 (1988).
138. M. J. Edirisinghe and J. R. G. Evans, *J. Mater. Sci.* 22: 2267 (1987).
139. N. J. Mills, *J. Mater. Sci. 17*: 558 (1982).
140. N. J. Mills, *Plast. Rubb. Proc. Appln. 3*: 181 (1983).
141. K. N. Hunt, J. R. G. Evans, N. J. Mills, and J. Woodthorpe, *J. Mater. Sci. 26*: 5229 (1991).
142. B. Kostic, T. Zhang, and J. R. G. Evans, *J. Amer. Ceram. Soc. 75*: 2773 (1992).
143. J. Jancar, J. Kucera, and P. Vesely, *J. Mater. Sci. Lett. 7*: 1377 (1988).
144. T. Zhang and J. R. G. Evans, *J. Euro. Ceram. Soc. 6*: 15 (1990).
145. T. Zhang and J. R. G. Evans, *J. Euro. Ceram. Soc. 7*: 155 (1991).
146. K. Iisaka and K. Shibayama, *J. Appl. Polym. Sci. 22*: 1321 (1978).
147. G. J. Howard and R. A. Shanks, *J. Macromol. Sci. Phys. B19*: 167 (1981).

148. T. K. Kwei, *J. Polym. Sci. A 3*: 3229 (1965).
149. A. R. Greenberg, *J. Mater. Sci. Lett. 6*: 78 (1987).
150. S. Sahu and L. J. Broutman, *Polym. Eng. Sci. 12*: 91 (1972).
151. E. P. Plueddeman and G. L. Stark, Proc. 3_.h Ann. Tech. Conf. S.P.E. 20B-1 (1980).
152. G. A. McFarren, T. F. Sanderson, and F. G. Schappell, *Soc. Plast. Eng. Tech. Props. 22*: 19 (1976).
153. N. H. Sung and A. Kaul, *Polym. Eng. Sci. 22*: 637 (1982).
154. A. Kaul and N. H. Sung, *Polym. Eng. Sci. 24*: 493 (1984).
155. B. Pukanszky, F. Tudos, J. Jancar, and J. Kolarik, *J. Mater. Sci. Lett. 8*: 1040 (1989).
156. M. J. Marmo, M. A. Mustafa, H. Jinnai, F. M. Fowkes, and J. A. Manson, *Ind. Eng. Chem. Prod. Res. Dev. 15*: 206 (1976).
157. J. A. Manson and J. T. Williams, *Org. Coat. Plast. Chem. 42*: 175 (1980).
158. J. A. Manson, J. S. Lin, and A. Tiburcio, *Org. Coat. Plast. Chem. 46*: 121 (1982).
159. J. A. Manson, *Pure Appl. Chem. 57*: 1667 (1985).
160. R. Wiech, U.S. Patent 4197118, April 8, 1980.
161. R. M. German, *Powder Injection Molding*. Metal Powder Industrial Federation, Princeton, New Jersey, 1990, p. 313, pp. 330–331.
162. P. Nahass, R. L. Pober, W. E. Rhine, W. L. Robbins, and H. K. Bowen, *J. Am. Ceram. Soc. 75*: 2373 (1992).
163. R. E. Wiech, Euro. Pat. 0032403, 14 Jan. 1981.
164. T. S. Wei and R. M. German, *Int. J. Powder Met. 24*: 327 (1988).
165. M. J. Cima, M. Dudziak, and J. A. Lewis, *J. Amer. Ceram. Soc. 72*: 1087 (1989).
166. D. J. Cumberland and R. J. Crawford, *The Packing of Particles*. Elsevier, Amsterdam, 1987, p. 60.
167. D. W. van Krevelen, *Properties of Polymers*. Elsevier, Amsterdam, 1972, p. 258.
168. J. K. Wright and J. R. G. Evans, *Ceram. Int. 17*: 79 (1991).
169. Y. Bao and J. R. G. Evans, *J. Euro. Ceram. Soc. 8*: 81 (1991).
170. Y. Bao and J. R. G. Evans, *J. Euro. Ceram. Soc. 8*: 95 (1991).
171. J. K. Wright, J. R. G. Evans, and M. J. Edirisinghe, *J. Am. Ceram. Soc. 72*: 1822 (1989).
172. J. K. Wright and J. R. G. Evans, *J. Mater. Sci. 26*: 4897 (1991).
173. G. C. Stangle and I. A. Aksay, *Chem. Eng. Sci. 45*: 1719 (1990).
174. D. S. Tsai, *A.I.Ch.E. J. 37*: 547 (1991).
175. P. Calvert and M. Cima, *J. Am. Ceram. Soc. 73*: 575 (1990).
176. J. R. G. Evans, M. J. Edirisinghe, J. K. Wright, and J. Crank, *Proc. Roy. Soc. A 432*: 321 (1991).
177. J. S. Vrentas and J. L. Duda, *Encyclopaedia of Polymer Science and Engineering 5*: 36 (1986).
178. S. Vrentas, *J. Polym. Sci. Polym. Phys. 15*: 403 (1977).
179. J. L. Duda, J. S. Vrentas, S. T. Ju, and H. T. Liu, *A.I.Ch. J. 28*: 279 (1982).
180. J. L. Duda, G. K. Kimmerly, W. L. Sigelko, and J. S. Vrentas, *Ind. Eng. Chem. Fundam. 12*: 133 (1973).

181. R. M. Barrer, in *Diffusion in Polymers* (J. Crank and G. S. Park, eds.). Academic Press, London, 1968, p. 165.
182. D. Bedeaux and R. Kapral, *J. Chem. Phys. 79*: 1783 (1983).
183. S. Sridharan and R. I. Cukier, *J. Phys. Chem. 91*: 2962 (1987).
184. J. R. G. Evans and M. J. Edirisinghe, *J. Mater. Sci. 26*: 2081 (1991).
185. C. L. Quackenbush, K. French, and J. T. Neil, *Ceram. Eng. Sci. Proc. 3*: 20 (1982).
186. A. Tiburcio and J. A. Manson, *Org. Coatings Plast. Chem. 51*: 539 (1984).
187. M. J. Edirisinghe, *J. Mater. Sci. Lett. 9*: 1039 (1990).
188. J. Woodthorpe, M. J. Edirisinghe, and J. R. G. Evans, *J. Mater. Sci. 24*: 1038 (1989).
189. A. Johnsson, E. Carlström, L. Hermansson, and R. Carlsson, *Proc. Brit. Ceram. Soc. 33*: 139 (1983).
190. K. Saito, T. Tanaka, and T. Hibino, U.S. Patent 4000110, Dec. 28, 1976. Assigned to Tokyo Shibaura Electric Co., Ltd.
191. S. J. Stedman, J. R. G. Evans, and J. Woodthorpe, *Ceramics International 16*: 107 (1990).
192. S. Chaudri, *Polymer 9*: 604 (1968).
193. R. H. Hansen, C. A. Russell, T. de Benedictus, W. M. Martin, and J. V. Pascale, *J. Polym. Sci. A 2*: 587 (1964).
194. S. Masia, P. D. Calvert, W. E. Rhine, and H. K. Bowen, *J. Mater. Sci. 24*: 1907 (1989).

Index

Printed in the United States
by Baker & Taylor Publisher Services